Probability Theory and Stochastic Modelling

Volume 70

The **Stochastic Modelling and Probability Theory** series is a merger and continuation of Springer's two well established series Stochastic Modelling and Applied Probability and Probability and Its Applications series. It publishes research monographs that make a significant contribution to probability theory or an applications domain in which advanced probability methods are fundamental. Books in this series are expected to follow rigorous mathematical standards, while also displaying the expository quality necessary to make them useful and accessible to advanced students as well as researchers. The series covers all aspects of modern probability theory including

- Gaussian processes
- Markov processes
- Random Fields, point processes and random sets
- Random matrices
- Statistical mechanics and random media
- Stochastic analysis

as well as applications that include (but are not restricted to):

- Branching processes and other models of population growth
- Communications and processing networks
- Computational methods in probability and stochastic processes, including simulation
- Genetics and other stochastic models in biology and the life sciences
- Information theory, signal processing, and image synthesis
- Mathematical economics and finance
- Statistical methods (e.g. empirical processes, MCMC)
- Statistics for stochastic processes
- Stochastic control
- Stochastic models in operations research and stochastic optimization
- Stochastic models in the physical sciences

For further volumes:
http://www.springer.com/series/13205

David Applebaum

Probability on Compact Lie Groups

Foreword by Herbert Heyer

David Applebaum
School of Mathematics and Statistics
University of Sheffield
Sheffield
UK

Foreword by
Herbert Heyer
University of Tübingen
Tübingen
Germany

ISSN 2199-3130 ISSN 2199-3149 (electronic)
ISBN 978-3-319-37579-3 ISBN 978-3-319-07842-7 (eBook)
DOI 10.1007/978-3-319-07842-7
Springer Cham Heidelberg New York Dordrecht London

Mathematics Subject Classification (2010): 60B15, 43A05, 43A77, 43A30

Printed on acid-free paper

Springer is part of Springer Science+Business Media (www.springer.com)

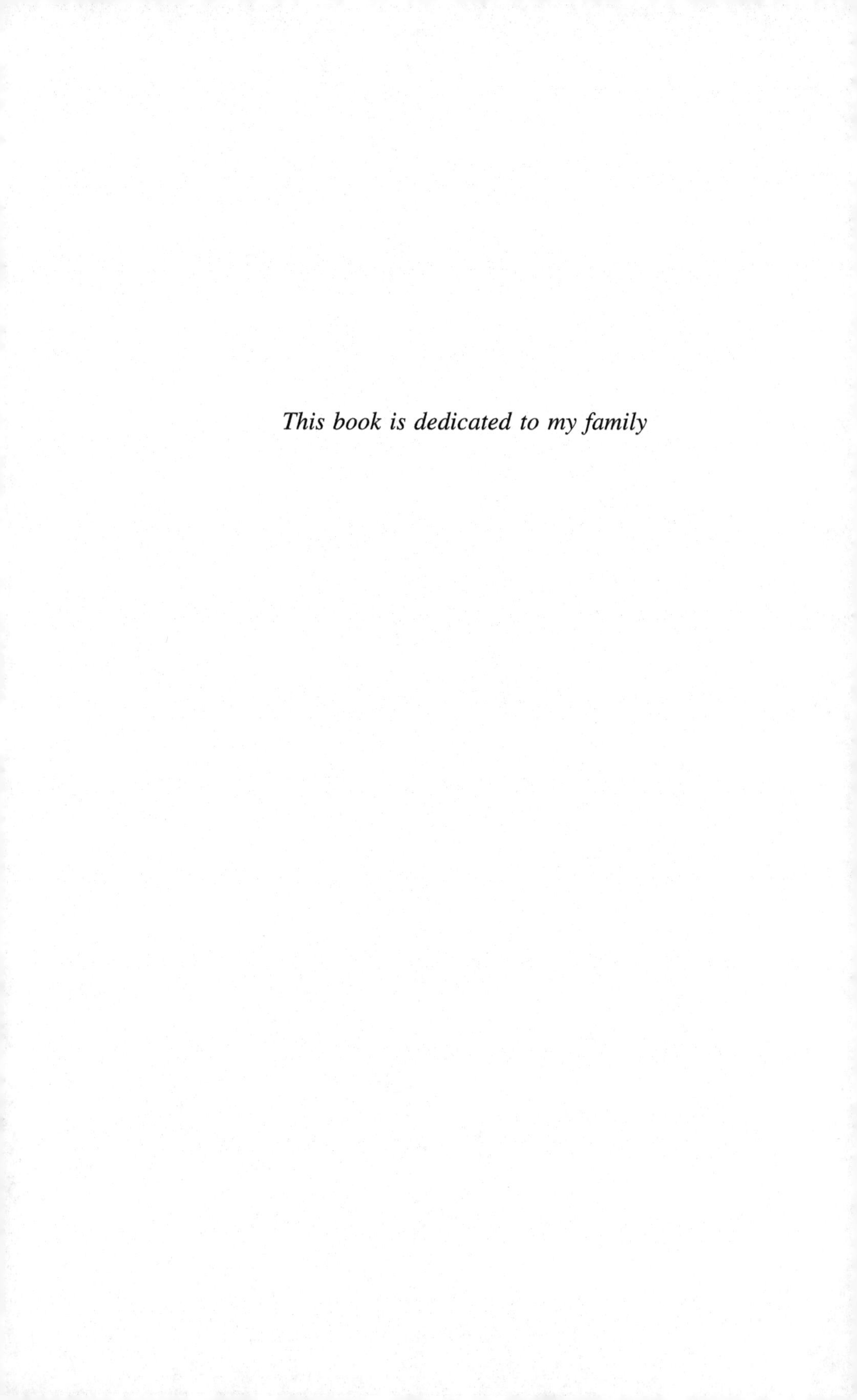

This book is dedicated to my family

Symmetry is a vast subject, significant in art and nature. Mathematics lies at its root, and it would be hard to find a better one on which to demonstrate the working of the mathematical intellect

Hermann Weyl, "Symmetry"

You should meditate often on the connection of all things in the universe and their relationship to each other

Marcus Aurelius, "Meditations"

Foreword

Since the trendsetting monographs of Grenander, Parthasarathy and Heyer the study of probability measures on algebraic-topological structures such as topological semigroups, groups and hypergroups has developed into a comprehensive area of mathematical research, with applications ranging from the dynamics of stochastic processes to statistical decision theory. A basic tool in advancing the theory was the abstract harmonic analysis of locally compact groups, in particular the technique of Fourier transformation based on unitary representations. While the initial studies mostly dealt with general classes of groups and measures, the emphasis during the last four decades has been concretization in two directions: to choose special classes of groups and to consider special types of measures. In the seminal books of Diaconis, Hazod and Siebert, and Liao innovative contributions about random walks on, and Lévy processes in groups were included up to the actual state of the art. Originally, there was a broad variety of problems of classical probability theory that called for generalization and standardization within the framework of locally compact groups. We mention selectively the embedding of infinitely divisible measure into convolution semigroups, the canonical decomposition of convolution semigroups, the central limit problem and arithmetical characterizations of prominent types of probability measures. All of these problems could be more easily approached once the underlying group was Abelian, although some of the analysis had already been extended beyond this special case at an early stage. The author of the present book chooses compact Lie groups in order to describe recent advances of the theory. On the basis of current sources he discusses for example results on positive definiteness of mappings on duals of compact Lie groups, on regularity of densities of probability measures, and on deconvolution. Part survey and part treatise, the presentation supplements the existing expository literature on the subject. It invites the reader to learn first about compact Lie groups and their representations before it leads him to topics in structurally oriented probability theory. A wealth of references accompanying the exposition opens the eye of the reader towards future research. It is a pleasant experience to see well-established results and recent progress in the theory of probability measures on compact Lie groups summarized in book form. Of course in a handy publication the choice of relevant topics can only be subjective. Yet the author's very own research and his overall picture of the field guarantee successful reading for all who wish to absorb research in probability theory from an algebraic-topological point of view.

Herbert Heyer

Acknowledgments

Most of this book was written in the academic year 2011–2012 when the author was granted study leave by the University of Sheffield. He would like to thank Tony Dooley and the University of New South Wales for hospitality in January–February 2012 where much progress on this project was made despite the attractions of Coogee Beach and cricket on the TV nearly every night! He would also like to thank Marta Sanz-Solé and the University of Barcelona for supporting a ten-day visit in May 2012, where he continued to make progress despite the rather different attractions of Catalan culture and tapas. Thanks are also due to many colleagues for discussions on topics relating to material in the book including Rodrigo Bañuelos, Fabrice Baudoin, Tony Dooley, Neil Dummigan, Herbert Heyer, Peter Kim, Fionntan Roukema and René Schilling. My profound thanks go to Michael Bingham, who read through the document with great care and made many extremely valuable (and sometimes rather deep) comments. The customary free copy of the published volume is a small reward for his labours. Thanks are also due to Ming Liao, Jonathan Manton and Markus Riedle, who each read and made helpful comments on particular chapters, to Christian Fonseca Mora, who very thoroughly worked through the whole draft manuscript and helped me remove many typos, and to Nick Bingham who provided me with a "full red-ink job", that enabled me to correct a host of grammatical inaccuracies. I would also like to thank two anonymous referees who gave me extremely valuable feedback that has led to many improvements. Thanks are also due a second time to Herbert Heyer for agreeing to write the Preface to this volume. Finally, it is a pleasure to thank my editor Marina Reizakis, and all the staff at Springer who contributed towards the production of the book.

Contents

Introduction

Probability on groups gives a mathematical context to the interplay of *chance* with *symmetry*. To study this subject involves investigating probability measures on groups and random variables, stochastic processes and random fields that are group-valued. At the very least this entails the interaction of (probabilistic) measure theory and group theory, but the range of mathematical tools required to develop the subject is much wider than this and includes representation theory, harmonic analysis, functional analysis and differential geometry, as well as stochastic analysis and stochastic differential geometry. Furthermore the subject is highly applicable, and in the first two decades of the twenty-first century there has been considerable interest in applications to engineering, such as signal processing, robotics and the measurement of biological molecules. There have also been complementary developments in statistical estimation and inference on groups and manifolds which have been strongly motivated by these applications.

We can study probability theory in, e.g. finite groups, algebraic groups, Lie groups, locally compact groups, loop groups and current groups, so the reader may well be asking—why *compact* Lie groups? There are a number of reasons for this. To begin with, this book is designed to be an *introduction*, and so it should avoid the temptations of generality. Indeed, the topic of probability theory on general locally compact groups is the theme of a classic monograph by Herbert Heyer [95], which although published in 1977 still remains highly relevant today. Compact Lie groups have the advantage that:

- They are the simplest class of continuous groups, beyond the abelian, to display the key feature of *noncommutativity* (in general) that leads to so much fascinating structure.
- They have a rather straightforward representation theory. In particular, their irreducible representations are all finite-dimensional. This leads to a *Fourier transform* (characteristic function) that is matrix-valued, while for more general groups this will operate in an infinite-dimensional space. Furthermore, the theory of highest weights enables us to carry out standard analytic operations (such as taking limits) in the "Fourier co-variable".
- Every sequence (or net) of probability measures on a compact group is tight.

- There are many important and interesting examples. These include the tori Π^n, the special orthogonal groups $SO(n)$ and the special unitary groups $SU(n)$. The group $SO(3)$ of rotations in three-dimensional space is of particular interest to engineers [48, 46, 47] and cosmologists [147]. The symmetry group of the standard model in elementary particle physics is the compact Lie group $U(1) \times SU(2) \times SU(3)$ [16].

Of course there is much that can be done with compact groups alone, but compact Lie groups admit a differential structure, and this enables us to exploit natural analytic/geometric structures such as the Laplacian and its associated heat semigroup and heat kernel. In fact, one can think of a Lie group informally as the natural context in which calculus meets symmetry, so one can, for example, describe symmetry transformations that preserve some invariant quantity, infinitesimally.

There are three key tools that are needed to study a probability measure μ or a process on a group G. The first is the Fourier transform as mentioned above, whose value at an irreducible representation π is $\widehat{\mu}(\pi) = \int_G \pi(g^{-1})\mu(dg)$. The second is functional analysis, which enters the picture when we consider convolution operators of the form $P_\mu f = f * \mu$ acting on various Banach spaces. Finally, we use stochastic analysis to study the interaction of systems with noise by means of stochastic differential, or stochastic partial differential equations. In this book we will almost entirely concentrate on the interplay between the first two of these tools and we will hardly use the third at all. Again there are some good reasons for this. First, this is an introductory book and I didn't want it to get too long. Secondly, from a pedagogic viewpoint, there is something to be said for imitating the process we go through as an undergraduate or first-year graduate student, where we take a course that covers topics like characteristic functions and the central limit theorem before we try to learn about stochastic integration. Thirdly, many topics that interest engineers and statisticians, such as deconvolution, seem to rely mainly on Fourier transform techniques. Finally, there was some intellectual curiosity on my own part—how far could I go without using one of my favourite tool kits? But the reader who gets to the end of the book will have encountered many places where stochastic analysis is needed to make further progress, and there are abundant references for further reading in this direction.

This book is specifically aimed at graduate students who have taken a standard course in measure-theoretic probability and have some training in functional analysis (particularly, Hilbert spaces) and a smattering of general topology, but do not necessarily have a strong background in group theory, representations and differential geometry. Of course it is also suitable for experts in Lie theory and associated harmonic analysis who want to find out what the probabilists are doing with their beautiful subject. Finally, I hope it will be accessible to the growing number of highly mathematically trained engineers, physicists and statisticians who are now working on applications of these ideas and who wish to fill in some gaps in their knowledge.

Guide to Reading the Book

Chapter 1 is a roller-coaster ride through the essentials of topological groups and Lie groups. Of course this is a huge topic and it cannot be learned in a hurry. I would encourage readers who are novices in this area to study an authoritative text as an accompaniment. Chapter 2 develops all the representation theory that we need for the book and proves one of the key results—the Peter–Weyl theorem. We also begin the important study of the Fourier transform of functions. Here I am highly indebted to Faraut's beautiful monograph [63], and large parts of the account are closely based on his approach. In the last part of the chapter, I introduce (in a somewhat condensed manner) the theory of weights and sketch the proofs of the wonderful character and dimension formulae of Hermann Weyl. Experts on Lie theory can skip these two chapters, but might want to glance at Sect. 2.3 to remind themselves of key properties of the Fourier transform.

The short Chap. 3 introduces the important tool of the Laplacian and the associated Sobolev spaces. In the second part of the chapter we utilise the one--to-one correspondence between irreducible representations and dominant weights to regard the Fourier transform as a "function" of the latter. Since weights live in a finite-dimensional vector space, we can now carry out standard analytic operations, such as taking limits, on the "Fourier co-variable". In particular, we present Sugiura's [200] far-reaching characterisation of smooth functions on the group in terms of decay properties at infinity of the Fourier transform.

In Chap. 4, we at last turn our attention to probability measures on groups and make a detailed study of characteristic functions (i.e. Fourier transforms of measures). In particular, we describe a compact-group analogue of the famous Lévy convergence theorem that is due to Kawada and Itô [114]. A key theme of this chapter is absolute continuity, and we begin to see the importance of convolution operators through a beautiful theorem of Raikov [164] that tells us that the probability measure μ is absolutely continuous if and only if the associated convolution operator P_μ is compact in the space of continuous functions on the group. Using Sugiura's techniques, we then find necessary and sufficient conditions for a measure to have a smooth density in terms of the decay of the characteristic function. In the last part of the chapter we study random walks on general (not necessarily compact) Lie groups, and give some potential-theoretic characterisations of recurrence due to Guivarc'h et al. [75].

Standard Gaussian measures on compact Lie groups are naturally described through their Fourier transform in terms of eigenvalues of the Laplacian (the Casimir spectrum). They have smooth densities which can be easily obtained from the heat kernel. In recent years there has been increasing interest in both theoretical work and modelling with non-Gaussian processes, and Lévy processes (which are essentially stochastic processes with stationary and independent increments) have been at the heart of this activity. In Chap. 5, we study the analogues of these processes on general Lie groups but at the level of measures. So we investigate convolution semigroups of probability measures and the associated C_0-semigroups of convolution operators. We obtain Hunt's important characterisation [103] of the generators of these semigroups, which can be seen as an analogue of the classical Lévy–Khintchine formula on Euclidean space. As a corollary we derive a less well-known and more direct generalisation of the Lévy–Khintchine formula using the Fourier transform. We also develop Breuillard's [35] presentation of the central limit theorem, itself originally due to Wehn. Subordination is an important technique for obtaining new examples of interesting convolution semigroups, and specialists in mathematical finance know that the most interesting cases are obtained by subordinating Brownian motion. The same is true for Lie groups, and we examine some important classes (such as analogues of the stable laws) and establish smoothness of densities.

Finally Chap. 6 is a short introduction to statistics on compact Lie groups. It focusses on work by Kim and Richards [118] on deconvolution in the context of estimating the distribution of a signal from observations made on an output that is distorted by an independent noise.

Each chapter finishes with a brief guide to further reading, and there are many references in the bibliography for readers to delve further into the literature. I should perhaps apologise to anyone whose work I have neglected to mention. The bibliography, like the book that it references, is not meant to be an exhaustive archive of contributions to this very wide area. It is rather a guide to the available literature, and in many cases I have referenced particular papers and books in the knowledge that the reader can find many more interesting references in the bibliography of that paper or book. Despite my best efforts, this book will surely contain many typos and (hopefully minor) errors. Please send these to me at d.applebaum@sheffield,ac.uk. They will be posted on my website at http://www.applebaum.staff.shef.ac.uk/books.html.

The book contains many proofs but in addition a large number of statements are included without proof, especially in Chap. 1. In particular I have tended to omit proofs that require a lot of structure theory, but include those that are more analytical, since I believe that this will make the volume more accessible for the typical reader, who is likely to be an analytically trained probabilist. Of course, where the proof is missing, there will be a reference to where it can be found in its full glory.

There are a number of appendices, and in particular, Appendices 1, 3, 5 and 6 give very heavily abridged accounts of those aspects of topology, differentiable manifolds, measures on locally compact spaces and compact operators in Hilbert space (respectively) that are needed. They can serve as a quick reminder of key facts, but are no substitute for systematic study of the topic in question. I have included references to standard texts for those who need them, but the reader can also find a fuller treatment of all of the above (and a great deal more, including a chapter on compact and locally compact groups, as well as representations of the former) in a two-volume work by Knapp [121, 122].

I hope the reader will find that the book contains a joyful and stimulating selection of topics. However one should be aware that compact groups are, though important, a restrictive class, and many important applications involve groups such as the Heisenberg group or the Euclidean group (the semi-direct product of translations with rotations). Analogues of some results obtained in this book are not yet available in these contexts, and there is a great deal of work still to be done. In particular, it is important that mathematicians, statisticians and engineers do not work in isolation as there is so much that they can learn from each other.

Guide to Notation and a Few Useful Facts

If S is a set, S^c denotes its complement. If T is another set then $S \setminus T := S \cap T^c$. If A is a finite set, then the number of elements in A is $\#A$. If A is a non-empty subset of a topological space then $\overline{A}$ is its closure.

$\mathbb{N}$, $\mathbb{Z}$, $\mathbb{Q}$, $\mathbb{R}$, $\mathbb{C}$ are the sets of natural numbers, integers, rational numbers, real numbers and complex numbers (respectively). $\mathbb{Z}_+ := \mathbb{N} \cup \{0\}$. In this short section we use F to denote $\mathbb{R}$ or $\mathbb{C}$.

If S is a topological space, then the *Borel* σ-algebra of S will be denoted $\mathcal{B}(S)$. It is the smallest σ-algebra of subsets of S which contains all the open sets. Sets in $\mathcal{B}(S)$ are called *Borel sets*. $\mathbb{R}$ and $\mathbb{C}$ will always be assumed to be equipped with their Borel σ-algebras, and measurable functions from $(S, \mathcal{B}(S))$ to $(F, \mathcal{B}(F))$ are sometimes called *Borel measurable*. Similarly, a measure defined on $(S, \mathcal{B}(S))$ is called a *Borel measure*.

If S is a locally compact Hausdorff space, then $B_b(S, F)$ is the linear space (with the usual pointwise operations of addition and scalar multiplication) of all bounded Borel measurable functions from S to F. It is an F-Banach space under the supremum norm $||f||_\infty := \sup_{x \in S} |f(x)|$ for $f \in B_b(S, F)$. The space of bounded continuous functions from S to F is denoted $C_b(S, F)$. It is a closed linear subspace of $B_b(S, F)$, and so an F-Banach space in its own right. A function f from S to F is said to *vanish at infinity* if given any $\epsilon > 0$ there exists a compact set K in S so that $|f(x)| < \epsilon$ whenever $x \in K^c$. The space $C_0(S, F)$ of all continuous F-valued functions on S which vanish at infinity is a closed linear subspace of $B_b(S, F)$ (and of $C_b(S, F)$), and so is also an F-Banach space in its own right. The *support* of an F-valued function f defined on S is the closure of the set $\{x \in S;\ f(x) \neq 0\}$ and it is denoted $\text{supp}\,(f)$. The linear space $C_c(S, F)$ of all continuous F-valued functions on S with compact support is a dense subspace of $C_0(S, F)$. When $F = \mathbb{C}$, we usually write $B_b(S) := B_b(S, \mathbb{C})$, $C_0(S) := C_0(S, \mathbb{C})$ etc. This will be the case throughout the book, except for Chap. 5 where $C_0(S)$ always means $C_0(S, \mathbb{R})$ etc.

Throughout this book "smooth" means "infinitely differentiable". If M is a smooth manifold and $p \in \mathbb{N}$, then $C^p(M, F)$ is the linear space of all p-times continuously differentiable functions from M to F, and the linear space of infinitely differentiable functions is $C^\infty(M, F) = \bigcap_{p \in \mathbb{N}} C^p(M, F)$. We write $C(M, F) := C^0(M, F)$ for the linear space of all continuous functions on M. If M is compact, then $C(M, F) = C_b(M, F) = C_0(M, F) = C_c(M, F)$. Note that the linear space of smooth functions of compact support $C_c^\infty(M,F) = C^\infty(M,F) \cap C_c(M,F)$ is dense in $C_0(M, F)$. If $A(S)$ is any function space of mappings from S to F, then $A(S)_+$ denotes the cone of non-negative functions in $A(S)$.

If $n \in \mathbb{N}$, then $M_n(F)$ is the F-algebra of all $n \times n$ matrices with values in F. The identity matrix in $M_n(F)$ is denoted by I_n. If $A \in M_n(F)$, then A^T denotes its transpose and $A^* = \overline{A^T}$ is its adjoint. The trace of a square matrix A, i.e. the sum of its diagonal entries, is denoted by tr(A), and its determinant is $\det(A)$. The Hilbert–Schmidt inner product on $M_n(F)$ is $\langle A,B\rangle_{\mathrm{HS}} := \mathrm{tr}(\mathrm{AB}^*)$, and the corresponding Hilbert–Schmidt norm is $||A||_{\mathrm{HS}} := \mathrm{tr}(\mathrm{AA}^*)^{\frac{1}{2}}$. The matrix A is said to be *non-negative definite* if $x^T A x \geq 0$ for all $x \in \mathbb{R}^n$, and *positive definite* if $x^T A x > 0$ for all $x \in \mathbb{R}^n \backslash \{0\}$.

If (S, Σ, μ) is a measure space and $f : S \to F$ is an integrable function, we often write the Lebesgue integral $\int_S f(x)\mu(dx)$ as $\mu(f)$. For $1 \leq p < \infty$, $L^p(S) := L^p(S, \Sigma, \mu; F)$ is the usual L^p space of equivalence classes of functions that agree almost everywhere with respect to μ for which

$$||f||_p = \left(\int_S |f(x)|^p \mu(dx)\right)^{\frac{1}{p}} < \infty$$

for all $f \in L^p(S)$. $L^p(S)$ is a Banach space with respect to the norm $|| \cdot ||_p$, and $L^2(S)$ is a Hilbert space with respect to the inner product

$$\langle f, g\rangle := \int_S f(x)\overline{g(x)}\mu(dx)$$

for $f, g \in L^2(S)$. Once again, $L^p(S)$ will mean $L^p(S, \mathbb{C})$ in the main part of the book (except in Chap. 5).

The *indicator function* 1_A of $A \in \Sigma$ is defined as follows:

$$1_A(x) = \begin{cases} 1 & \text{if } x \in A \\ 0 & \text{if } x \notin A. \end{cases}$$

If μ is σ-finite, and ν is a finite measure on (S, Σ), we write $\nu << \mu$ if ν is absolutely continuous with respect to μ, and $\frac{d\nu}{d\mu}$ is the corresponding Radon–Nikodym derivative.

If (Ω, F, P) is a probability space and $X : \Omega \to \mathbb{R}$ is a random variable (i.e. a measurable function from (Ω, F) to $(\mathbb{R}, B(\mathbb{R}))$) that is also integrable in that $\int_\Omega |X(\omega)|P(d\omega) < \infty$, then its *expectation* is defined to be

$$\mathbb{E}(X) := \int_\Omega X(\omega)P(d\omega).$$

If $T : V_1 \to V_2$ is a linear mapping between F-vector spaces V_1 and V_2, then Ker(T) is its kernel and Ran(T) is its range. If $T : H_1 \to H_2$ is a bounded linear operator between F-Hilbert spaces H_i having inner products $\langle\cdot,\cdot\rangle_i\,(i=1,2)$, its *adjoint* is the unique bounded linear operator $T^*: H_2 \to H_1$ for which

$$\langle T^*\psi, \phi\rangle_1 = \langle\psi, T\phi\rangle_2,$$

for all $\phi \in H_1$, $\psi \in H_2$. The bounded linear operator $U: H_1 \to H_2$ is said to be *unitary* if it is both an isometry and a co-isometry (i.e. U^* is also an isometry). Equivalently it is an isometric isomorphism for which $U^{-1} = U^*$. If H is a F-Hilbert space, then $\mathcal{L}(H)$ will denote the $*$-algebra of all bounded linear operators on H, where the involution is obtained by taking adjoints. $\mathcal{L}(H)$ is a Banach space (and in fact a C^*-algebra, but we won't need this) with respect to the operator norm

$$||T|| = \sup\{||Tx||; x \in H,\ ||x|| = 1\}.$$

Note that the algebra $M_n(F)$ may be realised as $\mathcal{L}(F^n)$. More generally, if H_1 and H_2 are distinct F-Hilbert spaces, then $\mathcal{L}(H_1, H_2)$ will denote the F-linear space of all bounded linear operators from H_1 to H_2 . If H is a Hilbert space and $x, y \in H$ are orthogonal so that $\langle x, y\rangle = 0$, we sometimes write $x \perp y$. The *complexification* of a real vector space V is the complex vector space $V_\mathbb{C}$ whose elements are of the form $x + iy$, with $x, y \in V$, with the addition law:

$$(x_1 + iy_1) + (x_2 + iy_2) = (x_1 + x_2) + i(y_1 + y_2),$$

for $x_i, y_i \in V (i = 1, 2)$ and scalar multiplication:

$$(\alpha + i\beta)(x + iy) = (\alpha x - \beta y) + i(\alpha y + \beta x),$$

for $x, y \in V$, $\alpha, \beta \in \mathbb{R}$. If V is equipped with an inner product $\langle\cdot,\cdot\rangle$, then we may define the inner product $\langle\cdot,\cdot\rangle_\mathbb{C}$ on $V_\mathbb{C}$ by the prescription

$$\langle x_1 + iy_1, x_2 + iy_2\rangle_\mathbb{C} = \langle x_1, x_2\rangle + \langle y_1, y_2\rangle + i(\langle y_1, x_2\rangle - \langle x_1, y_2\rangle).$$

In the main part of the book, we will always write $\langle\cdot,\cdot\rangle_\mathbb{C}$ as $\langle\cdot,\cdot\rangle$. It is perhaps worth emphasising that all inner products on complex vector spaces are linear on the left and conjugate-linear on the right, which is standard in mathematics (but not in physics). The algebraic dual of an F-linear space V will typically be denoted by V^*, so V^* is the F-linear space of all F-linear functionals from V to F.

If T is a linear operator defined on an F-Banach space E, its domain is written as $\mathrm{Dom}(T)$. The *restriction* of T to $D \subset \mathrm{Dom}(T)$ is denoted $T|_D$. We say that T is *densely defined* if $\mathrm{Dom}(T)$ is dense in E and *closed* if its graph G_T is closed in $E \times E$, where $G_T := \{(\psi, T\psi); \psi \in \mathrm{Dom}\,(T)\}$. The linear operator T is said to be *closable* if it has at least one closed extension, in which case its *closure* $\overline{T}$ is the smallest such extension (i.e. that which has the smallest domain).

Let G be a group. We always denote its neutral element by e. General elements are typically denoted by g, h, but in the last part of the book we often use σ, τ. The group operation is usually written multiplicatively. If G is a Lie group its Lie algebra is denoted by $\mathfrak{g}$. Linear or unitary representations of groups are generically written as π, and the bounded operator $\pi(g)$ then acts on the complex Hilbert space V_π for each $g \in G$. The identity operator in V_π is denoted I_π. If V_π is finite-dimensional, then $d_\pi := \dim(V_\pi)$. The unitary dual of G is the set of all equivalence classes of irreducible unitary representations (with respect to unitary conjugation) denoted by $\widehat{G}$. This is precisely the dual group of G (i.e. the character group) when G is abelian. If G is a Lie group, the derived representation of its Lie algebra $\mathfrak{g}$ is $d\pi$. If G is a compact, connected Lie group, then the highest weight of π is denoted λ_π, its character is χ_π and the Casimir spectrum is $\{-\kappa_\pi, \pi \in \widehat{G}\}$. Generically we will denote characters by χ, weights by λ and roots by α. The Laplace–Beltrami operator associated to a Riemannian manifold is always written as Δ.

If G is a locally compact Hausdorff group, its left Haar measure is m_L and its right Haar measure is m_R. If G is also compact, then $m := m_L = m_R$, and we generally normalise so that $m(G) = 1$. If $f \in L^1(G, F)$, then $\int_G f(\sigma)m(d\sigma)$ is always written as $\int_G f(\sigma)d\sigma$. The Fourier transform of such an f at the representation π is $\widehat{f}(\pi) := \int_G \pi(g^{-1})f(g)dg$. Similarly, if μ is a finite Borel measure on G, then its Fourier transform at π is $\widehat{\mu}(\pi) := \int_G \pi(g^{-1})\mu(dg)$. Convolution of functions f_1 and f_2 is denoted as $f_1 * f_2$, and of measures μ_1 and μ_2 by $\mu_1 * \mu_2$.

We occasionally use Landau notation, according to which $(o(n), n \in \mathbb{N})$ is any real-valued sequence for which $\lim_{n\to\infty}(o(n)/n) = 0$ and $(O(n), n \in \mathbb{N})$ is any non-negative sequence for which $\limsup_{n\to\infty}(O(n)/n) < \infty$. Functions $o(t)$ and $O(t)$ are defined similarly. If f and g are functions from $\mathbb{R}$ to $\mathbb{R}$, then $f(x) \sim g(x)$ as $x \to \infty$ means that $\lim_{x\to\infty} \frac{f(x)}{g(x)} = 1$.

Is S is a countable set, the *Kronecker delta* is defined for $i, j \in S$ by

$$\delta_{ij} = \begin{cases} 0 & \text{if i} \neq \text{j} \\ 1 & \text{if i} = \text{j}. \end{cases}$$

If $a, b \in \mathbb{R}$, then $a \wedge b := \min\{a, b\}$.

Chapter 1
Lie Groups

Abstract In this chapter we review the main concepts from the theory of topological groups and Lie groups that we need for the main part of the book. We define topological groups and present some important examples, including the classical groups. We briefly describe the role of covering groups and investigate properties of Haar measure on locally compact, Hausdorff groups. We define Lie groups and their Lie algebras, and introduce the important concepts of the exponential map and the adjoint representation. In the last part of the chapter we consider the Laplacian as an algebraic object belonging to the universal enveloping algebra of the group.

The purpose of this opening chapter is to give a brief overview of key concepts in topological and Lie groups that will be needed in later chapters when we develop probability theory therein. Of necessity it is somewhat fragmentary and incomplete and we include very few proofs. The only part of the chapter where we give a more detailed account of the theory is in Sect. 1.2, where we derive properties of Haar measure that are crucial for later developments. Readers needing more thorough exposure to these concepts and ideas are advised to study a basic text or monograph. In particular we recommend Faraut [63] (for both this and the next chapter), Fegan [66], Chevalley [45] and Knapp [120]. Indeed if a proof is missing from the text and you want to locate it, then it will certainly appear in at least one of these books. We also point out to the reader that some background material in differential geometry is included in Appendix A.3.

1.1 Topological Groups

1.1.1 Definition and Examples

Let G be a set that is equipped with a binary operation which we write multiplicatively, i.e. there is a mapping from $G \times G \to G$ which takes $(g, h) \in G \times G$ to $gh \in G$. We say that G is a *topological group* if it is both a topological space and also a

D. Applebaum, *Probability on Compact Lie Groups*, Probability Theory and Stochastic Modelling 70, DOI: 10.1007/978-3-319-07842-7_1,

group with respect to the given binary operation, and if the mapping from $G \times G$ to G given by $(g, h) \to gh^{-1}$ is continuous. The neutral or identity element of G will always be denoted as e. A topological group (or subgroup, as appropriate) is abelian, normal etc. if the underlying group is, and is connected, compact, Hausdorff etc. if the underlying topological space is. We may consider subgroups of G and of particular interest are those closed in the given topology. For example, the *centre* of G which we denote by $Z(G) := \{h \in G; gh = hg \text{ for all } g \in G\}$ and $\{e\}$ are closed subgroups of G when G is Hausdorff.

Examples of abelian groups are $\mathbb{R}^d$, $\mathbb{C}^d$, the d-torus Π^d (where $\Pi := \mathbb{R}/2\pi\mathbb{Z}$) Π^∞, the p-adic number fieldsand any topological vector space, where the binary operation is addition. For examples of non-abelian groups, we consider *matrix groups*, i.e. those that are subsets of $M_n(\mathbb{C})$ or $M_n(\mathbb{R})$. Examples include:

The general linear groups	$GL(n, \mathbb{C})$	$= \{X \in M_n(\mathbb{C}); \det(X) \neq 0\}$.
	$GL(n, \mathbb{R})$	$= GL(n, \mathbb{C}) \cap M_n(\mathbb{R})$.
The special linear groups	$SL(n, \mathbb{C})$	$= \{X \in GL(n, \mathbb{C}); \det(X) = 1\}$.
	$SL(n, \mathbb{R})$	$= SL(n, \mathbb{C}) \cap M_n(\mathbb{R})$.
The unitary groups	$U(n)$	$= \{X \in GL(n, \mathbb{C}); X^*X = I_n\}$.
The special unitary groups	$SU(n)$	$= U(n) \cap SL(n, \mathbb{C})$.
The orthogonal groups	$O(n)$	$= U(n) \cap GL(n, \mathbb{R})$.
The special orthogonal groups	$SO(n)$	$= O(n) \cap SL(n, \mathbb{R}) = U(n) \cap SL(n, \mathbb{R})$.

We also define the symplectic group $Sp(n)$ which is the group of all matrices with coefficients in the algebra of quaternions Q which preserve the symplectic product in Q^n. For details see Appendix A.4.1. The groups in the table above together with $Sp(n)$ are often referred to as the *classical groups* (see e.g. Chap. 1 of Chevalley [45]).[1]

The direct product $G_1 \times G_2 \times \cdots \times G_n$ of n topological groups $G_1, G_2, \ldots, G_n$ is itself a topological group with respect to the product topology, e.g. the group $U(1) \times SU(2) \times SU(3)$ plays an important role in the *standard model* of elementary particles [16].

For G a topological group and H a closed subgroup, we may form the *homogeneous space* of left cosets $G/H := \{gH; g \in G\}$. It is itself a topological space, where the *quotient topology* is induced from that in G by the requirement

[1] There is one additional classical group, the complex symplectic group $Sp(n, \mathbb{C})$, which we will not need to discuss in this volume.

that for each $A \subseteq G$, $O_A := \{gH; g \in A\}$ is open in G/H if and only if $\bigcup_{g \in A} \bigcup_{h \in H} \{gh\}$ is open in G. Important examples are the *n-spheres* $S^n \subseteq \mathbb{R}^{n+1}$ where $S^n := \{x \in \mathbb{R}^{n+1}; |x| = 1\}$. We may realise the n-spheres in a number of different ways as quotients of classical groups. In fact for $n \geq 2$,

$$S^{n-1} = O(n)/O(n-1) = SO(n)/SO(n-1),$$

$$S^{2n-1} = U(n)/U(n-1) = SU(n)/SU(n-1),$$

and

$$S^{4n-1} = Sp(n)/Sp(n-1).$$

The homogeneous space G/H is itself a topological group if and only if H is a normal subgroup of G. For example S^{n-1} is a group if and only if $n = 1, 2$ and 4. Note that in these cases S^{n-1} may be realised as the group formed by all unit vectors in the following normed division algebras: the real numbers ($n = 1$), complex numbers ($n = 2$) and quaternions ($n = 4$).[2]

If G is a topological group and $g \in G$ we define *left translation*[3] l_g, *right translation* r_g and *conjugation* c_g by

$$l_g(h) = g^{-1}h,\ r_g(h) = hg \text{ and } c_g(h) = ghg^{-1},$$

for all $h \in G$. These are all homeomorphisms of G with

$$r_g^{-1} = r_{g^{-1}},\ l_g^{-1} = l_{g^{-1}} \text{ and } c_g^{-1} = c_{g^{-1}},$$

and c_g is also an automorphism of G. Observe that for $g_1, g_2 \in G$,

$$l_{g_1 g_2} = l_{g_2} l_{g_1},\ r_{g_1 g_2} = r_{g_2} r_{g_1} \text{ and } c_{g_1 g_2} = c_{g_1} c_{g_2}.$$

We also define *inversion* in G by $I(g) = g^{-1}$ for each $g \in G$. The mapping I is also a homeomorphism of G and an anti-automorphism, in that it is bijective and $I(g_1 g_2) = I(g_2)I(g_1)$ for each $g_1, g_2 \in G$.

From now on we will always assume that G is both *locally compact* and *Hausdorff*.

We consider the function space $C_0(G)$ of all continuous complex-valued functions on G which vanish at infinity, i.e. given any $\epsilon > 0$ there exists a compact set K in G so that $|f(x)| < \epsilon$ whenever $x \in K^c$. It is a Banach space with respect

---

[2] It is interesting that these are the only real finite-dimensional division algebras. There is one further nonassociative real finite-dimensional division algebra and this is the octonions. The collection of unit vectors therein may be identified with the sphere S^7. Another fascinating fact is that the only spheres that have the geometric property of being *parallisable* are S^0, S^1, S^3 and S^7.

[3] The use of the inverse in left but not right translation makes life easier later on when we want the homomorphism property to hold for the pullback to various function spaces.

to the norm $\|f\|_\infty := \sup_{g\in G}|f(g)|$ for $f \in C_0(G)$. The linear space $C_c(G)$ of continuous complex-valued functions of compact support is dense in $C_0(G)$ (see Proposition A.1.1 in Appendix A.1 for the proof). For each $g \in G, f \in C_0(G)$, define $L_g f = f \circ l_g$, $R_g f = f \circ r_g$, $C(g)f = f \circ c(g)$ and $\iota(f) = f \circ I$. It is easy to see that each of these mappings is an isometric isomorphism of $C_0(G)$.

We say that a function $f : G \to \mathbb{C}$ is *left uniformly continuous* (respectively *right uniformly continuous*) if given any $\epsilon > 0$ there exists an open neighbourhood U of e such that $\|L_g f - f\|_\infty < \epsilon$ whenever $g \in U$ (respectively $\|R_g f - f\|_\infty < \epsilon$ whenever $g \in U$). Clearly any left or right uniformly continuous function is continuous. All complex-valued functions in $C_c(G)$ are both left and right uniformly continuous. As the proof is a little technical, we present it in Appendix A.2.

1.1.2 Covering Groups

This is a very brief account of some elegant mathematics, which is largely based on Chapter II of Chevalley [45] and Kosniowski [124] . Readers should be aware that these concepts, although very important, will not be used in the main part of the book other than incidentally. Heuristically the main idea that you need to pick up is that for a suitably nice connected group G, we can always find a group $\tilde{G}$ that is even nicer (in fact, simply connected) and which is locally indistinguishable from G itself.

Let X be a connected topological space. It is *path-connected* if for every $p, q \in X$ there exists a continuous mapping $h_{p,q} : [0, 1] \to X$ such that $h_{p,q}(0) = p$ and $h_{p,q}(1) = q$. A (not necessarily connected) topological space is *locally path-connected* if each point has a neighbourhood that itself contains a path-connected neighbourhood. If X is both connected and locally path-connected, then it is path-connected.

Let X and X_1 be path connected topological spaces. We say that (X_1, f) is a *covering space* of X if there exists a continuous surjective mapping $f : X_1 \to X$ such that each point $p \in X$ has a neighbourhood U_p so that $f^{-1}(U_p) = \bigcup_{\alpha\in I} V_\alpha$ for some collection $\{V_\alpha, \alpha \in I\}$ of mutually disjoint open subsets of X_1 and the restriction of f to V_α is a homeomorphism onto U_p for all $\alpha \in I$. We call f a *covering map*. Distinct covering spaces (X_1, f_1) and (X_2, f_2) are said to be isomorphic if there exists a homeomorphism $\natural : X_1 \to X_2$ so that $f_1 = f_2 \circ \natural$. The space X is said to be *simply connected* if for every point p in X, any closed loop based at p is continuously deformable to p. It is an important theorem in topology that a connected, locally path-connected space X has a simply connected covering space $(\tilde{X}, f)$ (which is called a *universal cover* of X) if and only if it is semilocally simply connected[4] (see e.g. Theorem 22.1 in Kosniowski [124], pp. 171–173). This universal cover is unique up to isomorphism. The *fundamental group* $\pi_1(X)$ of such a space X is the group comprising all homeomorphisms ϕ of $\tilde{X}$ with itself such that $f \circ \phi = f$. We say that

[4] This means that every point p in X has a neighbourhood U so that any closed loop in U that is based at p may be continuously deformed to p. Any connected finite-dimensional manifold has this property.

$\tilde{X}$ is an *n-covering* if $\#\pi_1(X) = n$. Note that $\pi_1(X)$ is not abelian in general, but the higher homotopy groups (which we will not require here) all are.

Let G be a connected locally path-connected group. We say that it has a *covering group* G_1 if (G_1, f) is a covering space for which G_1 is also a group, and the covering map f is a homomorphism. We say that $(\tilde{G}, f)$ is a *universal covering group* for G if it is a universal cover wherein $\tilde{G}$ is a covering group. If such a group exists (and it may not in general), then it is unique up to isomorphism.[5] If G is a connected Lie group (to be defined in Sect. 1.3), then it has a universal covering group $(\tilde{G}, f)$ wherein $\tilde{G}$ is also a Lie group. We then have that $\pi_1(G)$ is isomorphic to $\text{Ker}(f)$. Indeed, $\pi_1(G)$ is a discrete, abelian, normal subgroup of $Z(\tilde{G})$ and $G \cong \tilde{G}/\pi_1(G)$.

We consider some examples. $\mathbb{R}^n$, $SU(n)$ and $Sp(n)$ are simply connected for all $n \in \mathbb{N}$. The group $O(n)$ is not connected. It has two connected components corresponding to those matrices having determinant 1 (which is the closed, normal subgroup $SO(n)$) and -1. The covering group of Π^n is $\mathbb{R}^n$ and $\pi_1(\Pi^n) = \mathbb{Z}^n$. It is easy to see that $SO(2) \cong \Pi^1$. For $n \geq 3$, $SO(n)$ has a double cover which is called $\text{Spin}(n)$. In fact $\pi_1(SO(n)) = \mathbb{Z}_2$. In particular $\text{Spin}(3) \cong SU(2) \cong Sp(1)$. For more about the spin groups, see Appendix A.4.

1.2 Haar Measure

Let $\mathcal{B}(G)$ be the Borel σ-algebra of G, i.e. the smallest σ-algebra that contains all the open sets of G. We say that a (non-trivial) regular Borel measure[6] m_L on G is a *left Haar measure* if $m_L(gA) = m_L(A)$ for all $A \in \mathcal{B}(G), g \in G$, i.e. it is *left invariant*. Similarly we say that m_R is a *right Haar measure* if it is *right invariant* in that $m_R(Ag) = m_R(A)$ for all $A \in \mathcal{B}(G), g \in G$. A key theorem is the following:

Theorem 1.2.1 *Let G be locally compact and Hausdorff. Left Haar measure and right Haar measure exist and are unique up to a positive multiplicative constant. Furthermore, by regularity, each such measure assigns finite mass to compact sets in G.*

We will not give a proof of the theorem here, but in Sect. 1.4, we will sketch an argument using differential forms which establishes the result when G is a Lie group. The standard approach to proving the existence of (say) left Haar measure in general locally compact Hausdorff groups G is to construct a linear functional I on $C_c(G)$ such that $I(L_g f) = I(f)$ for all $f \in C_c(G), g \in G$ and then use the Riesz representation theorem (see Appendix A.5) to obtain the required measure; see e.g. Hewitt and Ross [91], pp.184–195 for details.[7] We remark that if G is also second

---

[5] In this case $\natural$ must also be a group homomorphism.

[6] See Appendix A.5 for background on regular Borel measures.

[7] An alternative approach (which is more in the spirit of Haar's pioneering 1933 paper) in which the measure is constructed directly, may be found in e.g. Cohn [50], Theorem 9.2.1, pp. 305–309 or Folland [68], Sect. 2.2.

countable, then both left and right Haar measures are σ-finite (see [50], Propositions A.2.3 and A.2.5, pp. 206–208). Haar measure may not exist if the group G fails to be locally compact (e.g. see Kuo [130] for the case where G is the additive group of an infinite-dimensional Hilbert space).

We may consider integration in the usual Lebesgue sense with respect to left and right Haar measures to form the integrals $\int_G f(x)m_L(dx)$ and $\int_G f(x)m_R(dx)$, whenever these exist. In particular we may form the Banach spaces

$L^p(G, \mathcal{B}(G), m_L; \mathbb{C})$ and $L^p(G, \mathcal{B}(G), m_R; \mathbb{C})$ for $1 \leq p \leq \infty$. We write these succinctly as $L^p(G, m_L)$ and $L^p(G, m_R)$ (respectively). So for $1 \leq p < \infty$, $L^p(G, m_L)$ comprises equivalence classes of functions that agree almost everywhere with respect to m_L and which satisfy

$$\|f\|_p = \left(\int_G |f(x)|^p m_L(dx) \right)^{\frac{1}{p}} < \infty.$$

Note that $L^2(G, m_L)$ is a Hilbert space with respect to the inner product $\langle f_1, f_2 \rangle = \int_G f_1(x)\overline{f_2}(x)m_L(dx)$. Finally $L^\infty(G, m_L)$ is the space of equivalence classes of functions that agree almost everywhere with respect to m_L and which satisfy[8]

$$\|f\|_\infty := \inf\{K > 0; |f(x)| \leq K \text{ a.e.}\} < \infty.$$

If G is second countable, then $L^p(G, m_L)$ and $L^p(G, m_R)$ are separable for $1 \leq p < \infty$.

Since the space $C_c(G)$ is dense in each of $L^p(G, m_L)$ and $L^p(G, m_R)$ for $1 \leq p < \infty$ (for the proof of a more general statement, see Proposition A.5.1 in Appendix A.5) we can, for each $g \in G$, restrict the action of L_g from $C_0(G)$ to $C_c(G)$ and then extend it to an action on $L^p(G, m_L)$. In fact it is easily verified that L_g is an isometric isomorphism, and is unitary when $p = 2$. Similarly R_g extends to an isometric isomorphism (unitary when $p = 2$) of $L^p(G, m_R)$.

The following continuity property will be useful for us (see also Folland [68], Proposition 2.41, p. 53). For $A, B \subseteq G$, we use the notation

$$AB := \{g \in G; g = ab, a \in A, b \in B\}.$$

Proposition 1.2.1 *For each $1 \leq p < \infty$ the mapping $g \to L_g f$ (respectively, $g \to R_g f$) is continuous from G to $L^p(G, m_L)$ (respectively G to $L^p(G, m_R)$).*

Proof Fix $1 \leq p < \infty$. By the isometry property, for a given $f \in L^p(G, m_L)$, it is sufficient to prove that given any $\epsilon > 0$, we can find a neighbourhood U of e (having compact closure) so that $g \in U \Rightarrow \|L_g f - f\|_p < \epsilon$. Now by density of $C_c(G)$ in

---

[8] We use the same notation for norms in $C_0(G)$ and $L^\infty(G, m_L)$, but it should be clear from the context which space we are in.

$L^p(G, m_L)$ we can find a function $\phi \in C_c(G)$ such that $\|f - \phi\|_p < \frac{\epsilon}{3}$. Using the triangle inequality and the fact that L_g is an isometry we have

$$\begin{aligned}\|L_g f - f\|_p &\leq \|L_g f - L_g\phi\|_p + \|L_g\phi - \phi\|_p + \|\phi - f\|_p \\ &< \|L_g\phi - \phi\|_p + \frac{2\epsilon}{3}.\end{aligned}$$

Since ϕ is left uniformly continuous (see Theorem A.2.1 in Appendix A.2), and choosing a smaller U if necessary, we find that $g \in U$ implies that

$$\|L_g\phi - \phi\|_\infty < \frac{\epsilon}{3m_L(U\text{supp}(\phi))^{\frac{1}{p}}},$$

and so $\|L_g\phi - \phi\|_p < \frac{\epsilon}{3}$, and the result follows. The result for right translation is proved similarly, or it may be deduced from the above by using the fact that $\iota \circ L_g = R_g \circ \iota$. □

1.2.1 The Modular Function

Let m_L be a given left Haar measure on G. Fix $\rho \in G$ and define

$$m_L^\rho(A) := m_L(A\rho)$$

for each $A \in \mathcal{B}(G)$. Then m_L^ρ is another left Haar measure on G. Hence by Theorem 1.2.1 there exists $\Delta(\rho) > 0$ such that

$$m_L(A\rho) = \Delta(\rho)m_L(A)$$

for all $A \in \mathcal{B}(G)$. We call the mapping $\rho \to \Delta(\rho)$ from G to $(0, \infty)$ the *modular function* of G.

Theorem 1.2.2 *The modular function is a continuous homomorphism from G to the multiplicative group* $(0, \infty)$.

Proof First we establish the homomorphism property. Choose a non-trivial $f \in C_c(G)$ with $f \geq 0$ and observe that for all $\rho_1, \rho_2 \in G$,

$$\int_G f(x\rho_2^{-1}\rho_1^{-1})m_L(dx) = \Delta(\rho_1\rho_2)\int_G f(x)m_L(dx).$$

But we also have

$$\begin{aligned}\int_G f(x\rho_2^{-1}\rho_1^{-1})m_L(dx) &= \Delta(\rho_2)\int_G f(x\rho_1^{-1})m_L(dx)\\ &= \Delta(\rho_2)\Delta(\rho_1)\int_G f(x)m_L(dx)\\ &= \Delta(\rho_1)\Delta(\rho_2)\int_G f(x)m_L(dx),\end{aligned}$$

and the result follows. For the continuity proof, choose f as above to have the additional property $\int_G f(x)m_L(dx) = 1$. Then $\Delta(\rho) = \int_G f(\rho^{-1}x)m_L(dx)$ for all $\rho \in G$ and we easily see that $|\Delta(\rho_1) - \Delta(\rho_2)| \leq \|L_{\rho_1}f - L_{\rho_2}f\|_1$. The result then follows by Proposition 1.2.1. □

We say that a group is *unimodular* if $\Delta \equiv 1$.

Corolarry 1.2.1 *If G is compact, then it is unimodular.*

Proof Since Δ is continuous and G is compact, $\text{Im}(G)$ is compact. But Δ is a homomorphism and so $\text{Im}(G)$ is a group. But $\{1\}$ is the only compact subgroup of $(0, \infty)$. □

Now let G be an arbitrary locally compact Hausdorff group and fix a left Haar measure m_L on G.

Theorem 1.2.3 *1. The prescription $\tilde{m}_R(A) := \int_A \Delta(x^{-1})m_L(dx)$ for $A \in \mathcal{B}(G)$ defines a right Haar measure $\tilde{m}_R$ on G.*
2. For all $f \in C_c(G)$ we have

$$\int_G f(x^{-1})m_L(dx) = \int_G f(x)\Delta(x^{-1})m_L(dx). \tag{1.2.1}$$

Proof 1. Let $f \in C_c(G)$. Then as Δ is a homomorphism, for all $g \in G$

$$\begin{aligned}\int_G f(xg)\tilde{m}_R(dx) &= \int_G f(xg)\Delta(x^{-1})m_L(dx)\\ &= \int_G f(x)\Delta(x^{-1})\Delta(g)m_L(dxg^{-1})\end{aligned}$$

$$= \int_G f(x)\Delta(x^{-1})\Delta(g)\Delta(g)^{-1}m_L(dx)$$

$$= \int_G f(x)\tilde{m}_R(dx).$$

By the Riesz representation theorem (see Appendix A.5) the measure $\tilde{m}_R$ is uniquely determined by its action on $C_c(G)$ and the result follows.

2. As $A \to m_L(A^{-1})$ defines a right Haar measure on G, we have

$$\int_G f(x^{-1})m_L(dx) = C\int_G f(x)\Delta(x^{-1})m_L(dx),$$

for some $C > 0$. Now replace f by the mapping $x \to f(x^{-1})\Delta(x)$ to obtain

$$\int_G f(x)\Delta(x^{-1})m_L(dx) = C\int_G f(x^{-1})m_L(dx).$$

Hence $C = 1$ and the result follows. □

Suppose that G is unimodular. Then by Theorem 1.2.3(1) there is a one-to-one correspondence between left and right Haar measures in this case. For unimodular groups we simply call any such measure a (bi-invariant) Haar measure and denote it as m. By Theorem 1.2.3(2) we have

$$\int_G f(x)m(dx) = \int_G f(x^{-1})m(dx).$$

We also write $L^p(G, m)$ as $L^p(G)$ for $1 \leq p \leq \infty$.

Note that when G is unimodular, $C(g)$ is an isometric isomorphism of $L^p(G)$ for each $g \in G$, $1 \leq p < \infty$. In particular we have seen that every compact group is unimodular, and in this case, by regularity (see Appendix A.5) $m(G) < \infty$. We define *normalised Haar measure* on the compact group G by

$$m_G(A) = \frac{m(A)}{m(G)},$$

so m_G is a probability measure which plays the role on G of the uniform distribution. Henceforth we will always write $\int_G f(x)m_G(dx) = \int_G f(x)dx$ for $f \in L^1(G)$.

Of course any abelian group is unimodular and a Haar measure on $\mathbb{R}^d$ is Lebesgue measure. Normalised Haar measure on Π^d is given by Lebesgue measure divided by the normalising factor $(2\pi)^d$. It can be shown that $GL(n, \mathbb{R})$ is unimodular and a Haar measure is $m(dx) = \dfrac{1}{\det(x)^n}\displaystyle\prod_{1\leq 1, j\leq n} dx_{ij}$ (see e.g. Faraut [63], p. 78).

1.3 Lie Groups and Lie Algebras

1.3.1 Lie Algebras

An F*-Lie algebra* $\mathcal{L}$ is a vector space over the field F (where $F = \mathbb{R}$ or $\mathbb{C}$) which is equipped with a bilinear mapping $[\cdot,\cdot] : \mathcal{L} \times \mathcal{L} \to \mathcal{L}$ such that

1. $[X, Y] = -[Y, X]$ for all $X, Y \in \mathcal{L}$;
2. (*The Jacobi Identity*) $[X, [Y, Z]] + [Y, [Z, X]] + [Z, [X, Y]] = 0$, for all $X, Y, Z \in \mathcal{L}$.

The mapping $[\cdot,\cdot]$ is called the *Lie bracket*. Note that by (1), we have $[X, X] = 0$ for all $X \in \mathcal{L}$.

In the case where $F = \mathbb{R}$, we will always refer to $\mathcal{L}$ as a *Lie algebra*, and when $F = \mathbb{C}$, we will use the terminology *complex Lie algebra*. If $\mathcal{L}$ is a Lie algebra with bracket $[\cdot,\cdot]$, its *complexification* is the usual complexification $\mathcal{L}_{\mathbb{C}}$ of $\mathcal{L}$ as a vector space,[9] which is equipped with the bracket

$$[X_1 + iY_1, X_2 + iY_2]_{\mathbb{C}} := [X_1, X_2] - [Y_1, Y_2] + i([X_1, Y_2] - [X_2, Y_1]),$$

for $X_1, X_2, Y_1, Y_2 \in \mathcal{L}$. Then $\mathcal{L}_{\mathbb{C}}$ is a complex Lie algebra.

A Lie algebra $\mathcal{L}$ is *finite dimensional* if it is a finite-dimensional vector space. It is *abelian* if $[X, Y] = 0$ for all $X, Y \in \mathcal{L}$. For example $\mathbb{R}^d$ becomes an abelian Lie algebra when all Lie brackets are defined to be zero. When $d = 3$ we can give $\mathbb{R}^3$ a different Lie algebra structure by taking the Lie bracket to be the cross product of the vectors. $M_n(\mathbb{C})$ (which we treat as a real $2n^2$ -dimensional vector space) becomes a non-abelian Lie algebra when we define $[X, Y] := XY - YX$. So in this case $[\cdot,\cdot]$ is the usual *commutator* of two matrices. The *direct sum* of two Lie algebras $\mathcal{L}_1$ and $\mathcal{L}_2$ having Lie brackets $[\cdot,\cdot]_1$ and $[\cdot,\cdot]_2$ (respectively) is the vector space direct sum $\mathcal{L}_1 \oplus \mathcal{L}_2$ equipped with the Lie bracket $[(X_1, Y_1), (X_2, Y_2)] := ([X_1, X_2]_1, [Y_1, Y_2]_2)$ for $X_i \in \mathcal{L}_1, Y_i \in \mathcal{L}_2$ $(i = 1, 2)$. We construct the direct sum of finitely many Lie algebras by the usual recursive approach.

A linear subspace $\mathcal{M}$ of $\mathcal{L}$ is a *Lie subalgebra* if $[X, Y] \in \mathcal{M}$ whenever $X, Y \in \mathcal{M}$, and it is an *ideal* if $[X, Y] \in \mathcal{M}$ whenever $X \in \mathcal{L}, Y \in \mathcal{M}$. A non-abelian Lie algebra is said to be *simple* if its only ideals are $\{0\}$ and $\mathcal{L}$, and *semisimple* if it can be written as a direct sum of finitely many simple ideals.

The following important examples are all Lie subalgebras of $M_n(\mathbb{C})$:

$\mathbf{sl(n}, \mathbb{C})$ - the space of all $n \times n$ matrices X for which $\text{tr}(X) = 0$.

$\mathbf{sl(n}, \mathbb{R}) = \mathbf{sl(n}, \mathbb{C}) \cap M_n(\mathbb{R})$.

$\mathbf{u(n)}$ - the space of all $n \times n$ skew-hermitian matrices, i.e. $X \in \mathbf{u(n)}$ if and only if $X + X^* = 0$.

$\mathbf{su(n)} = \mathbf{u(n)} \cap \mathbf{sl(n}, \mathbb{C})$.

[9] See the section "Guide to Notation and a Few Useful Facts", if you need background on the complexification of a real vector space.

$\mathbf{so(n)} = \mathbf{su(n)} \cap M_n(\mathbb{R})$.

$\mathbf{sp(n)} = \{X \in \mathbf{u(2n)}; JX + X^T J = 0\}$ where $J = \begin{pmatrix} 0 & I_n \\ -I_n & 0 \end{pmatrix}$.

There is a famous classification due to Wilhelm Killing (1847–1923) and Elie Cartan (1869–1951) of simple **complex** Lie algebras. The real case is more complicated but some examples include $\mathbf{sl(n, \mathbb{R})}$, $\mathbf{u(n)}$, $\mathbf{su(n)}$, $\mathbf{so(n)}$ and $\mathbf{sp(n)}$.

A *Lie algebra homomorphism* f is a linear mapping from a Lie algebra $\mathcal{L}_1$ to a Lie algebra $\mathcal{L}_2$ for which $f([X, Y]_1) = [f(X), f(Y)]_2$ for all $X, Y \in \mathcal{L}_1$ where $[\cdot, \cdot]_i$ is the Lie bracket in $\mathcal{L}_i$ for $i = 1, 2$. The mapping f is a Lie algebra isomorphism if it is both a Lie algebra homomorphism and a bijection. If $\mathcal{L}_1$ and $\mathcal{L}_2$ are isomorphic Lie algebras, then we will write $\mathcal{L}_1 \simeq \mathcal{L}_2$.

A *derivation* of a Lie algebra is a linear map $\delta : \mathcal{L} \to \mathcal{L}$ for which

$$\delta([X, Y]) = [\delta(X), Y] + [X, \delta(Y)] \tag{1.3.2}$$

for all $X, Y \in \mathcal{L}$.

1.3.2 Lie Groups and Their Lie Algebras

In this section we will make frequent use of differential geometric ideas as summarised in Appendix A.3.

A *Lie group* is a group G that is also a finite-dimensional C^∞ real manifold for which the map $(g, h) \to gh^{-1}$ from $G \times G$ to G is C^∞ (i.e. infinitely differentiable).[10] It follows that every Lie group is locally compact, Hausdorff and second countable. All the examples of topological groups mentioned in Sect. 1.1 are in fact Lie groups except for Π^∞, the p-adic number fields and the infinite-dimensional topological vector spaces.[11] The direct product of finitely many Lie groups is again a Lie group. The dimension of a generic Lie group will be denoted by d. In the sequel, we will always assume that homomorphisms between two Lie groups are also C^∞.

We will now describe the relationship between Lie groups and Lie algebras. Let X be a C^∞ vector field defined on a Lie group G. We say that it is *left invariant* if $dl_{g^{-1}} X_h = X_{gh}$ for all $g, h \in G$. Right invariant vector fields are defined similarly. The linear space of all left invariant vector fields forms a Lie algebra $\mathcal{L}(G)$ under the usual vector field bracket operations. We can similarly form the Lie algebra of all right invariant vector fields on G, and the linear mapping dI is a Lie algebra

[10] In some books, both instances of C^∞ in the definition of a Lie group are replaced by the stronger condition of "real analyticity". In fact, these two seemingly different ways of defining a Lie group give rise to exactly the same objects. This is a consequence of the solution of Hilbert's Fifth Problem; see the classic work by Montomery and Zippin [151], and for more modern treatments, Chapter H in Stroppel [199] or Chapter II of Kaplansky [112].

[11] The group Π^∞ is compact, and this is a consequence of Tychonoff's theorem (see Appendix A.1). The p-adic number fields are locally compact, abelian groups (see Folland [68], pp. 34–36).

isomorphism between them. We obtain a linear isomorphism between $T_e(G)$ (the tangent space to G at e) and $\mathcal{L}(G)$ by mapping each $X \in T_e(G)$ to the left invariant vector field whose value at $g \in G$ is $dl_{g^{-1}}X$. We can use this mapping to define a Lie bracket on $T_e(G)$ by the prescription:

$$[X, Y] := dl_g[dl_{g^{-1}}(X), dl_{g^{-1}}(Y)],$$

for each $X, Y \in T_e(G), g \in G$. Then $T_e(G)$ and $\mathcal{L}(G)$ are isomorphic Lie algebras and in future we will often identify them. In particular we call $T_e(G)$ the *Lie algebra of the Lie group* G and denote it by $\mathfrak{g}$. It is helpful to think of $\mathfrak{g}$ as giving an *infinitesimal* or *local* description of the group G.

Let G and H be Lie groups with Lie algebras $\mathfrak{g}$ and $\mathfrak{h}$ (respectively). It follows from the properties of differentials (see Appendix A.3) that if $\Phi : G \to H$ is a homomorphism, then $d\Phi : \mathfrak{g} \to \mathfrak{h}$ is a Lie algebra homomorphism. Conversely, it can be shown that if G is simply connected and $f : \mathfrak{g} \to \mathfrak{h}$ is a Lie algebra homomorphism, then $f = d\Phi$ for some unique homomorphism $\Phi : G \to H$.

If $(\tilde{G}, f)$ is a covering Lie group of G and $\tilde{G}$ has Lie algebra $\tilde{\mathfrak{g}}$, then df is an isomorphism between $\tilde{\mathfrak{g}}$ and $\mathfrak{g}$. This follows from the fact that $\tilde{G}$ and G are locally indistinguishable topologically. If $G = G_1 \times G_2$ is a direct product of Lie groups G_1 and G_2 having Lie algebras $\mathfrak{g}_1$ and $\mathfrak{g}_2$, then its Lie algebra is the direct sum $\mathfrak{g}_1 \oplus \mathfrak{g}_2$. A Lie group is said to be *simple/semisimple* if its Lie algebra is.

A submanifold (H, ϕ) of G is said to be a *Lie subgroup* of G if H is a Lie group, and $\phi : H \to G$ is a homomorphism. Note that H is a closed subgroup of G only if the range of ϕ is closed. If (H, ϕ) is a Lie subgroup of G, then the Lie algebra $\mathfrak{h}$ of H is a Lie subalgebra of G. If H is also a normal subgroup, then $\mathfrak{h}$ is an ideal of $\mathfrak{g}$. Hence a simple Lie group has no non-trivial normal Lie subgroups (of positive dimension).

We give a table of some of the classical Lie groups, their Lie algebras and their dimensions (as real manifolds). Note that the dimension of the group is easily calculated from that of its Lie algebra.

Lie group	Lie algebra	Dimension
$\mathbb{R}^d$	$\mathbb{R}^d$	d
Π^d	$\mathbb{R}^d$	d
$GL(n, \mathbb{C})$	$M_n(\mathbb{C})$	$2n^2$
$GL(n, \mathbb{R})$	$M_n(\mathbb{R})$	n^2
$SL(n, \mathbb{C})$	$\mathbf{sl(n}, \mathbb{C})$	$2n^2 - 1$
$SL(n, \mathbb{R})$	$\mathbf{sl(n}, \mathbb{R})$	$n^2 - 1$
$U(n)$	$\mathbf{u(n)}$	n^2
$SU(n)$	$\mathbf{su(n)}$	$n^2 - 1$
$O(n)$	$\mathbf{so(n)}$	$\frac{1}{2}n(n-1)$
$SO(n)$	$\mathbf{so(n)}$	$\frac{1}{2}n(n-1)$
$Sp(n)$	$\mathbf{sp(n)}$	$2n^2 + n$

We give one example of how to calculate the dimension. If $X \in \mathbf{so(n)}$, then all matrix entries are real. There are n^2 of these, but $X + X^T = 0$ implies firstly that all diagonal entries are zero, and secondly that those above the main diagonal are determined by those below it. This leaves $\frac{1}{2}(n^2 - n)$ linearly independent components.

Since compactness plays a vital role in this book it is worth mentioning that the compact Lie groups in the above list are Π^d, $U(n)$, $SU(n)$, $O(n)$, $SO(n)$ and $Sp(n)$. All but $O(n)$ are connected. Indeed, as pointed out in Sect. 1.1.2, $O(n)$ has two connected components corresponding to matrices with determinant $+1$ and -1, respectively. The first of these is $SO(n)$, which contains the neutral element, and hence the Lie algebra of $O(n)$ is $T_e(O(n)) = \mathbf{so(n)}$.

Another important class of compact connected Lie groups is the spin groups, described in Appendix A.4. If G is a non-trivial compact, connected, abelian Lie group, then $G \cong \Pi^d$ for some $d \in \mathbb{N}$.

Any Lie group G is orientable. To see this choose a non-zero element ω_0 of $\Lambda^d T_e^*(G)$ and let $\omega = dl_{g^{-1}}^*(\omega_0)$ be its left translate, where * is the pullback as defined in Appendix A.3. Then ω is a non-vanishing left-invariant d-form in that $dl_h^*\omega = \omega$ for all $h \in G$ and hence G is orientable. From now on we assume that G is oriented, and that the form ω is positive (if it were not, we would just replace ω_0 with by $-\omega_0$). We can then obtain left Haar measure m_L on G as the induced regular Borel measure associated to ω, as described in Appendix A.3. In particular, it is then immediate that $m_L(G) < \infty$ when G is compact. Right Haar measure is obtained by taking right translates of ω_0.

1.3.3 The Exponential Map

Let $X \in \mathfrak{g}$. Although we usually identify tangent vectors with the left invariant vector fields they generate, we will in this section write X_L for the left invariant vector field associated to X, so that $X_L(g) := dl_{g^{-1}}X$ for each $g \in G$. Consider the differential equation in G given by

$$\frac{d\psi(t)}{dt} = X_L(\psi(t))$$

with initial condition $\psi(0) = e$. In local co-ordinates, if we write

$$X_L(g) = \sum_{i=1}^{d} a_i(g)\frac{\partial}{\partial x_i},$$

for $g = (x_1, \ldots, x_d)$, and for each $t \in \mathbb{R}$, $\psi(t) = (\psi_1(t), \ldots, \psi_d(t))$, then we have

$$\frac{d\psi_i(t)}{dt} = a_i(\psi(t))$$

for $1 \leq i \leq d$. This system of differential equations has a unique solution, and we write it as $\psi(t) = \exp(tX)$. Indeed, if G is a matrix group, then

$$\exp(tX) = e^{tX} = \sum_{n=0}^{\infty} \frac{t^n X^n}{n!}$$

and the series converges (uniformly on finite intervals) in any matrix norm.

Let us return to general Lie groups. The mapping $t \to \exp(tX)$ is a continuous homomorphism from $\mathbb{R}$ into G. This means that $\{\exp(tX), t \in \mathbb{R}\}$ forms a *one-parameter subgroup* of G. Indeed for each $s, t \in \mathbb{R}$,

$$\exp((s+t)X) = \exp(sX)\exp(tX) = \exp(tX)\exp(sX),$$

and

$$\exp(tX)^{-1} = \exp(-tX).$$

The map from $\mathfrak{g} \to G$ given by $X \to \exp(X)$ is called the *exponential map*. There exists a neighbourhood V of the origin in $\mathfrak{g}$ which is mapped diffeomorphically by exp to a neighbourhood N of e in G. Furthermore, if we fix a basis $\{X_1, \ldots, X_d\}$ of $\mathfrak{g}$, then there exist smooth mappings $x_i : N \to \mathbb{R}$, $(1 \leq i \leq d)$ called *canonical co-ordinates* (with respect to the given basis) such that for each $1 \leq i \leq d$,

$$x_i\left(\exp\left(\sum_{j=1}^{d} a_j X_j\right)\right) = a_i,$$

whenever $\sum_{j=1}^{d} a_j X_j \in V$.

If G is both compact and connected, then $\exp : \mathfrak{g} \to G$ is onto.

Finally the following can be quite useful. If $f \in C^{\infty}(G)$ then for all $g \in G$,

$$X_L f(g) = \left.\frac{d}{dt} f(g\exp(tX))\right|_{t=0} \text{ and } X_R f(g) = \left.\frac{d}{dt} f(\exp(tX)g)\right|_{t=0}. \tag{1.3.3}$$

1.3.4 Ad, ad and the Killing Form

Let $g \in G$ and recall the Lie group automorphism $c_g(h) = ghg^{-1}$ for each $h \in G$. Now define a Lie algebra automorphism $\mathrm{Ad}(g) : \mathfrak{g} \to \mathfrak{g}$ by $\mathrm{Ad}(g) = dc_g$. Ad is called the *adjoint representation* of G on $\mathfrak{g}$.[12] Note that (by properties of differentials, see Appendix A.3)

[12] The reason for the word "representation" here will become clear in Chap. 2.

$$\mathrm{Ad}(g_1g_2) = \mathrm{Ad}(g_1) \circ \mathrm{Ad}(g_2)$$

for all $g_1, g_2 \in G$. For example, if G is a matrix group, then $\mathrm{Ad}(g)\mathrm{X} = g\mathrm{X}g^{-1}$ for all $g \in G, X \in \mathfrak{g}$. It will be helpful later on to know that when G is compact, we can always define an Ad-invariant inner product on $\mathfrak{g}$, i.e. an inner product $\langle\langle\cdot,\cdot\rangle\rangle$ for which $\mathrm{Ad}(g)$ acts isometrically on $\mathfrak{g}$ for each $g \in G$. To see that this is the case, choose any inner product $\langle\cdot,\cdot\rangle$ on $\mathfrak{g}$ and define

$$\langle\langle X, Y\rangle\rangle := \int_G \langle \mathrm{Ad}(g)\mathrm{X}, \mathrm{Ad}(g)\mathrm{Y}\rangle dg,$$

for each $X, Y \in \mathfrak{g}$.

If G is a connected Lie group, then (as is shown in e.g. Faraut [63], Proposition 5.5.4, p. 88) its modular function is given by $\Delta(g) = \det(\mathrm{Ad}(g))$ for each $g \in G$. In particular if G is also semisimple, then it is unimodular (see e.g. Faraut [63], Corollary 5.5.5, pp. 88–89).

Now fix $X \in \mathfrak{g}$ and consider the linear mapping $\mathrm{ad(X)} : \mathfrak{g} \to \mathfrak{g}$ given by $\mathrm{ad(X)(Y)} = [\mathrm{X}, \mathrm{Y}]$ for all $Y \in \mathfrak{g}$. Of course (subject to a choice of basis in $\mathfrak{g}$) we can always regard $\mathrm{ad}(X)$ as a $d \times d$ matrix. A straightforward application of the Jacobi identity shows that $\mathrm{ad}(X)$ is a derivation of $\mathfrak{g}$ (recall (1.3.2)). We now state a beautiful relationship between Ad, ad and exp:

$$\mathrm{Ad(exp(X))} = \mathrm{e}^{\mathrm{ad(X)}}, \tag{1.3.4}$$

for all $X \in \mathfrak{g}$, where the right hand side is understood in the sense of exponentials of matrices (see e.g. Helgason [88], p. 128 for a proof). We can gain insight into this result by considering the case $G = U(n)$. Then for $X, Y \in \mathbf{u(n)}$ we have $\mathrm{Ad(exp(X))Y} = \mathrm{exp(X)Y\,exp(-X)}$ and if you formally expand the exponentials on the right hand side and collect together terms you will find the beginning of the series expansion for $e^{\mathrm{ad(X)}}Y$.

Another interesting relationship between Ad and exp is

$$g\exp(X)g^{-1} = \exp(\mathrm{Ad}(g)\mathrm{X}), \tag{1.3.5}$$

for all $g \in G, X \in \mathfrak{g}$ (see e.g. Helgason [88], p. 127).

The *Killing form* B is the bilinear symmetric mapping B from $\mathfrak{g} \times \mathfrak{g}$ to $\mathbb{R}$ defined for all $X, Y \in \mathfrak{g}$ by

$$B(X, Y) = \mathrm{tr(ad(X)ad(Y))} \tag{1.3.6}$$

where the right hand side is the trace of the product of the two $d \times d$ matrices.

Lemma 1.3.1 *For all $X, Y \in \mathfrak{g}, g \in G$,*

$$\mathrm{B(Ad}(g)\mathrm{X}, \mathrm{Ad}(g)\mathrm{Y}) = \mathrm{B(X, Y)}.$$

Proof First observe that since $\mathrm{Ad}(g)$ is a Lie algebra homomorphism we have the identity

$$\mathrm{ad}(\mathrm{Ad}(g)\mathrm{X}) = \mathrm{Ad}(g) \circ \mathrm{ad}(\mathrm{X}) \circ \mathrm{Ad}(g^{-1}),$$

indeed

$$\begin{aligned}\mathrm{ad}(\mathrm{Ad}(g)\mathrm{X})(\mathrm{Ad}(g)\mathrm{Y}) &= [\mathrm{Ad}(g)\mathrm{X}, \mathrm{Ad}(g)\mathrm{Y}]\\ &= \mathrm{Ad}(g)[\mathrm{X}, \mathrm{Y}]\\ &= \mathrm{Ad}(g)\mathrm{ad}(\mathrm{X})(\mathrm{Y}).\end{aligned}$$

But then (treating $\mathrm{Ad}(g)$ as a $d \times d$ matrix), we have

$$\begin{aligned}\mathrm{B}(\mathrm{Ad}(g)\mathrm{X},\mathrm{Ad}(g)\mathrm{Y}) &= \mathrm{tr}(\mathrm{Ad}(g)\mathrm{ad}(\mathrm{X})\mathrm{ad}(\mathrm{Y})\mathrm{Ad}(g^{-1}))\\ &= B(X, Y),\end{aligned}$$

using commutativity under the trace, and the fact that $\mathrm{Ad}(g^{-1}) = \mathrm{Ad}(g)^{-1}$. □

The following theorem is important for the study of semisimple Lie groups:

Theorem 1.3.1 *1. The Lie algebra* $\mathfrak{g}$ *is semisimple if and only if* B *is non-degenerate, i.e. if* $Y \in \mathfrak{g}$ *is such that* $B(X, Y) = 0$ *for all* $X \in g$*, then* $Y = 0$.
2. A semisimple Lie group is compact if and only if B *is negative definite, i.e.* $B(X, X) < 0$ *for all* $X \neq 0$.

We will not include a proof of this result here (see e.g. Fegan [66], Theorem 5.8, pp. 35–36), but the following argument gives some insight into why compactness of G is related to negative-definiteness of B. Let G be compact and recall the Ad-invariant inner product $\langle\langle\cdot, \cdot\rangle\rangle$. Then for each $g \in G$, $\mathrm{Ad}(g)$ acts orthogonally on $(\mathfrak{g}, \langle\langle\cdot, \cdot\rangle\rangle)$ and so by (1.3.4), $\mathrm{ad}(\mathrm{X})$ is a skew-symmetric matrix for each $X \in \mathfrak{g}$. If we write $\mathrm{ad}(\mathrm{X}) = (\mathrm{x_{ij}})$ relative to some basis in $\mathfrak{g}$, then

$$B(X, X) = \sum_{i,j=1}^{d} x_{ij}x_{ji} = -\sum_{i,j=1}^{d} x_{ij}^2 \leq 0.$$

From Lemma 1.3.1 and Theorem 1.3.1 we see that $-B$ is an Ad-invariant inner product on the Lie algebra of a compact, semisimple Lie group G. Furthermore we obtain a Riemannian metric ρ on G by the prescription

$$\rho(X_g, Y_g) = -B(dl_g(X_g), dl_g(Y_g)), \tag{1.3.7}$$

for each $X_g, Y_g \in T_g(G)$. In the sequel, we will use the terminology "Ad-invariant metric" on a Lie group G to mean a Riemannian metric on G that is induced by an Ad-invariant inner product on $\mathfrak{g}$, as in (1.3.7).

1.3.5 The Universal Enveloping Algebra and Laplacians

The Lie algebra $\mathfrak{g}$ of a Lie group G is not an associative algebra, and we seek the smallest such algebra into which $\mathfrak{g}$ is naturally embedded as a linear space. To be precise we define the *universal enveloping algebra* of $\mathfrak{g}$ to be $\mathcal{U}(\mathfrak{g}) = \mathcal{T}(\mathfrak{g})/\mathcal{J}(\mathfrak{g})$ where $\mathcal{T}(\mathfrak{g})$ is the *tensor algebra* over $\mathfrak{g}$ and $\mathcal{J}(\mathfrak{g})$ is the two-sided ideal of $\mathcal{T}(\mathfrak{g})$ that is generated by tensors of the form $X \otimes Y - Y \otimes X - [X, Y]$ for all $X, Y \in \mathfrak{g}$. Recall that the tensor algebra $\mathcal{T}(V)$ of a real vector space V is the direct sum $\bigoplus_{k=0}^{\infty} V^{\otimes^k}$, where $V^{\otimes^0} = \mathbb{R}$, $V^{\otimes^1} = V$ and for $k \geq 2$,

$$V^{\otimes^k} = \underbrace{V \otimes V \otimes \cdots \otimes V}_{k \text{ copies}}.$$

Elements of $\mathcal{T}(V)$ are formal sequences $(v) = (v_0, v_1, v_2, \ldots)$ where $v_k \in V^{\otimes^k}$, $k \in \mathbb{Z}_+$. Finally, given two sequences (v) and (w) in $\mathcal{T}(V)$ their product is the sequence whose kth term is $\sum_{i=0}^{k} v_i \otimes w_{k-i}$.

From an analytic point of view, a very useful characterisation of $\mathcal{U}(\mathfrak{g})$ is given by the celebrated *Poincaré-Birkhoff-Witt theorem*:

Theorem 1.3.2 (Poincaré-Birkhoff-Witt) *Let $\{X_1, \ldots, X_d\}$ be a basis for $\mathfrak{g}$. Then a basis for $\mathcal{U}(\mathfrak{g})$ is given by the set of all monomials of the form $X_{i_1}^{j_1} \cdots X_{i_r}^{j_r}$ where $i_1 < \cdots < i_r$, $\{i_1, \ldots, i_r\} \subseteq \{1, \ldots, d\}$ and $j_1, \ldots, j_r \in \mathbb{N}$.*

For a proof, see e.g. Knapp [120], pp. 217–221.

We can use Theorem 1.3.2 to extend an automorphism ϕ of $\mathfrak{g}$ to an automorphism of $\mathcal{U}(\mathfrak{g})$ via its action on the basic monomials, i.e.

$$\phi(X_{i_1}^{j_1} \cdots X_{i_r}^{j_r}) = \phi(X_{i_1})^{j_1} \cdots \phi(X_{i_r})^{j_r}.$$

A typical example that we will need in the sequel is $\phi = \mathrm{Ad}(g)$ for some $g \in G$.

We typically will want to consider elements of $\mathcal{U}(\mathfrak{g})$ as linear operators defined on $C^\infty(G)$ where the basis vectors $X_1, \ldots, X_d$ are realised as left invariant differential operators. For example, let us equip G with a Riemannian metric ρ. Indeed, we can always construct such a metric by choosing an inner product on $\mathfrak{g}$, and then transporting to an arbitrary tangent space using left translation, as at the end of Sect. 1.3.4. Define $\rho_{ij} := \rho(X_i, X_j)$ for each $1 \leq i, j \leq d$ where we use the basis for $\mathfrak{g}$ from Theorem 1.3.2. The matrix $R = (\rho_{ij})$ is positive definite and hence non-singular with inverse $R^{-1} = (\rho_{ij}^{-1})$. We define the *Laplace-Beltrami operator* or *Laplacian* on (G, ρ) to be the element $\Delta \in \mathcal{U}(\mathfrak{g})$ given by

$$\Delta = \sum_{i,j=1}^{d} \rho_{ij}^{-1} X_i X_j. \tag{1.3.8}$$

Sometimes we will consider $X = (X_1, \ldots, X_d)^T$ as a column vector in the Euclidean space $\mathfrak{g}^d$ and then we have the condensed notation:

$$\Delta = X^T R^{-1} X. \tag{1.3.9}$$

It is convenient to define the *dual basis* $\{\tilde{X^1}, \ldots, \tilde{X^d}\}$ by $\tilde{X^i} = \sum_{j=1}^d \rho_{ij}^{-1} X_j$ for $1 \leq i \leq d$, so that $\rho(\tilde{X}_i, X_j) = \delta_{ij}$ for all $1 \leq i, j \leq d$. Then we may rewrite (1.3.8) as

$$\Delta = \sum_{i=1}^d X_i \tilde{X^i}. \tag{1.3.10}$$

Note that if $\{X_1, \ldots, X_d\}$ is an orthonormal basis for $\mathfrak{g}$ (with respect to ρ) so that $\rho(X_i, X_j) = \delta_{ij}$ for all $1 \leq i, j \leq d$, then

$$\Delta = \sum_{i=1}^d X_i^2.$$

Proposition 1.3.1 *The Laplacian Δ is independent of the choice of basis in $\mathfrak{g}$.*

Proof Let $\{Y_1, \ldots, Y_d\}$ be another basis for $\mathfrak{g}$ and write $Y = (Y_1, \ldots, Y_d)^T \in \mathfrak{g}^d$. Then there exists a non-singular matrix $A = (a_{ij}) \in M_d(\mathbb{R})$ such that $Y_i = \sum_{j=1}^d a_{ij} X_j$ for all $1 \leq i \leq d$. We find it convenient to write this as $Y = AX$. Let $S = (s_{ij})$ be defined by $s_{ij} = \rho(Y_i, Y_j)$ for $1 \leq i, j \leq d$. Then

$$\begin{aligned} s_{ij} &= \sum_{k,l=1}^d \rho(a_{ik} X_k, a_{jl} X_l) \\ &= \sum_{k,l=1}^d a_{ik} \rho_{kl} a_{jl}, \end{aligned}$$

and so $S = ARA^T$. If we write Δ' for the Laplacian written with respect to the basis $\{Y_1, \ldots, Y_d\}$, then by (1.3.9)

$$\begin{aligned} \Delta' &= Y^T S^{-1} Y \\ &= X^T A^T.(A^T)^{-1} R^{-1} A^{-1}.AX \\ &= X^T R^{-1} X \\ &= \Delta, \end{aligned}$$

as was required. □

Let $Z(\mathfrak{g})$ be the *centre* of $\mathcal{U}(\mathfrak{g})$, i.e. $Z(\mathfrak{g}) := \{X \in \mathcal{U}(\mathfrak{g}); XY = YX \text{ for all } Y \in \mathcal{U}(\mathfrak{g})\}$. It follows from Theorem 1.3.2 that if $XY = YX$ for all $Y \in \mathfrak{g}$ then $X \in Z(\mathfrak{g})$.

Theorem 1.3.3 *If G is a compact Lie group equipped with an Ad-invariant metric ρ, then*

1. $\mathrm{Ad}(g)\Delta = \Delta$ *for all* $g \in G$,
2. $\Delta \in Z(\mathfrak{g})$.

Proof 1. Fix $g \in G$. Since $\mathrm{Ad}(g)$ is a Lie algebra automorphism, we may construct a new basis $\{Y_1, \ldots, Y_d\}$ for $\mathfrak{g}$ with $Y_i = \mathrm{Ad}(g)\mathrm{X_i}$ for $1 \leq i \leq d$. But then

$$\begin{aligned}\mathrm{Ad}(g)\Delta &= \sum_{i,j=1}^{d} \rho^{-1}(X_i, X_j)(\mathrm{Ad}(g)\mathrm{X_i})(\mathrm{Ad}(g)\mathrm{X_j})\\ &= \sum_{i,j=1}^{d} \rho^{-1}(\mathrm{Ad}(g)\mathrm{X_i},\mathrm{Ad}(g)\mathrm{X_j})(\mathrm{Ad}(g)\mathrm{X_i})(\mathrm{Ad}(g)\mathrm{X_j})\\ &= \sum_{i,j=1}^{d} \rho^{-1}(Y_i, Y_j)Y_iY_j\\ &= \Delta,\end{aligned}$$

by Proposition 1.3.1.

2. For all $t \in \mathbb{R}$, $X \in \mathfrak{g}$ we have by (1) and (1.3.4) that

$$\Delta = \mathrm{Ad}(\exp(tX))\Delta = e^{t\mathrm{ad}(X)}\Delta.$$

Now differentiate both sides of this equation with respect to t and then put $t = 0$ to find that $\mathrm{ad}(X)\Delta = 0$, and the result follows. □

We say that $T \in \mathcal{U}(\mathfrak{g})$ is *bi-invariant* if (as linear operators on $C^\infty(G)$):

$$L_gT = TL_g \text{ and } R_gT = TR_g$$

for all $g \in G$.

Theorem 1.3.4 *If G is a compact Lie group equipped with an Ad-invariant metric ρ, then the Laplacian Δ is bi-invariant.*

Proof As each element of $\mathfrak{g}$ is left-invariant, it commutes with L_g for all $g \in G$. It follows that Δ also has this property. For $g \in G$, consider the action of dr_g on $\mathfrak{g}$. Since ρ is Ad-invariant, it is also dr_g invariant since for each $X, Y \in \mathfrak{g}$,

$$\begin{aligned}\rho(X, Y) &= \rho(\mathrm{Ad}(g^{-1})\mathrm{X}, \mathrm{Ad}(g^{-1})\mathrm{Y})\\ &= \rho(dr_g \circ dl_g(X), dr_g \circ dl_g(Y))\\ &= \rho(dr_g(X), dr_g(Y)).\end{aligned}$$

It follows from Proposition 1.3.1 that $dr_g\Delta = \Delta$, and the result follows. □

We will return to the study of the Laplacian from an analytic perspective in Sect. 3.1.

We usually consider elements of $\mathcal{U}(\mathfrak{g})$ acting on $C^\infty(G)$, but the following useful result (due to Sugiura [200], pp. 42–43) gives a valuable characterisation of lower-order differentiability.

Theorem 1.3.5 *Let G be a connected Lie group. Fix $p \in \mathbb{N}$ and let f be an arbitrary real-valued function defined on G. The following are equivalent.*

1. *The mapping $f \in C^p(G, \mathbb{R})$.*
2. *For all $X_1, X_2, \ldots, X_p \in \mathfrak{g}$, the mapping from G to $\mathbb{R}$ given by $g \to X_1X_2\cdots X_pf(g)$ is well-defined and continuous.*

1.4 Notes and Further Reading

Lie groups were created by the Norwegian mathematician Sophus Lie (1842–1899). He was partly motivated by the need to extend Galois theory from algebraic to differential equations and hence saw the need to focus on "continuous" rather than discrete groups for this endeavour. Among his great discoveries was what we now call the Lie algebra as a means of describing the group infinitesimally. Lie algebras were independently discovered by Wilhelm Karl Joseph Killing (1847–1923) whose life's work was the classification of the simple Lie algebras over the complex field. Some aspects of this programme were completed in the doctoral thesis of Elie Cartan (1869–1951). Although it does not play any role in this book, we list the classification here for the reader's benefit. The majority of these Lie algebras fall into four infinite families, which are historically denoted as $A_n (n \geq 1)$, $B_n (n \geq 2)$, $C_n (n \geq 3)$ and $D_n (n \geq 4)$. In modern terminology $A_n = \mathbf{sl(n+1}, \mathbb{C})$, $B_n = \mathbf{so(2n+1}, \mathbb{C})$, $C_n = \mathbf{sp(2n}, \mathbb{C})$ and $D_n = \mathbf{so(2n}, \mathbb{C})$. Killing also discovered five additional simple Lie algebras, the *exceptional Lie algebras* which are denoted G_2, F_4, E_6, E_7 and E_8 and which have dimensions 14, 52, 78, 133 and 248 respectively. After many years of relative neglect, the exceptional Lie algebras have begun to see some interesting mathematical development (and even speculative applications to elementary particle physics) in the early part of the twenty-first century—see e.g. Sect. 4 of Baez [17]. A scholarly account of the early history of Lie group theory is Hawkins [83]. Yaglom [218] is a shorter book on this same theme which emphasises the significance of the friendship between Lie and Felix Klein (1849–1925). Of course group theory lay at the heart of Klein's famous *Erlangen programme*.[13] A concise account of the history of Lie theory (on which this paragraph has borrowed heavily) can be found in the on-line lecture notes by Varadarajan [208]. I also recommend Chap. 10 of Stewart [194].

[13] "Geometry is the science which studies the properties of figures preserved under the transformations of a certain group of transformations, or, as one also says, the science which studies the invariants of a group of transformations," see Yaglom [218], p. 115.

If you want to give yourself a thorough immersion in the theory of topological groups and Lie groups with a view to doing research in the field, then Chevalley [45] remains an incomparable guide. It will need to be followed up by a more modern treatment, and for this I recommend Knapp [120], with Helgason [88] on the side (where in particular you will find a thorough treatment of the classification problem). Some nice introductory material on Lie groups can be found in Carter et. al. [40]. Hall [77] presents a very thorough single-volume treatment of both Lie groups and representation theory, covering most of the material (and some additional topics) found in both Chaps. 1 and 2 of this book. Varadarajan [206] is an older monograph that covers similar ground. A very concise and informative introduction to topological groups and Haar measure may be found in Chapter 2 of Folland [68], and since the theme of that book is harmonic analysis on locally compact groups, it provides a lot of valuable background for the topic of the current volume. The standard source for a purely algebraic treatment of Lie algebras (without Lie groups) is Jacobson [108].

If you want to learn Lie groups quickly with a view to rapidly assimilating all the material that you'll need to be able to navigate the current book, then I recommend Fegan [66], Simon [188] (Chap. 7 onwards) or Sepanski [183], all three of which are specifically dedicated to compact Lie groups. For structure of more general locally compact groups, see volume one of Hewitt and Ross [91] or Stroppel [199], while Hofmann and Morris [99] gives a monograph treatment of compact groups. If you are interested in applications of Lie groups to physics, then Sternberg [193] is highly recommended (see also Jones [110]) and for engineering applications, see Chirikjian and Kyatkin [48] and the two volume work by Chirikjian [46, 47].

Chapter 2
Representations, Peter-Weyl Theory and Weights

Abstract Representation theory is a deep and beautiful subject. Our goal in this chapter is to develop those concepts and results that we need for applications to Fourier analysis on compact groups and hence to probability theory. In the first part, we give a self-contained account of key aspects of the representation theory of compact groups, including proofs of Schur's lemma, the Schur orthogonality relations and the Peter-Weyl theorem. We also introduce the Fourier transform for suitable functions on the group and establish some of its elementary properties. In the second part of the chapter, we introduce weights and roots and sketch proofs of the Weyl character and Weyl dimension formulae. This part is far less rigorous and many proofs are omitted. The key point that readers need to absorb is that irreducible representations are in one-to-one correspondence with highest weights and, as we will see in later chapters, this enables us to carry out a finer analysis of functions and measures in "Fourier space". Finally we illustrate the abstract theory by finding all the irreducible representations of $SU(2)$. (We use a lot of elementary Hilbert space ideas in this chapter. Readers requiring a reminder of key concepts should consult a standard text such as Debnath and Mikusinski [55] or Reed and Simon [166]).

2.1 Unitary Group Representations: Basic Concepts, Examples and Schur's Lemma

Let G be a topological group. A *(linear) representation* of G is a strongly continuous homomorphism π from G into the group of all bounded invertible operators on some topological vector space V_π. Note that strong continuity in this context means that for each fixed $\psi \in V_\pi$ the mapping $g \to \pi(g)\psi$ is continuous from G to V_π. A representation is said to be *unitary* if V_π is a complex separable Hilbert space and $\pi(g)$ is a unitary operator in V_π for each $g \in G$. Henceforth we will be mainly interested in the unitary case and so we drop the qualifier "unitary" (and when we

D. Applebaum, *Probability on Compact Lie Groups*, Probability Theory and Stochastic Modelling 70, DOI: 10.1007/978-3-319-07842-7_2,

do consider the more general case, we will emphasise the qualifier "linear"). We summarise key properties of representations below:

- $\pi(gh) = \pi(g)\pi(h)$,
- $\pi(g^{-1}) = \pi(g)^{-1} = \pi(g)^*$,
- $\pi(e) = I_\pi$,

for all $g, h \in G$, where I_π is the identity operator acting in V_π.

Example 1 Every group G has the *trivial representation* π_0 where $V_{\pi_0} = \mathbb{C}$ and $\pi_0(g) = 1$, for all $g \in G$.

Example 2 Let Π^d be the d-torus. For each $n \in \mathbb{Z}^d$ we obtain a representation π_n of G in $V_\pi = \mathbb{C}$ by defining

$$\pi_n(x) = e^{in\cdot x}, \text{ for each } x \in \Pi^d.$$

Example 3 For each $y \in \mathbb{R}^d$ we obtain a representation π_y of $\mathbb{R}^d$ in $V_\pi = \mathbb{C}$ by defining $\pi_y(x) = e^{iy\cdot x}$, for each $x \in \mathbb{R}^d$.

Example 4 The identity representation of $U(n)$ is just the mapping that sends each $X \in U(n)$ to itself. Here we have $V_\pi = \mathbb{C}^n$.

Example 5 Let G be a Lie group and equip its Lie algebra with an Ad-invariant inner product. Then Ad is a representation of G on $V_\pi = \mathfrak{g}_\mathbb{C}$. If G is semi-simple and compact, then we may take $-B$ to be the inner product.

Example 6 Let G be locally compact and Hausdorff and choose a left Haar measure m_L on G. Then the mapping $g \rightarrow L_g$ is a representation called the *left regular representation* of G on $V_\pi = L^2(G, m_L)$. Similarly if G is equipped with a right Haar measure m_R, then $g \rightarrow R_g$ defines the *right regular representation* on $V_\pi = L^2(G, m_R)$.

Suppose that π_i is a representation of G acting on the Hilbert space V_i ($i = 1, 2$). Then we can form new representations called the *direct sum* $\pi_1 \oplus \pi_2$ and the *tensor product* $\pi_1 \otimes \pi_2$ on $V_1 \oplus V_2$ and $V_1 \otimes V_2$ (respectively). The direct sum is defined by

$$(\pi_1 \oplus \pi_2)(g)\psi = (\pi_1(g)\psi_1, \pi_2(g)\psi_2),$$

for each $g \in G$, $\psi = (\psi_1, \psi_2) \in V_1 \oplus V_2$. The tensor product is defined by continuous linear extension of the prescription:

$$(\pi_1 \otimes \pi_2)(g)\psi = \pi_1(g)\psi_1 \otimes \pi_2(g)\psi_2,$$

for each $g \in G$ and $\psi = \psi_1 \otimes \psi_2 \in V_1 \otimes V_2$. The construction of finite and countable direct sums of representations and of finite tensor products proceeds analogously. If π is a representation acting in V_π, we may also define the *conjugate* representation

$\overline{\pi}$ acting in the dual space V_π^*. Recall that there is a natural conjugate isomorphism J between V_π and V_π^* (given by the Riesz lemma), which is such that $\langle J\phi, J\psi\rangle = \langle\psi, \phi\rangle$ for all $\phi, \psi \in V_\pi$. We then define

$$\overline{\pi}(g) = J\pi(g)J^{-1} \tag{2.1.1}$$

for all $g \in G$. If we write $\overline{\phi} = J\phi$ for all $\phi \in V_\pi$ then we deduce that

$$\langle\overline{\pi}(g)\overline{\phi}, \overline{\psi}\rangle = \langle\psi, \pi(g)\phi\rangle, \tag{2.1.2}$$

for all $\phi, \psi \in V_\pi, g \in G$.

Let π be a representation of G on V_π and W_π be a closed linear subspace of V_π. We say that W_π is an *invariant subspace* for π if $\pi(g)(W_\pi) \subseteq W_\pi$ for each $g \in G$. In this case the restriction of π to W_π defines a representation of G called a *subrepresentation* of π. For example let $\pi_1 \oplus \pi_2$ be the direct sum representation that we introduced previously. Define $\widetilde{\pi_1}$ to be the representation of G on $V_1 \oplus \{0\}$ defined by $\widetilde{\pi_1}(g)\psi = (\pi_1(g)\psi_1, 0)$ for $\psi = (\psi_1, 0)$. Then $\widetilde{\pi_1}$ is a subrepresentation of $\pi_1 \oplus \pi_2$.

Recall that if W_π is a closed linear subspace of V_π, then so is its orthogonal complement $W_\pi^\perp$ and we have the direct sum decomposition $V_\pi = W_\pi \oplus W_\pi^\perp$.

Lemma 2.1.1 *If π is a representation of G on V_π and W_π is an invariant subspace for π, then so is $W_\pi^\perp$. We have the direct sum decomposition $\pi = \pi_W \oplus \pi_W^\perp$ where π_W and $\pi_W^\perp$ are the subrepresentations obtained by restricting π to W_π and $W_\pi^\perp$ (respectively).*

Proof Let $\psi \in W_\pi$ be arbitrary. Then for all $g \in G, \phi \in W_\pi^\perp$,

$$\langle\pi(g)\phi, \psi\rangle = \langle\pi(g^{-1})^*\phi, \psi\rangle = \langle\phi, \pi(g^{-1})\psi\rangle = 0,$$

and so $\pi(g)\phi \in W_\pi^\perp$. It follows that $W_\pi^\perp$ is invariant for π and the rest follows easily. □

We say that a representation π is *irreducible* if $V_\pi \neq \{0\}$ and the only closed invariant subspaces of π are V_π and $\{0\}$. The representations described in Examples 1–3 above are all irreducible. Let π_1 and π_2 be representations of G in V_1 and V_2, respectively. We say that a bounded linear operator T from V_1 to V_2 is an *intertwining operator* for π_1 and π_2 if $T\pi_1(g) = \pi_2(g)T$ for all $g \in G$. In the case where $\pi_1 = \pi_2$ and $V_1 = V_2$, we will simply say that T is *intertwining*. Two representations π_1 and π_2 are said to be *equivalent* if they can be intertwined by a unitary operator (in which case the Hilbert spaces V_1 and V_2 are isomorphic). It is not difficult to verify that equivalence of representations is an equivalence relation. We define $\widehat{G}$ to be the set of all equivalence classes of *irreducible* representations of G. We call $\widehat{G}$ the *unitary dual* of G. The complete description of $\widehat{G}$ for groups of interest is one of the main goals of representation theory. We will often be quite cavalier and fail to make an explicit distinction between the equivalence class $[\pi]$ and a typical representative element π of that class. A representation π is called *completely reducible* if (up to equivalence) it has a direct sum decomposition $\pi = \bigoplus_{n=1}^{\infty} \pi_n$ where π_n is irreducible for each $n \in \mathbb{N}$. We say that a representation π is *finite-dimensional* if $d_\pi := \dim(V_\pi) < \infty$. As we will see later in Corollary 2.2.2, for compact groups every finite-dimensional representation is either irreducible or completely reducible. But this is not true for more general locally compact groups. A representation π is said to be *faithful* if the mapping π is injective from G to $\mathcal{L}(V_\pi)$. If G is a compact Lie group, then it always has a finite-dimensional faithful representation (see e.g. Theorem 4.1 in Bröcker and tom Dieck [36] pp. 136–137 for a proof).

The next result is very important.

Theorem 2.1.1 (Schur's Lemma)

1. *Let π_1 and π_2 be irreducible finite-dimensional representations of G acting in the complex Hilbert spaces V_1 and V_2 (respectively) and T be an intertwining operator. Then either T is a linear isomorphism between V_1 and V_2 or $T = 0$.*
2. *Let π be an irreducible finite-dimensional representation of G on the complex Hilbert space V_π and let $T \in \mathcal{L}(V_\pi)$ be such that $T\pi(g) = \pi(g)T$ for all $g \in G$. Then there exists $\lambda \in \mathbb{C}$ such that $T = \lambda I_\pi$.*

Proof

1. Either $\text{Ker}(T) = \{0\}$ or $\text{Ker}(T) \neq \{0\}$. We first assume that $\text{Ker}(T) = \{0\}$ and $T \neq 0$, so that T is injective. $\text{Ran}(T)$ is a closed linear subspace of V_2, and we will show that it is invariant. If $\psi \in \text{Ran}(T)$, then $\psi = T\phi$ for some $\phi \in V_1$. But then for all $g \in G, \pi_2(g)\psi = \pi_2(g)T\phi = T\pi_1(g)\phi \in \text{Ran}(T)$. But π_2 is irreducible and $T \neq 0$ so $\text{Ran}(T) = V_2$. It follows that T is a linear isomorphism. Now assume that the closed linear subspace $\text{Ker}(T) \neq \{0\}$. $\text{Ker}(T)$ is invariant, for if $\psi \in \text{Ker}(T)$, for all $g \in G$ we have $T\pi_1(g)\psi = \pi_2(g)T\psi = 0$. But π_1 is irreducible, and so $\text{Ker}(T) = V_1$, i.e. $T = 0$.
2. T has at least one eigenvalue $\lambda \in \mathbb{C}$ (with corresponding eigenvector ϕ), and it follows from the assumption on T that $(T - \lambda I_\pi)\pi(g) = \pi(g)(T - \lambda I_\pi)$ for all $g \in G$. The operator $(T - \lambda I_\pi)$ cannot be an isomorphism as $\phi \in \text{Ker}(T - \lambda I_\pi)$, so $T - \lambda I_\pi = 0$ by (1). □

NB. For Schur's lemma to hold, it is essential that it is applied to representations acting in **complex, finite-dimensional** Hilbert spaces. Readers are counseled not to attempt to use it outside this context. However, note that Theorem 21.30 in Hewitt and Ross [91] pp. 324–345 yields a direct generalisation of Theorem 2.1.1(2) for irreducible representations acting in infinite-dimensional complex Hilbert spaces.

2.2 Representations of Compact Groups and Locally Compact Abelian Groups

2.2.1 Compact Groups: Schur Orthogonality and the Peter-Weyl Theorem

Throughout this subsection we take G to be a compact Hausdorff group, equipped with its normalised Haar measure. It's worth pointing out that in this case, any linear representation acting in a Hilbert space can be made into a unitary one, as the following demonstrates:

Proposition 2.2.1 *If π is a linear representation of a compact group G acting in a complex Hilbert space V_π, then there exists an inner product on V_π for which $\pi(g)$ is unitary for all $g \in G$.*

Proof We simply define a new inner product on V_π by

$$\langle\langle \phi, \psi \rangle\rangle = \int_G \langle \pi(h)\phi, \pi(h)\psi \rangle dh,$$

for each $\phi, \psi \in G$. First note that the integral is well-defined, indeed by compactness we have

$$|\langle\langle \phi, \psi \rangle\rangle| \leq \sup_{h \in G} ||\pi(h)\phi||. \sup_{h \in G} ||\pi(h)\psi|| < \infty.$$

It is easily verified that $\langle\langle \cdot, \cdot \rangle\rangle$ is indeed an inner product on V_π. By left invariance of Haar measure we deduce that for all $g \in G$, $\pi(g)$ is an isometry. From this we find that $\pi(g^{-1}) = \pi(g)^{-1} = \pi(g)^*$. Hence $\pi(g)^*$ is also an isometry and so $\pi(g)$ is unitary as required. □

For the remainder of this subsection, we follow the excellent account in Faraut [63] Chap. 6 which exploits the theory of compact linear operators in Hilbert space (see Appendix A.6 for relevant background, if required) to analyse the properties of representations of compact groups.

Let π be a representation of G on V_π and fix $\psi \in V_\pi$. We introduce the linear operator $Q_\psi : V_\pi \to V_\pi$ defined by

$$Q_\psi \phi = \int_G \langle \pi(g^{-1})\phi, \psi \rangle \pi(g)\psi dg, \tag{2.2.3}$$

for all $\phi \in V_\pi$ (where the integral is to be understood in the Bochner sense, for which see e.g. Cohn [50] Appendix E, pp. 350–354). It is easily verified that Q_ψ is bounded and self-adjoint.

Lemma 2.2.1 *The operator Q_ψ is compact and intertwining.*

Proof We will show that Q_ψ is in fact Hilbert-Schmidt, then it is compact (see Appendix A.6). Let $(e_n, n \in \mathbb{N})$ be any complete orthonormal basis for V_π. Then by Pythagoras' theorem in Hilbert space,

$$\sum_{n=1}^{\infty} ||Q_\psi e_n||^2 = \sum_{n=1}^{\infty} \int_G \int_G \langle e_n, \pi(g)\psi\rangle\langle\pi(h)\psi, e_n\rangle\langle\pi(g)\psi, \pi(h)\psi\rangle dg dh$$
$$= \int_G \int_G |\langle\pi(g)\psi, \pi(h)\psi\rangle|^2 dg dh$$
$$= \int_G \int_G |\langle\pi(h^{-1}g)\psi, \psi\rangle|^2 dg dh$$
$$= \int_G |\langle\pi(g)\psi, \psi\rangle|^2 dg \leq ||\psi||^2.$$

The interchange of summation and integration is justified by use of the Cauchy-Schwarz inequality for series, the Plancherel theorem and Fubini's theorem.

To see that Q_ψ is intertwining: if $h \in G$, then for all $\phi \in V_\pi$,

$$Q_\psi \pi(h)\phi = \int_G \langle\pi(h)\phi, \pi(g)\psi\rangle\pi(g)\psi dg$$
$$= \int_G \langle\phi, \pi(h^{-1}g)\psi\rangle\pi(g)\psi dg.$$

Now make the change of variable $g \to hg$ to see that

$$Q_\psi \pi(h)\phi = \int_G \langle\phi, \pi(g)\psi\rangle\pi(h)\pi(g)\psi dg = \pi(h) Q_\psi \phi,$$

by continuity. □

Lemma 2.2.1 immediately yields two valuable results:

Corollary 2.2.1 *Every irreducible representation of a compact Hausdorff group is finite dimensional.*

Proof Let π be an irreducible representation of a compact Hausdorff group G. By Lemma 2.2.1 the linear operator Q_ψ is compact and self-adjoint on V_π. It is not difficult to show that Q_ψ is non-zero if $\psi \neq 0$, and we assume this from now on. It follows that Q_ψ has a real non-zero eigenvalue and the corresponding eigenspace W_π is finite-dimensional. But we showed in Lemma 2.2.1 that Q_ψ is intertwining, hence W_π is invariant. Since π is irreducible, $W_\pi = V_\pi$ and we are done. □

Corollary 2.2.2 *Every finite-dimensional representation of a compact Hausdorff group is either irreducible or completely reducible.*

Proof Assume that π is a representation acting on a finite-dimensional space V_π and that it is not irreducible. By Lemma 2.1.1, we can find a non-trivial subrepresentation π_S such that $\pi = \pi_S \oplus \pi_S^\perp$. Now if both π_S and $\pi_S^\perp$ are irreducible we are done. So suppose that at least one of these is not and iterate the application of Lemma 2.1.1. By finite dimensionality of V_π, this procedure must eventually terminate. □

If π_1 and π_2 be finite-dimensional representations of a compact Hausdorff group G, then $\pi_1 \otimes \pi_2$ is also finite-dimensional. So by Corollary 2.2.2, we have uniquely (up to equivalence),

$$\pi_1 \otimes \pi_2 = \bigoplus_{\pi \in S} m_\pi \pi. \tag{2.2.4}$$

Here $S \subset \widehat{G}$ is finite and for each $\pi \in S$, m_π is a non-negative integer which gives the multiplicity with which the irreducible representation π occurs in the decomposition, i.e. the direct sum decomposition of $V_{\pi_1 \otimes \pi_2}$ contains
$\underbrace{V_\pi \oplus \cdots \oplus V_\pi}_{m_\pi \text{ times}}$.

Note that if π_1 and π_2 are irreducible, then $\pi_1 \otimes \pi_2$ will not in general be so, and then its decomposition into irreducibles is given by (2.2.4). In that case, the natural numbers m_π are called *Clebsch-Gordon coefficients.*[1]

Using (2.1.1) we see that the conjugate representation $\overline{\pi}$ of G is irreducible if and only if π is. Indeed, W is an invariant subspace for π if and only if JW is invariant for $\overline{\pi}$. Note however that these two representations are typically not unitarily equivalent (in which case they determine different elements of $\widehat{G}$).

The result of the next theorem contains the important *Schur orthogonality relations* as a special case, as we will see later.

Theorem 2.2.1 *If π acts irreducibly on V_π, then for all $\phi_i, \psi_i \in V_\pi (i = 1, 2)$,*

$$\int_G \langle \pi(g)\phi_1, \psi_1 \rangle \langle \psi_2, \pi(g)\phi_2 \rangle dg = \frac{1}{d_\pi} \langle \phi_1, \phi_2 \rangle \langle \psi_2, \psi_1 \rangle. \tag{2.2.5}$$

Proof We showed in Lemma 2.2.1 that the operator Q_ψ intertwines π. It follows by Schur's lemma that $Q_\psi = \lambda_\psi I_\pi$ for some $\lambda_\psi \in \mathbb{C}$. Taking inner products with ϕ on both sides of (2.2.3) yields

$$\langle Q_\psi \phi, \phi \rangle = \int_G |\langle \pi(g)\psi, \phi \rangle|^2 dg = \lambda_\psi ||\phi||^2.$$

[1] These numbers have many important applications to quantum physics, see e.g. Jones [110] pp. 109–118.

However, on interchanging the roles of ψ and ϕ, and recalling that normalised Haar measure is invariant under the mapping $g \to g^{-1}$, we find that

$$\begin{aligned} \langle Q_\phi \psi, \psi \rangle &= \int_G |\langle \pi(g)\phi, \psi \rangle|^2 dg \\ &= \int_G |\langle \phi, \pi(g^{-1})\psi \rangle|^2 dg \\ &= \int_G |\langle \pi(g)\psi, \phi \rangle|^2 dg = \langle Q_\psi \phi, \phi \rangle. \end{aligned}$$

Hence we deduce that $\lambda_\psi ||\phi||^2 = \lambda_\phi ||\psi||^2$. But ψ and ϕ were chosen arbitrarily, and it follows that for all $\psi \in V_\pi$, $\lambda_\psi = c||\psi||^2$ for some fixed $c \in \mathbb{R}$. Now choose an arbitrary orthonormal basis $\{e_1, \ldots, e_{d_\pi}\}$ for V_π. By the Plancherel identity for Hilbert spaces, we have for all $g \in G$,

$$||\phi||^2 = \sum_{i=1}^{d_\pi} |\langle \pi(g)\phi, e_i \rangle|^2.$$

Integrate both sides of this expression to obtain

$$\begin{aligned} ||\phi||^2 &= \sum_{i=1}^{d_\pi} \int_G |\langle \pi(g)\phi, e_i \rangle|^2 dg \\ &= \sum_{i=1}^{d_\pi} \langle Q_\phi e_i, e_i \rangle \\ &= \sum_{i=1}^{d_\pi} \langle Q_{e_i} \phi, \phi \rangle \\ &= \sum_{i=1}^{d_\pi} \lambda_{e_i} ||\phi||^2 \\ &= c \sum_{i=1}^{d_\pi} ||\phi||^2 = c d_\pi ||\phi||^2. \end{aligned}$$

It follows that $c = \frac{1}{d_\pi}$ and hence for all $\phi, \psi \in V_\pi$,

$$\int_G |\langle \pi(g)\psi, \phi \rangle|^2 dg = \frac{1}{d_\pi} ||\phi||^2 ||\psi||^2.$$

The identity (2.2.5) may be deduced from this one by use of the polarisation identity. □

We now turn our attention to the analysis of the complex Hilbert space $L^2(G)$. For each irreducible representation π we define the linear subspace $\mathcal{M}_\pi$ to be the linear span of all mappings of the form $g \to \langle \pi(g)\psi, \phi \rangle$, where $\psi, \phi \in V_\pi$. It is an easy exercise to show that if π_1 and π_2 are equivalent, then $\mathcal{M}_{\pi_1} = \mathcal{M}_{\pi_2}$, and so we need only consider the spaces $\mathcal{M}_\pi$ with $\pi \in \widehat{G}$.

Theorem 2.2.2 *If π and π' are distinct elements of $\widehat{G}$, then $\mathcal{M}_\pi$ and $\mathcal{M}_{\pi'}$ are orthogonal.*

Proof Let $T \in \mathcal{L}(V_\pi, V_{\pi'})$ and consider the linear operator S_T defined for all $\phi \in V_\pi$ as the (vector-valued) Lebesgue integral

$$S_T\phi = \int_G \pi'(g^{-1})T\pi(g)\phi dg.$$

It is easily verified that $S_T \in \mathcal{L}(V_\pi, V_{\pi'})$ and, by using right invariance of Haar measure, that S_T intertwines π and π'. Hence by Schur's lemma, since π and π' are inequivalent, $S_T = 0$. Now choose $T = T_{\phi,\phi'}$ to be the rank-one operator $T_{\phi,\phi'}\psi = \langle \psi, \phi \rangle \phi'$, where $\phi \in V_\pi$ and $\phi' \in V_{\pi'}$ are fixed and $\psi \in V_\pi$. Then for all $\psi' \in V_{\pi'}$ we have

$$0 = \langle S_{T_{\phi,\phi'}}\psi, \psi' \rangle = \int_G \langle \pi(g)\psi, \phi \rangle \overline{\langle \pi'(g)\psi', \phi' \rangle} dg.$$

Since this identity holds for all $\phi \in V_\pi$ and all $\phi' \in V_{\pi'}$, the result follows. □

Corollary 2.2.3 (Schur Orthogonality Relations) *If $\pi_1, \pi_2 \in \widehat{G}$, then for all $\phi_i, \psi_i \in V_{\pi_i}, i = 1, 2$,*

$$\int_G \langle \pi_1(g)\phi_1, \psi_1 \rangle \langle \psi_2, \pi_2(g)\phi_2 \rangle dg = \begin{cases} 0 & \text{if } \pi_2 \neq \pi_1 \\ \dfrac{1}{d_{\pi_1}} \langle \phi_1, \phi_2 \rangle \overline{\langle \psi_1, \psi_2 \rangle} & \text{if } \pi_2 = \pi_1 \end{cases}$$

Proof This is an immediate consequence of Theorems 2.2.1 and 2.2.2. □

Now let $\mathcal{E}(G)$ be the smallest linear subspace of $L^2(G)$ which contains all the $\mathcal{M}_\pi$'s, i.e. $\mathcal{E}(G)$ is the linear span of the mappings $\{g \to \langle \pi(g)\psi, \phi \rangle, \psi, \phi \in V_\pi, \pi \in \widehat{G}\}$. The next result is the famous *Peter-Weyl theorem*:

Theorem 2.2.3 (Peter-Weyl 1) *$\mathcal{E}(G)$ is dense in $L^2(G)$.*

Proof Let $W = \mathcal{E}(G)^\perp$. The result will follow if we can show that $W = \{0\}$, so we suppose that $W \neq \{0\}$ and seek a contradiction. Recall that the right regular representation $h \to R_h$ of G acts unitarily on $L^2(G)$. In particular, for each $\pi \in \widehat{G}$, it sends the mapping $g \to \langle\pi(g)\psi, \phi\rangle\psi$ to $g \to \langle\pi(g)\pi(h)\psi, \phi\rangle\psi$, which is another mapping of the same form. Hence it preserves the space $\mathcal{E}(G)$ and also (by Lemma 2.1.1) the space W. Either it acts irreducibly on W or W contains a closed proper subspace of the form $V_{\pi'}$ for some $\pi' \in \widehat{G}$. Let us assume the latter (in the former case, W is itself of the form $V_{\pi'}$). Let $f \in V_{\pi'}$ with $f \neq 0$, and for each $g \in G$, define $\Gamma_f(g) := \langle R_g f, f\rangle = \langle\pi'(g)f, f\rangle$. Then $\Gamma_f \in \mathcal{M}_{\pi'}$. Let $\pi \in \widehat{G}$ be arbitrary and $u, v \in V_\pi$. We have

$$\begin{aligned}\int_G \Gamma_f(g)\overline{\langle\pi(g)u, v\rangle}dg &= \int_G \langle R_g f, f\rangle\overline{\langle\pi(g)u, v\rangle}dg \\ &= \int_G\int_G f(hg)\overline{f(h)\langle\pi(g)u, v\rangle}dhdg \\ &= \int_G \overline{f(h)}\left(\int_G f(g)\overline{\langle\pi(g)u, \pi(h)v\rangle}dg\right)dh,\end{aligned}$$

where we have made the change of variable $g \to h^{-1}g$ and used Fubini's theorem to obtain the last line. The inner integral is zero as $f \in W$ and the mapping $g \to \langle\pi(g)u, \pi(h)v\rangle$ is in $\mathcal{E}(G)$. But then $\Gamma_f \in W$, and since we've already observed that $\Gamma_f \in \mathcal{M}_{\pi'} \subseteq W^\perp$, we conclude that $\Gamma_f = 0$. But $||f||^2 = \Gamma_f(e)$ and so $f = 0$ (a.e.). This yields our desired contradiction and the result follows. □

Fix once and for all a basis $\{e_i^{(\pi)}, 1 \leq i \leq d_\pi\}$ in each $V_\pi, \pi \in \widehat{G}$ and define, relative to these bases, the *co-ordinate functions* $\pi_{ij}(g) = (\pi(g))_{ij}$ for each $g \in G, 1 \leq i, j \leq d_\pi, \pi \in \widehat{G}$, where $(\pi(g))_{ij}$ is the (i, j)th entry of the matrix $\pi(g)$, i.e. $(\pi(g))_{ij} := \langle\pi(g)e_i^{(\pi)}, e_j^{(\pi)}\rangle$.

Corollary 2.2.4 (Peter-Weyl 2) *The set* $\{d_\pi^{\frac{1}{2}}\pi_{ij}, 1 \leq i, j, \leq d_\pi, \pi \in \widehat{G}\}$ *is a complete orthonormal basis for* $L^2(G)$.

Proof Orthonormality is a direct consequence of (2.2.5). For completeness, suppose that ψ is orthogonal to every vector in this set. Then by linearity, ψ is orthogonal to $\mathcal{M}_\pi$ for each $\pi \in \widehat{G}$, and so ψ is orthogonal to $\mathcal{E}(G)$. It then follows by Theorem 2.2.3 that $\psi = 0$, as required. □

Corollary 2.2.5 *If G is second countable then the set $\widehat{G}$ is countable.*

Proof If G is second countable, then $L^2(G)$ is separable.[2] It follows that the orthonormal basis that we described in Corollary 2.2.4 is countable, and the result follows. □

---

[2] To see this, note that by Proposition 3.4.5 in Cohn [50], p. 110, we need only show that the σ-algebra $\mathcal{B}(G)$ is countably generated. But if G is second countable, then it has a countable basis for its topology which generates $\mathcal{B}(G)$.

In particular, Corollary 2.2.5 tells us that $\widehat{G}$ is countable whenever G is a compact Lie group.

We recall that a family $\mathcal{C}$ of complex-valued functions defined on a topological space S *separates points* if for all $x, y \in S$ with $x \neq y$ there exists $f \in \mathcal{C}$ so that $f(x) \neq f(y)$. The celebrated *Stone-Weierstrass theorem* says that if S is compact and if $\mathcal{A}$ is a sub-algebra of $C(S)$ that is closed under complex conjugation, contains all the constant functions and separates points, then $\mathcal{A}$ is dense in $C(S)$ (see e.g. Ruzhansky and Turunen [172] pp. 63–65).

Theorem 2.2.4 (Peter-Weyl 3) *$\mathcal{E}(G)$ is dense in $C(G)$ (equipped with the topology of uniform convergence).*

Proof We show that the conditions of the Stone-Weierstrass theorem are satisfied. First we prove that $\mathcal{E}(G)$ is an algebra. Let $\pi_1, \pi_2 \in \widehat{G}$ and $\psi_i, \phi_i \in V_{\pi_i}, i = 1, 2$. Then for all $g \in G$, by (2.2.4)

$$\begin{aligned}\langle \pi_1(g)\psi_1, \phi_1\rangle\langle \pi_2(g)\psi_2, \phi_2\rangle &= \langle(\pi_1 \otimes \pi_2)(g)\psi_1 \otimes \psi_2, \phi_1 \otimes \phi_2\rangle\\ &= \sum_{s\in S} m_s\langle \pi_s(g)\psi_s, \phi_s\rangle,\end{aligned}$$

where S is a finite subset of $\widehat{G}$, $\{m_s, s \in S\}$ are the Clebsch-Gordon coefficients and $\phi_s, \psi_s \in V_{\pi_s}$ for each $s \in S$. This verifies the algebra property.

To see that $\mathcal{E}(G)$ contains all constant functions it suffices to take π to be trivial, and to see that it is closed under conjugation, observe that for all $\pi \in \widehat{G}, \psi, \phi \in V_\pi, g \in G, \overline{\langle \pi(g)\psi, \phi\rangle} = \langle \overline{\pi}(g)\overline{\phi}, \overline{\psi}\rangle$ by (2.1.2), and recall that $\overline{\pi}$ is irreducible whenever π is. To complete the proof, we must show that $\mathcal{E}(G)$ separates points, and we will demonstrate this within a separate lemma (Lemma 2.3.1), at the end of the next section. □

The algebra $\mathcal{E}(G)$ is sometimes called the *coefficient algebra* of G, as it contains the matrix entries (or matrix "coefficients") of all irreducible representations of G. It can be given the structure of a Hopf algebra and so yields the prototype for a *quantum group* (see e.g. Cartier [41] Sect. 3.1).[3]

2.2.2 Irreducible Representations of Locally Compact Abelian Groups

Let G be a locally compact abelian group. A *character* of G is a continuous homomorphism χ from G to the one-torus $\mathbb{T}^1$. So if $g, h \in G$,

$$\chi(gh) = \chi(g)\chi(h) = \chi(h)\chi(g); \quad \chi(g^{-1}) = \overline{\chi(g)}; \quad \chi(e) = 1.$$

It is clear that a character defines a one-dimensional representation of G. Conversely we have

[3] Of course, in a generic quantum group, there is no underlying group G, only the Hopf algebra structure.

Theorem 2.2.5 *If G is a compact Hausdorff abelian group, then every irreducible representation of G is one-dimensional and is given by a character of G.*

Proof Suppose that $\pi : G \to V_\pi$ is an irreducible representation of G. Then V_π is finite-dimensional by Corollary 2.2.1. Fix an arbitrary $g \in G$. Then since G is abelian, $\pi(g)\pi(h) = \pi(h)\pi(g)$ for all $h \in G$. But then by Schur's lemma (Theorem 2.1.1 (2)) $\pi(g) = \chi(g)I_\pi$ where $\chi : G \to \mathbb{C}$. Then by irreducibility we have $\dim(V_\pi) = 1$. We show that χ is a character. Firstly, since π is a representation, it follows that χ is a continuous homomorphism. The fact that $\pi(g^{-1}) = \pi(g)^*$ implies that $\chi(g^{-1}) = \overline{\chi(g)}$ and $\pi(g)\pi(g)^* = 1$ tells us that $|\chi(g)| = 1$, for all $g \in G$. Hence χ takes values in $\mathbb{T}^1$, as required. □

Note Theorem 2.2.5 extends to locally compact abelian Hausdorff groups by using Theorem 21.30 in Hewitt and Ross [91] pp. 324–325.

Hence if G is locally compact, Hausdorff and abelian we may identify the unitary dual $\widehat{G}$ with the set of all characters of G. In fact, in this case $\widehat{G}$ is itself a locally compact, Hausdorf abelian group, called the *dual group* of G. It is easily checked that the set of all characters forms an abelian group. Indeed, if χ_1 and χ_2 are characters of G we define the composition law by $(\chi_1.\chi_2)(g) = \chi_1(g)\chi_2(g)$ for each $g \in G$, the neutral element is the character that maps each $g \in G$ to 1, and the inverse χ^{-1} of the character χ is given by $\chi^{-1}(g) = \overline{\chi(g)}$. The dual group $\widehat{G}$ is a locally compact, Hausdorff topological group with respect to the compact-open topology[4] (see Hewitt and Ross [91] pp. 360–362). We present some brief examples. Details can be found in [91] pp. 366–368.

Example 7 The torus $\mathbb{T}^1$. In this case, the character group is the set of integers $\mathbb{Z}$ regarded as a *discrete group* (i.e. a group equipped with the discrete topology). The action of $\mathbb{Z}$ on $\mathbb{T}^1$ is given by $e^{i\theta} \to e^{in\theta}$ for each $n \in \mathbb{Z}, \theta \in [0, 2\pi)$. In this case the Peter-Weyl theorem (specifically Corollary 2.2.4) reduces to the well-known fact that $\{e_n, n \in \mathbb{Z}\}$ is a complete orthonormal basis for $L^2(\mathbb{T}^1)$, where $e_n(\theta) = e^{in\theta}$ for each $n \in \mathbb{Z}, \theta \in [0, 2\pi)$. Similarly we find that the dual group of the d-torus $\mathbb{T}^d$ is $\mathbb{Z}^d$.

Example 8 The integers $\mathbb{Z}$. The dual group of the discrete group of integers is the torus $\mathbb{T}^1$ with the action given by $n \to e^{in\theta}$ for each $\theta \in [0, 2\pi), n \in \mathbb{Z}$. Similarly the dual group of $\mathbb{Z}^d$ is $\mathbb{T}^d$.

Example 9 The real line $\mathbb{R}$. The dual group of $\mathbb{R}$ is $\mathbb{R}$ itself with the action $x \to e^{ixy}$ for each $y \in \mathbb{R}$ (as dual group) and $x \in \mathbb{R}$. Similarly the dual group of $\mathbb{R}^d$ is $\mathbb{R}^d$.

[4] A basis $\mathcal{U}_e$ for a neighbourhood of the neutral element in $\widehat{G}$ is given by the sets $\{\chi \in \widehat{G}; |\chi(g) - 1| < \epsilon$ for all $g \in K\}$ where K is a compact subset of G and $\epsilon > 0$. A basis for the topology of $\widehat{G}$ is then given by the collection of all left translates of sets in $\mathcal{U}_e$.

It can be shown that the dual group of a compact group is always a discrete group and the dual group of a discrete group is always a compact group (see Theorem 23.17 on p. 362 of Hewitt and Ross [91]), and Examples 1 and 2 illustrate this result. Our last topic in this section concerns the process of iterating duality of abelian groups. We write $\widehat{\widehat{G}}$ for the dual of the dual group. For each $g \in G$ we define a character $\widehat{g}$ of $\widehat{G}$ by the prescription $\widehat{g}(\chi) = \chi(g)$ for all $\chi \in \widehat{G}$. The celebrated *Pontryagin duality theorem* tells us that these are the only characters on $\widehat{G}$—in fact the mapping $g \to \widehat{g}$ is a topological isomorphism of G with $\widehat{\widehat{G}}$. For a proof of this key structural theorem in the study of locally compact abelian groups, see e.g. Hewitt and Ross [91] pp. 376–380. Once again, Examples 1 and 2 are illuminating.

2.3 The Non-commutative Fourier Transform

For the remainder of this chapter, G is always a compact, Hausdorff group (unless otherwise stated). In this section, we begin the development of harmonic analysis on such groups, which is one of the main themes of this book. We begin by introducing some preliminary notions. Firstly we recall that the Hilbert-Schmidt norm on the matrix algebra $M_n(\mathbb{C})$ is defined by

$$||A||_{\mathrm{HS}} := [\mathrm{tr}(\mathrm{AA}^*)]^{\frac{1}{2}},$$

for each $A \in M_n(\mathbb{C})$. The associated inner product is $\langle A, B\rangle_{\mathrm{HS}} := \mathrm{tr}(\mathrm{AB}^*)$ for $A, B \in M_n(\mathbb{C})$. We will also need the operator norm of a matrix, $||A||_{\mathrm{op}} := \sup\{||A\psi||; \psi \in \mathbb{C}^n, ||\psi|| = 1\}$. Secondly, let G be a compact, Hausdorff, second countable group, and define the set

$$\mathcal{M}(\widehat{G}) := \bigcup_{\pi \in \widehat{G}} M_{d_\pi}(\mathbb{C}).$$

We say that a mapping $F : \widehat{G} \to \mathcal{M}(\widehat{G})$ is *compatible* if $F(\pi) \in M_{d_\pi}(\mathbb{C})$ for each $\pi \in \widehat{G}$. The set of all compatible mappings forms a complex linear space, which we write $\mathcal{L}(\widehat{G})$, under the pointwise operations

$$(\lambda F + G)(\pi) := \lambda F(\pi) + G(\pi),$$

for each $\lambda \in \mathbb{C}$, $F, G \in \mathcal{L}(\widehat{G})$, $\pi \in \widehat{G}$. Let $\mathcal{H}_2(\widehat{G})$ be the linear subspace of all $F \in \mathcal{L}(\widehat{G})$ for which $|||F|||_2^2 := \sum_{\pi \in \widehat{G}} d_\pi ||F(\pi)||_{HS}^2 < \infty$. Then $\mathcal{H}_2(\widehat{G})$ is a complex Hilbert space with inner product

$$\langle\langle F, G\rangle\rangle := \sum_{\pi \in \widehat{G}} d_\pi \langle F(\pi), G(\pi)\rangle_{HS},$$

for $F, G \in \mathcal{H}_2(\widehat{G})$ and associated norm $|||\cdot|||_2$. If F is a compatible mapping, we sometimes use the terminology *compatible matrices* for the set $\{F(\pi), \pi \in \widehat{G}\}$.

Next for each $f \in L^1(G)$ we introduce the *non-commutative Fourier transform*[5]

$$\widehat{f}(\pi) := \int_G \pi(g^{-1})f(g)dg, \tag{2.3.6}$$

for each $\pi \in \widehat{G}$, which we regard as a matrix-valued integral, so for each $1 \leq i, j \leq d_\pi$,

$$\widehat{f}(\pi)_{ij} = \int_G \pi_{ij}(g^{-1})f(g)dg.$$

We can immediately verify that

$$||\widehat{f}(\pi)||_{\text{op}} \leq ||f||_1.$$

It is also a straightforward exercise to show that for all $g \in G$,

$$\widehat{R_g f}(\pi) = \pi(g)\widehat{f}(\pi) \quad \text{and} \quad \widehat{L_g f}(\pi) = \widehat{f}(\pi)\pi(g)^*.$$

For example, if $G = \Pi^n$, so that $\widehat{G} = \mathbb{Z}^n$, we have that (at $\pi = n$) $\widehat{f}(n)$ is the usual Fourier coefficient: $\widehat{f}(n) = \frac{1}{(2\pi)^n}\int_{\Pi^n} f(x)e^{-in\cdot x}dx$. More generally, if G is a locally compact Hausdorff abelian group, then $\widehat{G}$ is the dual group of characters, and $\widehat{f}(\chi) = \int_G f(g)\overline{\chi(g)}dg$ for each $\chi \in \widehat{G}$ defines a mapping from $\widehat{G}$ to $\mathbb{C}$. This case is developed extensively in Chap. 1 of Rudin [171].

We return to our main framework where G is a more general (not necessarily abelian) compact group and we let $\mathcal{F} : L^1(G) \to \mathcal{L}(\widehat{G})$ be the *Fourier transformation* mapping given by $\mathcal{F}(f) = \widehat{f}$ for each $f \in L^1(G)$. It is clearly linear. Now let us consider its restriction to $L^2(G)$, which we continue to denote by $\mathcal{F}$. Using the Peter-Weyl theorem and elementary Hilbert space techniques we get the following important result:

Theorem 2.3.1 *1. Fourier expansion. For all $f \in L^2(G)$,*

$$f = \sum_{\pi \in \widehat{G}} d_\pi \text{tr}(\widehat{f}(\pi)\pi), \tag{2.3.7}$$

(where convergence is in the L^2-sense).

---

[5] In the representation theory literature, it is quite common to meet the alternative notation $\pi(f) := \widehat{\widehat{f}}(\pi) = \int_G f(g)\pi(g)dg$.

2. *Parseval-Plancherel Identity. The operator $\mathcal{F}$ is an isometry from $L^2(G)$ into $\mathcal{H}_2(\widehat{G})$ so that for all $f, f_1, f_2 \in L^2(G)$,*

$$\int_G |f(g)|^2 dg = \sum_{\pi\in\widehat{G}} d_\pi ||\widehat{f}(\pi)||^2_{\mathrm{HS}} \tag{2.3.8}$$

$$\text{and} \quad \int_G f_1(g)\overline{f_2}(g) dg = \sum_{\pi\in\widehat{G}} d_\pi \langle \widehat{f_1}(\pi), \widehat{f_2}(\pi)\rangle_{\mathrm{HS}} \tag{2.3.9}$$

Proof 1. Using Corollary 2.2.4 together with the usual Fourier expansion in Hilbert space we find that

$$\begin{aligned} f &= \sum_{\pi\in\widehat{G}} d_\pi \sum_{i,j=1}^{d_\pi} \langle f, \pi_{ij}\rangle \pi_{ij} \\ &= \sum_{\pi\in\widehat{G}} d_\pi \sum_{i,j=1}^{d_\pi} \left(\int_G f(\sigma)\overline{\pi_{ij}(\sigma)} d\sigma \right) \pi_{ij} \\ &= \sum_{\pi\in\widehat{G}} d_\pi \sum_{i,j=1}^{d_\pi} \left(\int_G f(\sigma)\pi_{ji}(\sigma^{-1}) d\sigma \right) \pi_{ij} \\ &= \sum_{\pi\in\widehat{G}} d_\pi \sum_{i,j=1}^{d_\pi} \widehat{f}(\pi)_{ji}\pi_{ij} \\ &= \sum_{\pi\in\widehat{G}} d_\pi \mathrm{tr}(\widehat{f}(\pi)\pi). \end{aligned}$$

2. The usual Plancherel theorem in Hilbert space yields

$$||f||^2_{L^2(G)} = \sum_{\pi\in\widehat{G}} d_\pi \sum_{i,j=1}^{d_\pi} |\langle f, \pi_{i,j}\rangle|^2,$$

and (2.3.8) follows from this since

$$\begin{aligned} \sum_{i,j=1}^{d_\pi} |\langle f, \pi_{i,j}\rangle|^2 &= \sum_{i,j=1}^{d_\pi} \left(\int_G f(\sigma)\overline{\pi_{ij}(\sigma)} d\sigma \right) \left(\int_G \overline{f(\sigma)}\pi_{ij}(\sigma) d\sigma \right) \\ &= \sum_{i,j=1}^{d_\pi} \widehat{f}(\pi)_{ji}\widehat{f}(\pi)^*_{ij} = \mathrm{tr}(\widehat{f}(\pi)\widehat{f}(\pi)^*). \end{aligned}$$

Here we have used the fact that

$$\widehat{f}(\pi)^* = \int_G \overline{f(\sigma)}\pi(\sigma^{-1})^* d\sigma = \int_G \overline{f(\sigma)}\pi(\sigma)d\sigma.$$

Finally (2.3.9) follows from (2.3.8) by using the polarisation identity. □

We may generalise the space $\mathcal{H}_2(\widehat{G})$ as follows. Let $1 \leq p < \infty$ and define $\mathcal{H}_p(\widehat{G})$ to be the (complex) linear space of all $A \in \mathcal{M}(\widehat{G})$ for which

$$|||A|||_p = \left(\sum_{\pi \in \widehat{G}} d_\pi \mathrm{tr}([A(\pi)^* A(\pi)]^{\frac{p}{2}}) \right)^{\frac{1}{p}} < \infty.$$

Then $\mathcal{H}_p(\widehat{G})$ is a Banach space with respect to the norm $||| \cdot |||_p$. In Edwards [61] (Theorem 2.14.5, p. 148) we meet the following extension of the classical Hausdorff-Young theorem:

Theorem 2.3.2 *If* $1 \leq p \leq 2$ *and* $f \in L^p(G)$*, then* $\widehat{f} \in \mathcal{H}_q(\widehat{G})$*, where* $\dfrac{1}{p} + \dfrac{1}{q} = 1$ *and*

$$|||\widehat{f}|||_q \leq ||f||_p \tag{2.3.10}$$

Note that in the case $p = 2$, we have equality in (2.3.10). Indeed this is precisely the Plancherel theorem (2.3.8).

The case $p = 1$ is of considerable interest. Let $\mathcal{A}(G)$ be the subspace of $L^1(G)$ comprising mappings f for which $|||\widehat{f}|||_1 < \infty$. Then $\mathcal{A}(G)$ is a Banach space (in fact a commutative Banach algebra where the product is convolution) with respect to the norm $|| \cdot ||_{\mathcal{A}(G)}$, where $||f||_{\mathcal{A}(G)} := |||\widehat{f}|||_1$. Then the Fourier transformation $f \to \widehat{f}$ is an isometric isomorphism of $\mathcal{A}(G)$ into $\mathcal{H}_1(\widehat{G})$. The algebra $\mathcal{A}(G)$ is of interest as it is a (proper, when G is nonabelian) subset of the space of continuous functions on G that have an absolutely convergent Fourier series. When G is abelian, these two spaces coincide. Readers may also be interested in the discussion on pp. 546–550 of Ruzhansky and Turunen [172] concerning related spaces to $\mathcal{H}_p(\widehat{G})$.

We complete this section by filling in the gap (as promised) in the proof of the Peter-Weyl theorem (Theorem 2.2.4).

Lemma 2.3.1 *The algebra* $\mathcal{E}(G)$ *separates the points of* G.

Proof We follow the proof given in Faraut [63] (Lemma 6.4.4. p. 113). In fact we prove a slightly stronger result, namely if $g, h \in G$ with $g \neq h$, then there exists $\pi \in \widehat{G}$ so that $\pi(g) \neq \pi(h)$. Equivalently, if $g \in G$, with $g \neq e$, there exists $\pi \in \widehat{G}$ so that $\pi(g) \neq I_\pi$. Now suppose that there exists $g_0 \in G$ with $g_0 \neq e$ so that $\pi(g_0) = I_\pi$ for all $\pi \in \widehat{G}$. We choose an arbitrary non-zero $f \in C(G)$ and define $\Phi_f \in C(G)$ by $\Phi_f = R_{g_0} f - f$. Since $\widehat{R_{g_0} f}(\pi) = \pi(g_0)\widehat{f}(\pi)$, we see that

$\widehat{\Phi_f}(\pi) = \pi(g_0)\widehat{f}(\pi) - \widehat{f}(\pi) = 0$ for all $\pi \in \widehat{G}$. Then by (2.3.8) and continuity we deduce that Φ_f is identically zero, and so $R_{g_0} f = f$. In particular, $f(g_0) = f(e)$ and since f was arbitrary and $C(G)$ separates the points of G, we have our desired contradiction. □

2.4 Characters and Central Functions

Let f be a complex-valued function defined on G. It is said to be *central* (or a *class function*) if

$$f(hgh^{-1}) = f(g),$$

(or equivalently $f(gh) = f(hg)$) for all $g, h \in G$. The closed subspace of $L^2(G)$ comprising square-integrable functions that are central a.e., in the sense that for any $h \in G$, $f(hgh^{-1}) = f(g)$ for almost all $g \in G$, is denoted by $L^2_c(G)$. It is clearly non-trivial as all measurable functions that are almost everywhere constant lie in $L^2_c(G)$.

Theorem 2.4.1 *If $f \in L^2(G)$ then $f \in L^2_c(G)$ if and only if for all $\pi \in \widehat{G}$, $\widehat{f}(\pi) = c_\pi I_\pi$ for some $c_\pi \in \mathbb{C}$.*

Proof First suppose that $f \in L^2_c(G)$. Then for all $h \in G, \pi \in \widehat{G}$ we have

$$\begin{aligned}
\pi(h)\widehat{f}(\pi) &= \int_G f(g)\pi(hg^{-1})dg \\
&= \int_G f(gh)\pi(g^{-1})dg \\
&= \int_G f(hg)\pi(g^{-1})dg \\
&= \int_G f(g)\pi(g^{-1}h)dg \\
&= \widehat{f}(\pi)\pi(h),
\end{aligned}$$

and the result follows by Schur's lemma (Theorem 2.1.1 (2)). To establish the converse, assume that for each $\pi \in \widehat{G}$ there exists $c_\pi \in \mathbb{C}$ so that $\widehat{f}(\pi) = c_\pi I_\pi$. Then for given $h \in G$

$$\begin{aligned}
\widehat{f}(\pi) &= \pi(h^{-1})\widehat{f}(\pi)\pi(h) \\
&= \int_G f(g)\pi(h^{-1}g^{-1}h)dg
\end{aligned}$$

$$= \int_G f(hgh^{-1})\pi(g^{-1})dg.$$

It follows that $\int_G (f(g) - f(hgh^{-1}))\pi(g^{-1})dg = 0$. Now let $\phi, \psi \in V_\pi$ be arbitrary. Then we see that $\int_G \overline{(f(g) - f(hgh^{-1}))}\langle \phi, \pi(g^{-1})\psi \rangle dg = 0$. From this we deduce that the equivalence class of mappings $g \rightarrow \overline{f(g) - f(hgh^{-1})}$ is orthogonal to $\mathcal{M}_\pi$ for each $\pi \in \widehat{G}$, and the result then follows from Theorem 2.2.3. □

Let π be an arbitrary finite-dimensional representation of a compact group G. Define the *character* χ_π of the representation π by the prescription

$$\chi_\pi(g) = \mathrm{tr}(\pi(g)) \tag{2.4.11}$$

for each $g \in G$, i.e. $\chi_\pi(g) = \sum_{i=1}^{d_\pi} \pi_{ii}(g)$. Note that $\chi_\pi(e) = d_\pi$. Clearly χ_π is a continuous function on G. It is also central, since for each $g, h \in G$

$$\chi_\pi(gh) = \mathrm{tr}(\pi(gh)) = \mathrm{tr}(\pi(g)\pi(h)) = \mathrm{tr}(\pi(h)\pi(g)) = \chi_\pi(hg).$$

Theorem 2.4.2 *Let π_1 and π_2 be finite-dimensional representations of a compact group G.*

(i) $\chi_{\pi_1 \oplus \pi_2} = \chi_{\pi_1} + \chi_{\pi_2}$,
(ii) $\chi_{\pi_1 \otimes \pi_2} = \chi_{\pi_1}\chi_{\pi_2}$,
(iii) $\chi_{\overline{\pi}} = \overline{\chi_\pi}$,
(iv) $\chi_{\pi_1 \otimes \pi_2} = \sum_{\pi \in S} m_\pi \chi_\pi$, *where we use the notation of* (2.2.4).

Proof (i) is straightforward and left as an exercise for the reader.

(ii) Let $(e_i, 1 \leq i \leq d_{\pi_1})$ be an orthonormal basis for V_{π_1} and $(f_j, 1 \leq j \leq d_{\pi_2})$ be an orthonormal basis for V_{π_2}. Then $(e_i \otimes f_j, 1 \leq i \leq d_{\pi_1}, 1 \leq j \leq d_{\pi_2})$ is an orthonormal basis for $V_{\pi_1} \otimes V_{\pi_2}$ and for all $g \in G$,

$$\begin{aligned}(\chi_{\pi_1 \otimes \pi_2})(g) &= \sum_{i=1}^{d_{\pi_1}} \sum_{j=1}^{d_{\pi_2}} \langle (\pi_1(g) \otimes \pi_2(g))(e_i \otimes f_j), e_i \otimes f_j \rangle \\ &= \left(\sum_{i=1}^{d_{\pi_1}} \langle \pi_1(g)e_i, e_i \rangle \right) \left(\sum_{j=1}^{d_{\pi_2}} \langle \pi_2(g)f_j, f_j \rangle \right) \\ &= \chi_{\pi_1}(g)\chi_{\pi_2}(g).\end{aligned}$$

(iii) Let $(e_i, 1 \leq i \leq d_\pi)$ be an orthonormal basis for V_π. Then $(Je_i, 1 \leq i \leq d_\pi)$ is an orthonormal basis for V_π^*. Now for all $g \in G$, using (2.1.2)

$$\chi_{\overline{\pi}}(g) = \sum_{i=1}^{d_\pi} \langle \overline{\pi}(g)Je_i, Je_i \rangle$$

$$= \sum_{i=1}^{d_\pi} \langle e_i, \pi(g)e_i \rangle$$
$$= \overline{\chi_\pi(g)}.$$

(iv) follows immediately from (2.2.4) and (i). □

We observe that χ is well-defined on $\widehat{G}$, for if π and π' are equivalent irreducible representations of G, then there exists a unitary isomorphism U between V_π and $V_{\pi'}$ so that $\pi'(g) = U\pi(g)U^{-1}$ for all $g \in G$. It follows that

$$\chi_{\pi'}(g) = \mathrm{tr}(U\pi(g)U^{-1}) = \chi_\pi(g).$$

When G is abelian, the characters $\{\chi_\pi, \pi \in \widehat{G}\}$ are precisely as discussed in Sect. 2.2.2.

Theorem 2.4.3 *The set $\{\chi_\pi, \pi \in \widehat{G}\}$ is a complete orthonormal basis for $L^2_c(G)$.*

Proof Since $\chi_\pi \in \mathcal{M}_\pi$ it follows from Theorem 2.2.2 that χ_π and $\chi_{\pi'}$ are orthogonal when $\pi \neq \pi'$. To see that they have the required normalisation, use (2.2.5) to obtain

$$||\chi_\pi||^2 = \int_G |\chi_\pi(g)|^2 dg$$
$$= \sum_{i,j=1}^{d_\pi} \int_G \langle \pi(g)e_i, e_i \rangle \overline{\langle \pi(g)e_j, e_j \rangle} dg$$
$$= \frac{1}{d_\pi} \sum_{i,j=1}^{d_\pi} |\langle e_i, e_j \rangle|^2$$
$$= \frac{1}{d_\pi} \sum_{i,j=1}^{d_\pi} \delta_{ij} = 1.$$

To show completeness, let $f \in L^2_c(G)$. Then by Theorem 2.4.1, for all $\pi \in \widehat{G}$, $\widehat{f}(\pi) = c_\pi I_\pi$ for some $c_\pi \in \mathbb{C}$. Then we have

$$d_\pi c_\pi = \mathrm{tr}(\widehat{f}(\pi)) = \int_G f(g)\overline{\chi_\pi(g)} dg,$$

and so $c_\pi = \dfrac{1}{d_\pi} \langle f, \chi_\pi \rangle$. Hence

$$\mathrm{tr}(\widehat{f}(\pi)\widehat{f}(\pi)^*) = \frac{1}{d_\pi} |\langle f, \chi_\pi \rangle|^2.$$

Then by the Parseval-Plancherel identity (2.3.8)

$$||f||^2 = \int_G |f(g)|^2 dg = \sum_{\pi \in \widehat{G}} |\langle f, \chi_\pi \rangle|^2,$$

and completeness follows from the Hilbert space version of Plancherel's theorem. □

If f is an L^1-central function on G, then we may define its *(central) Fourier transform* to be

$$\widehat{f}(\chi_\pi) = \int_G f(g)\chi_\pi(g^{-1})dg, \tag{2.4.12}$$

for all $\pi \in \widehat{G}$ and this is a more appropriate tool than (2.3.6) in this context. Note that the Fourier expansion of Theorem 2.3.1(1) now takes the form

$$f = \sum_{\pi \in \widehat{G}} \widehat{f}(\chi_\pi)\chi_\pi,$$

for $f \in L^2_c(G)$.

When G is locally compact and abelian, (2.3.6) and (2.4.12) define the same object. In this case $L^1(G)$ is a commutative Banach algebra (where the product operation is convolution), and the Fourier transform is precisely the Gelfand transform (see e.g. Folland [68] p. 93).

2.5 Weights

2.5.1 The Derived Representation

Let $\mathcal{L}$ be a (finite-dimensional) Lie algebra. A *representation* of $\mathcal{L}$ is a homomorphism of $\mathcal{L}$ into the Lie algebra of all skew-hermitian matrices $\mathbf{u(n)}$ on some finite-dimensional complex vector space.[6] So if $X, Y \in \mathcal{L}$ and $\alpha \in \mathbb{R}$ we have

$$\rho(\alpha X + Y) = \alpha\rho(X) + \rho(Y) \text{ and } \rho([X, Y]) = [\rho(X), \rho(Y)].$$

Let G be a Lie group with Lie algebra $\mathfrak{g}$, and π be a finite-dimensional representation of G. We obtain a representation of $\mathfrak{g}$, called the *derived representation* and denoted $d\pi$, by the prescription

[6] This is more restrictive than necessary, but is all that we need here. In general we might want to consider infinite-dimensional representations.

$$\pi(\exp(X)) = e^{d\pi(X)}, \tag{2.5.13}$$

for all $X \in \mathfrak{g}$. Indeed, to see that $d\pi(X)$ exists and is skew-adjoint we note that $\{\pi(\exp(tX)), t \in \mathbb{R}\}$ is a strongly continuous one-parameter group of unitary operators in V_π, and so $d\pi(X)$ is its infinitesimal generator as given by Stone's theorem (see Theorem A.7.4 in Appendix A.7). We can extract the action of $d\pi(X)$ by strong differentiation on vectors $\psi \in V_\pi$ by

$$d\pi(X)\psi = \frac{d}{dt}\pi(\exp(tX))\psi\bigg|_{t=0} \tag{2.5.14}$$

The next result gives a useful application of these ideas.

Theorem 2.5.1 *1. Let π be a finite-dimensional representation of a connected Lie group acting in the complex vector space V_π. For all $\phi, \psi \in V_\pi$, the mappings $g \to \langle\pi(g)\phi, \psi\rangle$ are C^∞.*
2. If G is a compact, connected Lie group, then the mappings $g \to \pi_{i,j}(g)$ are C^∞ for all $1 \leq i, j \leq d_\pi, \pi \in \widehat{G}$.

Proof 1. Define $f^\pi_{\phi,\psi}(g) = \langle\pi(g)\phi, \psi\rangle$ for each $g \in G$. Then for all $X_1, \ldots, X_n \in \mathfrak{g}, n \in \mathbb{N}$, repeated application of (2.5.14) yields

$$\begin{aligned}
&X_1 \cdots X_n f^\pi_{\phi,\psi}(g) \\
&= \frac{\partial^n}{\partial u_1 \cdots \partial u_n}\langle\pi(g\exp(u_1X_1)\cdots\exp(u_nX_n))\phi, \psi\rangle\bigg|_{u_1=\cdots=u_n=0} \\
&= \langle\pi(g)d\pi(X_1)\cdots d\pi(X_n)\phi, \psi\rangle
\end{aligned}$$

So the mappings $g \to X_1 \cdots X_n f^\pi_{\phi,\psi}(g)$ are well-defined and continuous, and the result follows by Theorem 1.3.5.
2. This follows from (1) and Corollary 2.2.1. □

From now on in this section, G will be a compact, connected Lie group. A *maximal torus* in G is a maximal abelian subgroup $\mathcal{T}$. So if $\mathcal{T}$ exists, then it has the property that if $\mathcal{T}'$ is any other abelian subgroup of G, then $\mathcal{T} \subseteq \mathcal{T}' \Rightarrow \mathcal{T}' = \mathcal{T}$. Any maximal torus is isomorphic to $\Pi^r = (\mathbb{R}/(2\pi\mathbb{Z}))^r$, and r is called the *rank* of the group G.

Example 10 ($G = SU(n)$.) Here we have $\mathcal{T} = \text{diag}(e^{i\theta_1}, \ldots, e^{i\theta_n})$ with $\sum_{i=1}^n \theta_i = 2\pi m$ where $m \in \mathbb{Z}$. So $r = n - 1$ in this case.

Here are some key facts about maximal tori in compact, connected Lie groups (see e.g. Simon [188] Theorem VIII.1.1 pp. 165–179 or Fegan [66] pp. 43–59 for proofs):

- Maximal tori always exist,
- Any $g \in G$ lies on some maximal torus,

- Any two maximal tori $\mathcal{T}_1$ and $\mathcal{T}_2$ are conjugate, i.e. there exists $h \in G$ such that $h\mathcal{T}_1 h^{-1} = \mathcal{T}_2$.

Later on we will require the notion of a *regular* point. This is precisely an element of G that lies on only one maximal torus.

The Lie algebra $\mathfrak{t}$ of $\mathcal{T}$ is a Lie subalgebra of $\mathfrak{g}$ called a *Cartan subalgebra.* It is a maximal abelian subalgebra of $\mathfrak{g}$ and so $\mathfrak{t}$ is isomorphic to $\mathbb{R}^r$. Let π be a finite-dimensional representation of G. The matrices $\{d\pi(X), X \in \mathfrak{t}\}$ are mutually commuting, and so simultaneously diagonalisable, i.e. there exists a non-singular matrix Q such that

$$Qd\pi(X)Q^{-1} = \mathrm{diag}(i\lambda_1(X), \ldots, i\lambda_{d_\pi}(X)). \tag{2.5.15}$$

It is clear that for each $1 \leq i \leq d_\pi$, λ_i is a real linear mapping from $\mathfrak{t}$ to $\mathbb{R}$, i.e. $\lambda_i \in \mathfrak{t}^*$—the algebraic dual of $\mathfrak{t}$. The distinct linear functionals λ_j are called the *weights* of the representation π. We say that $v \in V_\pi$ is a *weight vector* corresponding to the weight λ if $v \neq 0$ and

$$d\pi(X)v = i\lambda(X)v \tag{2.5.16}$$

for all $X \in \mathfrak{t}$. The linear subspace V_λ of V_π comprising all weight vectors corresponding to λ, together with the zero vector, is called the *weight space* of λ.

We complete this subsection with some useful facts about dual spaces that we will need in the sequel. Suppose that $\mathfrak{t}$ is equipped with an inner product $(\cdot, \cdot)$. Then for each $\gamma \in \mathfrak{t}^*$, there exists $H_\gamma \in \mathfrak{t}$ so that for all $X \in \mathfrak{t}$,

$$\gamma(X) = (X, H_\gamma).$$

The mapping $\gamma \to H_\gamma$ is a linear isomorphism of $\mathfrak{t}^*$ with $\mathfrak{t}$, and we may define an inner product in $\mathfrak{t}^*$ (which we also denote by $(\cdot, \cdot)$) so that for all $\gamma, \delta \in \mathfrak{t}^*$,

$$(\gamma, \delta) = \gamma(H_\delta) = \delta(H_\gamma) = (H_\gamma, H_\delta). \tag{2.5.17}$$

For more details, see e.g. Knapp [120] p. 260 or Varadarajan [120] p. 274.

2.5.2 Casimir Operators

Let ρ be an Ad-invariant metric on G and let $\{X_1, \ldots, X_d\}$ be an orthonormal basis for $\mathfrak{g}$. Recall that the Laplacian on G is the element $\Delta \in \mathcal{U}(\mathfrak{g})$ defined by $\Delta = \sum_{i=1}^d X_i^2$. Now let π be a finite-dimensional representation of G so that for each $X \in \mathfrak{g}$, $d\pi(X)$ is a $d_\pi \times d_\pi$ matrix acting in V_π. The matrix

$$\Omega_\pi := \sum_{i=1}^d d\pi(X_i)^2$$

is called a *Casimir operator* corresponding to the representation π. It is useful to think of Ω_π as "$d\pi(\Delta)$". Note that when π is the trivial representation, $\Omega_\pi = 0$.

Lemma 2.5.1 *If π is a non-trivial finite-dimensional representation of G, then Ω_π is a hermitian matrix and $-\Omega_\pi$ is strictly positive.*

Proof We easily deduce that Ω_π is hermitian from the fact that $d\pi(X)$ is skew-adjoint for each $X \in \mathfrak{g}$. For the positivity, observe that for $\psi \in V_\pi$, $\psi \neq 0$,

$$\begin{aligned}\langle \Omega_\pi \psi, \psi \rangle &= \sum_{i=1}^{d} \langle d\pi(X_i)^2 \psi, \psi \rangle \\ &= \sum_{i=1}^{d} \langle d\pi(X_i)\psi, d\pi(X_i)^* \psi \rangle \\ &= -\sum_{i=1}^{d} ||d\pi(X_i)\psi||^2 < 0.\end{aligned}$$

The next result will be very important in the next chapter. □

Theorem 2.5.2 *If $\pi \in \widehat{G}$, then there exists $\kappa_\pi \geq 0$ such that $\Omega_\pi = -\kappa_\pi I_\pi$.*

Proof It follows by Theorem 1.3.3(2) that for all $X \in \mathfrak{g}$,

$$d\pi(X)\Omega_\pi = \Omega_\pi d\pi(X).$$

By iteration, $d\pi(X)^n \Omega_\pi = \Omega_\pi d\pi(X)^n$, for all $n \in \mathbb{N}$ and hence

$$e^{d\pi(X)}\Omega_\pi = \Omega_\pi e^{d\pi(X)},$$

or equivalently by (2.5.13),

$$\pi(\exp(X))\Omega_\pi = \Omega_\pi \pi(\exp(X)).$$

But since G is compact and connected, every $g \in G$ can be written $\exp(X)$ for some $X \in \mathfrak{g}$. The required result now follows by Schur's lemma (Theorem 2.1.1 (2)) and Lemma 2.5.1. □

In the sequel we will refer to the non-negative numbers $\{\kappa_\pi, \pi \in \widehat{G}\}$ as the *Casimir spectrum* of G. It follows from the work of this subsection that

$$\kappa_{\pi_0} = 0 \text{ and } \kappa_\pi > 0 \text{ if } \pi \neq \pi_0.$$

2.5.3 The Lattice of Weights

Let V be a finite-dimensional real vector space with $d := \dim(V)$. A *lattice* in V is an additive subgroup $\mathbb{L}$ that is closed (with respect to the vector topology), spans V and is discrete in that 0 has a neighbourhood N for which $N \cap \mathbb{L} = \{0\}$. A key fact about lattices is that we can always find $\{v_1, \ldots, v_d\} \subseteq V$ such that if $l \in \mathbb{L}$, then $l = \sum_{i=1}^{d} n_i v_i$ with $n_1, \ldots, n_d \in \mathbb{Z}$. If $\mathbb{L}$ is a lattice in V and V^* is the algebraic dual of V, then there exists a lattice $\mathbb{L}^*$ called the *dual lattice* to $\mathbb{L}$ and defined by $\mathbb{L}^* := \{\alpha \in V^*; \alpha(l) \in \mathbb{Z} \text{ for all } l \in \mathbb{L}\}$.

Now we return to the spaces $\mathfrak{t}$ and $\mathfrak{t}^*$. We obtain a lattice $\mathbb{I}$ in $\mathfrak{t}$ which is called the *integer lattice* by the prescription

$$\mathbb{I} = \{X \in \mathfrak{t}; \exp(2\pi X) = e\}.$$

Let $\mathbb{I}^*$ be the dual lattice in $\mathfrak{t}^*$. It is called the *lattice of weights*; the next result motivates this nomenclature.

Lemma 2.5.2 *Let G be a compact connected Lie group and ξ be a finite-dimensional representation of G. If λ is a weight of ξ, then $\lambda \in \mathbb{I}^*$.*

Proof Let $X \in \mathbb{I}$ so that $\xi(\exp(2\pi X)) = I_\xi$. Then by (2.5.13)

$$\exp(2\pi d\xi(X)) = I_\xi;$$

hence

$$\exp(2\pi Q d\xi(X) Q^{-1}) = Q \exp(2\pi d\xi(X)) Q^{-1} = I_\xi.$$

By (2.5.15),

$$\exp(2\pi i \operatorname{diag}(\lambda_1(X), \ldots, \lambda_{d_\xi}(X))) = I_\xi,$$

and so $\lambda_j(X) \in \mathbb{Z}$ for each $1 \leq j \leq d_\xi$, as required. □

2.5.4 Roots

Let $\mathfrak{g}_\mathbb{C}$ be the complexification of $\mathfrak{g}$ and Ad be the adjoint representation of G on $\mathfrak{g}_\mathbb{C}$. We can and will choose an Ad-invariant inner product $(\cdot, \cdot)$ on $\mathfrak{g}_\mathbb{C}$. The derived representation of Ad is ad (see (1.3.4)). The weights of the adjoint representation acting on $\mathfrak{g}$ equipped with $(\cdot, \cdot)$ are called the *roots* of G. The weight space corresponding to the root α is called the *root space* and denoted g_α. So

$$g_\alpha = \{X \in \mathfrak{g}_\mathbb{C}; \operatorname{ad}(H)X = i\alpha(H)X \text{ for all } H \in \mathfrak{t}\}.$$

Let $\mathcal{P}$ be the set of all roots of G. We list some useful facts about roots (for proofs see e.g. Helgason [88] Theorem 4.2 pp. 166–168 or Knapp [120] pp. 140–142)[7]:

- If $\alpha, \beta \in \mathcal{P}$ with $\alpha + \beta \neq 0$, then $\alpha + \beta \in \mathcal{P}$. Indeed this is a straightforward consequence of the Jacobi identity.
- If $\alpha \in \mathcal{P}$, then $-\alpha \in \mathcal{P}$.
- $\dim(g_\alpha) = 1$ for all $\alpha \in \mathcal{P}$.
- We have the direct sum decomposition:

$$\mathfrak{g}_{\mathbb{C}} = \mathfrak{t}_{\mathbb{C}} \oplus \bigoplus_{\alpha \in \mathcal{P}} g_\alpha. \tag{2.5.18}$$

In fact $\mathfrak{t}_{\mathbb{C}}$ is orthogonal to g_α for all $\alpha \in \mathcal{P}$ and g_α is orthogonal to g_β if $\alpha+\beta \neq 0$. For all $\alpha \in \mathcal{P}$, we choose vectors $E_\alpha \in g_\alpha$ with $(E_\alpha, E_{-\alpha}) = 1$, and call these *root vectors*. Recall the vectors $\{H_\gamma, \gamma \in \mathfrak{t}^*\}$ that were defined at the end of Sect. 2.5.1. It can be shown that for each $\alpha \in \mathcal{P}$,

$$[H_\alpha, E_\alpha] = i\alpha(H_\alpha)E_\alpha, \tag{2.5.19}$$

and

$$[E_\alpha, E_{-\alpha}] = iH_\alpha \tag{2.5.20}$$

(see e.g. Knapp [120], p. 143). Note that $E_{-\alpha}$ is the dual basis vector to E_α (in the sense of Sect. 1.3.5) with respect to the decomposition (2.5.18).

We adopt a convention for giving roots a sign as follows. Pick $v \in \mathfrak{t}$ such that $\mathcal{P} \cap \{\eta \in \mathfrak{t}^*; \eta(v) = 0\} = \emptyset$. Now define $\mathcal{P}_+ = \{\alpha \in \mathcal{P}; \alpha(v) > 0\}$ and $\mathcal{P}_- = \{\alpha \in \mathcal{P}; \alpha(v) < 0\}$, so that $\mathcal{P} = \mathcal{P}_+ \cup \mathcal{P}_-$. We call $\mathcal{P}_+$ (respectively $\mathcal{P}_-$) the set of *positive* (respectively *negative*) roots. We can always find a subset $\mathcal{Q} \subset \mathcal{P}_+$ so that $\mathcal{Q}$ forms a basis for $\mathfrak{t}^*$, and every $\alpha \in \mathcal{P}$ is a linear combination of elements of $\mathcal{Q}$ with integer coefficients, all of which are either nonnegative or nonpositive (see e.g. Knapp [120], Proposition 2.4.1, p. 155 or Simon [188], Theorem VIII.6.1, p. 184). The elements of $\mathcal{Q}$ are called *simple* or *fundamental* roots. A key role in the sequel will be played by the famous *half sum of positive roots* which is defined as

$$\rho := \frac{1}{2} \sum_{\alpha \in \mathcal{P}_+} \alpha. \tag{2.5.21}$$

The study of roots is very important for some aspects of Lie theory, particularly the classification of simple complex Lie algebras. Our main interest in roots is in enabling us to understand the general pattern of weights, and the next result is key in this respect.

[7] Note that both Helgason and Knapp work with complex semisimple Lie algebras and utilise the (non-degenerate) Killing form instead of an Ad-invariant inner product, but the proofs of those facts that we need transfer easily to our context.

Lemma 2.5.3 *Let π be a finite-dimensional representation of G, λ be a weight of π and α be a root. Then either $\lambda + \alpha = 0$ or $\lambda + \alpha$ is a weight of π.*

Proof Let $v \in V_\lambda$ and $Y \in g_\alpha$. Then for all $H \in \mathfrak{t}$,

$$\begin{aligned} d\pi(H)d\pi(Y)v &= d\pi(Y)d\pi(H)v + d\pi([H, Y])v \\ &= i(\lambda(H) + \alpha(H))d\pi(Y)v, \end{aligned}$$

and so either $\lambda + \alpha = 0$, or $\lambda + \alpha$ is a weight of π with weight vector $d\pi(Y)v$. □

2.5.5 The Highest Weight

The next result is important as it enables us to begin the process of identifying a specific weight that can characterise an irreducible representation (up to equivalence).

Theorem 2.5.3 *Let π be an irreducible representation of G. Then there exists a weight λ of π which is such that all other weights of π take the form*

$$\mu = \lambda - \sum_{\alpha \in Q} n_\alpha \alpha$$

where each n_α is a non-negative integer, with at least one n_α non-zero.

Proof (Sketch based on Simon [188] Theorem IX.4.3, p. 216). We may choose a weight λ such that λ has maximal norm amongst all the weights and $(\lambda, \alpha) \geq 0$ for all $\alpha \in Q$. Let $\mu \in \mathfrak{t}^*$ be of the form $\lambda + \sum_{\alpha \in Q} n_\alpha \alpha$, where each $n_\alpha \geq 0$ and at least one $n_\alpha > 0$. Then

$$\begin{aligned} |\mu|^2 &= |\lambda|^2 + 2 \sum_{\alpha \in Q} n_\alpha (\lambda, \alpha) + \left| \sum_{\alpha \in Q} n_\alpha \alpha \right|^2 \\ &> |\lambda|^2, \end{aligned}$$

and so μ cannot be a weight of π.

Now let $v \in V_\lambda$. We can extend the representation $d\pi$ of $\mathfrak{g}$ to an irreducible representation of the universal enveloping algebra $\mathcal{U}(\mathfrak{g})$. Clearly $\mathcal{U}(\mathfrak{g})v$ is an invariant subspace for $\mathcal{U}(\mathfrak{g})$ and so by irreducibility, $V_\pi = \mathcal{U}(\mathfrak{g})v$. By the Poincaré-Birkhoff-Witt theorem (Theorem 1.3.2) and (2.5.18), it can be shown that a basis for $\mathcal{U}(\mathfrak{g})$ consists of all possible monomials of the form

$$E_{-\alpha_1}^{m_1} E_{-\alpha_2}^{m_2} \cdots E_{-\alpha_k}^{m_k} H_1^{n_1} H_2^{n_2} \cdots H_r^{n_r} E_{\alpha_1}^{p_1} E_{\alpha_2}^{p_2} \cdots E_{\alpha_k}^{p_k},$$

where $H_1, \ldots, H_r$ is a basis for t, $k = \frac{1}{2}(d - r)$ and each $m_i, n_j, p_l \in \mathbb{Z}_+$. If we imitate the proof of Lemma 2.5.3 and first replace Y therein by the monomial $E_{\alpha_1}^{p_1} E_{\alpha_2}^{p_2} \cdots E_{\alpha_k}^{p_k}$, we obtain $\lambda + \sum_{i=1}^{k} p_i \alpha_i$, which we have just seen cannot be a weight. It is easy to see that replacing Y by any element of t leaves V_λ invariant. Finally, the action of $E_{-\alpha_1}^{m_1} E_{-\alpha_2}^{m_2} \cdots E_{-\alpha_k}^{m_k}$ yields $\lambda - \sum_{i=1}^{k} m_i \alpha_i$ and this is either zero or a weight by Lemma 2.5.3. □

The weight λ which appears in Theorem 2.5.3 is called the *highest weight* of π. If λ is the highest weight, then it can be shown that $\dim(V_\lambda) = 1$ (see e.g. Knapp [120], Proposition 5.11(b), pp. 284–285). By (2.5.15) it is clear that weights are invariant under intertwining, and so there is a one-to-one correspondence between $\widehat{G}$ and the space D of highest weights of all irreducible representations of G. We can thus parameterise $\widehat{G}$ by D, and this is a key step for Fourier analysis on nonabelian compact Lie groups. Indeed, the lattice structure on the weights means that we can effectively represent each element of $\widehat{G}$ by a string of integers.

2.5.6 Weights, Roots and the Casimir Spectrum

Recall the root space decomposition (2.5.18) and the discussion of the properties of the root vectors that followed this. Choose an orthonormal basis $\{T_i, 1 \leq i \leq r\}$ for t. Using (1.3.10), (2.5.18) and (2.5.20) we may then write the Laplacian as

$$\begin{aligned}\Delta &= \sum_{i=1}^{r} T_i^2 + \sum_{\alpha \in \mathcal{P}_+} (E_{-\alpha} E_\alpha + E_\alpha E_{-\alpha}) \\ &= \sum_{i=1}^{r} T_i^2 + \sum_{\alpha \in \mathcal{P}_+} (2E_{-\alpha} E_\alpha + i H_\alpha).\end{aligned}$$

Now let π_λ be an irreducible representation of G with highest weight λ. Let Ω_λ be the corresponding Casimir operator, so that $\Omega_\lambda = -\kappa_\lambda I_\pi$ and $\{\kappa_\lambda, \lambda \in D\}$ is the Casimir spectrum.

Theorem 2.5.4 *For each* $\lambda \in D$,

$$\kappa_\lambda = |\lambda + \rho|^2 - |\rho|^2, \tag{2.5.22}$$

where ρ *is the half-sum of positive roots.*

Proof We follow the approach of Lemma 1.1 (2) in Sugiura [200]. Relative to the decomposition of the Laplacian that we have just derived, we have

$$\Omega_\lambda = \sum_{i=1}^{r} d\pi_\lambda(T_i)^2 + \sum_{\alpha \in \mathcal{P}_+} (2d\pi_\lambda(E_{-\alpha}) d\pi_\lambda(E_\alpha) + i d\pi_\lambda(H_\alpha)).$$

Choose $v_\lambda \in V_\lambda$ with $v_\lambda \neq 0$. Then $\Omega_\lambda v_\lambda = -\kappa_\lambda v_\lambda$. Since λ is the highest weight, $\lambda + \alpha = 0$ for $\alpha \in \mathcal{P}_+$ by Lemma 2.5.3, and so $d\pi_\lambda(E_\alpha)v_\lambda = 0$. It follows by (2.5.16) that

$$\Omega_\lambda v_\lambda = \sum_{i=1}^{r} d\pi_\lambda(T_i)^2 v_\lambda + i \sum_{\alpha \in \mathcal{P}_+} d\pi_\lambda(H_\alpha) v_\lambda$$

$$= -\sum_{i=1}^{r} \lambda(T_i)^2 v_\lambda - \sum_{\alpha \in \mathcal{P}_+} \lambda(H_\alpha) v_\lambda.$$

But $\sum_{i=1}^{r} \lambda(T_i)^2 = |\lambda|^2$ and for each $\alpha \in \mathcal{P}$, $\lambda(H_\alpha) = (\lambda, \alpha)$ by (2.5.17). Hence

$$\kappa_\lambda = |\lambda|^2 + 2(\lambda, \rho)$$

and (2.5.22) follows. □

The following corollary gives useful estimates that we will utilise later.

Corollary 2.5.1 *For all $\lambda \in D$,*

$$|\lambda|^2 \leq \kappa_\lambda \leq C(1 + |\lambda|^2), \tag{2.5.23}$$

where $C = \max\{2, |\rho|^2\}$.

Proof By (2.5.22), $\kappa_\lambda = |\lambda|^2 + 2(\lambda, \rho)$ and since $(\lambda, \rho) \geq 0$, the left hand inequality is immediate. For the right hand one, we apply the Cauchy-Schwarz inequality to obtain

$$\kappa_\lambda \leq |\lambda|^2 + 2|\lambda|.|\rho| \leq 2|\lambda|^2 + |\rho|^2,$$

and the result follows. □

2.5.7 The Weyl Group and Dominant Weights

Let $\mathcal{T}$ be a maximal torus of G. The *Weyl group* of G is defined to be

$$W = N(\mathcal{T})/\mathcal{T},$$

where $N(\mathcal{T})$ is the *normaliser* of $\mathcal{T}$; $N(\mathcal{T}) := \{g \in G; g\mathcal{T}g^{-1} = \mathcal{T}\}$. For each $s \in W$, there exists $x \in G$ so that $s(h) = xhx^{-1}$, for all $h \in \mathfrak{t}$. Since each element of $\mathcal{T}$ is of the form $\exp(X)$ for some $X \in \mathfrak{t}$, we can consider $s \in W$ as acting on $\mathfrak{t}$ by the formula $s(\exp(X)) = \exp(s(X))$, and we also obtain an action on $\mathfrak{t}^*$ by duality: $s(\alpha)(X) = \alpha(s(X))$ for $\alpha \in \mathfrak{t}^*$.[8] Proofs of the following facts about the Weyl group can be found in Fegan [66] pp. 71–75 or Simon [188] Sect. VIII.8, pp. 192–196.

[8] Of course, there is some abuse of notation here.

Let $\alpha \in \mathcal{P}$, and consider the hyperplane $Y_\alpha := \{\beta \in \mathfrak{t}^*; (\alpha, \beta) = 0\}$. It is a fact that W is a finite group that is generated by the reflections $\{s_\alpha, \alpha \in \mathcal{P}\}$ in these hyperplanes. To be precise, for each $\alpha, \beta \in \mathcal{P}$

$$s_\alpha(\beta) = \beta - 2\frac{(\alpha, \beta)}{(\alpha, \alpha)}\alpha.$$

For future reference we define for each $s \in W$,

$$(-1)^s = \begin{cases} 1 & \text{if } s \text{ is an even product of reflections,} \\ -1 & \text{if } s \text{ is an odd product of reflections.} \end{cases}$$

The connected components of $\mathfrak{t}^* \setminus \bigcup_{\alpha \in \mathcal{P}} Y_\alpha$ are called *Weyl chambers.* It can be shown that W permutes the Weyl chambers. We pick once and for all a Weyl chamber C_0, and let C be its closure. C is called the *dominant chamber* in $\mathfrak{t}^*$, and any weight that lies in C is said to be *dominant.* It is a fact that there is a one-to-one correspondence between the highest weights and the dominant weights, and so between dominant weights and elements of $\widehat{G}$. Thus we can and will identify the sets $C \cap \mathbb{I}^*$, D and $\widehat{G}$ in the sequel.

Example: SU(n). In this case,

$$\mathfrak{t} = \{\text{diag}(i\theta_1, \ldots, i\theta_n); \theta_1 + \cdots + \theta_n = 0\},$$

and so $r = n - 1$. Let E_{ij} be the $n \times n$ matrix that is 1 in its (i, j)th position and 0 elsewhere. An easy computation shows that for all $X \in \mathfrak{t}$,

$$\text{ad}(X)(E_{ij}) = i(\theta_i - \theta_j)E_{ij},$$

and we thus deduce that the roots of $SU(n)$ are $\mathcal{P} = \{\pm(\theta_i^* - \theta_j^*); i < j\}$, where $\theta_i^*(\theta_k) = \delta_{ik}, 1 \leq i, k \leq n$. It follows that $\mathcal{P}_+ = \{\theta_i^* - \theta_j^*; i < j\}$. Moreover the simple roots are precisely those of the form $\theta_i^* - \theta_{i+1}^*$, for $1 \leq i \leq n - 1$. The Weyl group W is the symmetric group on n letters which acts on $\mathfrak{t}$ as follows: for each $w \in W$,

$$\text{diag}(i\theta_1, \ldots, i\theta_n) \to \text{diag}(i\theta_{w(1)}, \ldots, i\theta_{w(n)}).$$

Every weight is of the form $\sum_{i=1}^n m_i\theta_i^*$, where $m_1, \ldots, m_n \in \mathbb{Z}$ with $m_1 + \cdots + m_n = 0$, and the dominant weights are precisely those for which $m_1 \geq m_2 \geq \cdots \geq m_n$. To simplify even further, consider $SU(2)$. Here there are only two roots: $\theta_1^* - \theta_2^*$ and $\theta_2^* - \theta_1^*$. The Weyl group is $W = \{e, w\}$, where w is the reflection in the hyperplane $\theta_1^* = \theta_2^*$. The dominant weights are precisely those of the form $m(\theta_1^* - \theta_2^*)$, where $m \geq 0$, so that we have a bijection between the sets $\widehat{G}$ and $\mathbb{Z}_+$. We will have much more to say about the irreducible representations of $SU(2)$ in Sect. 2.6.

2.5.8 The Weyl Formulae

Let $X \in \mathfrak{t}$ and $\alpha \in \mathfrak{t}^*$. In the sequel we will use the convenient notation $e_\alpha(X) := e^{2\pi i\alpha(X)}$. A key role in this section will be played by the linear functionals A_α defined on $\mathfrak{t}$ by

$$A_\alpha(X) = \sum_{s \in W} (-1)^s e_{s(\alpha)}(X) \tag{2.5.24}$$

for all $X \in \mathfrak{t}$ and $\alpha \in \mathfrak{t}^*$,

We will be particularly interested in the case where $\alpha = \rho$ (the famous "half sum of roots" as defined in (2.5.21)). In particular it can be shown that (see Simon [188] Theorem IX.2.8, p. 213 for the first two of these):

- $A_\rho = \prod_{\alpha \in \mathcal{P}_+} (e_{\frac{\alpha}{2}} - e_{-\frac{\alpha}{2}})$,
- A_ρ^2 extends to a linear functional on $\mathcal{T}$ by the prescription

$$A_\rho^2(\exp(X)) := A_\rho(X)^2,$$

- $$\int_{\mathcal{T}} A_\rho^2(h)dh = \#W. \tag{2.5.25}$$

  Indeed, this follows directly from (2.5.24) and Plancherel's theorem on the r-torus $\mathcal{T}$.

We state but do not prove the following celebrated result (see e.g. Simon [188] pp. 214–215 for the proof).

Theorem 2.5.5 (Weyl's Integral Formula) *Let f be a bounded measurable central function on G. Then for any maximal torus $\mathcal{T}$ in G,*

$$\int_G f(g)dg = \frac{1}{\#W} \int_{\mathcal{T}} f(h)A_\rho^2(h)dh.$$

A key technique of the proof of Theorem 2.5.5 is to use the mapping $J : \mathcal{T} \times G/\mathcal{T} \to G$ given by $J([g], h) = hgh^{-1}$, to rewrite the integral over G as a double integral over $\mathcal{T} \times G/\mathcal{T}$, with respect to the product of Haar measure on $\mathcal{T}$ and the induced normalised measure on $G/\mathcal{T}$. The absolute value of the Jacobian of the transformation is A_ρ^2, and the corresponding integrand has no $G/\mathcal{T}$ dependence, and so the integral over that space is simply 1.

Now let χ_π be a character of the irreducible representation π. We recall that characters are central functions, and so they are determined by their values on a given maximal torus $\mathcal{T}$. If $X \in \mathfrak{t}$, we often abuse notation to write $\chi_\pi(X) := \chi_\pi(\exp(X))$. Furthermore, if π has highest weight λ we will write χ_λ and d_λ interchangeably, with χ_π and d_π (respectively), where we recall that d_π is the dimension of the representation space V_π. We sketch proofs of the next two results, but we omit many of the details. The proofs are based on those given in Simon [188] Chapter IX, and the reader who wants a thorough treatment is directed there (see also Chap. 9 of Fegan [66] and Sects. A.6.3 and A.6.4 of Hall [77]).

Theorem 2.5.6 (Weyl's Character Formula) *Let χ_λ be the character of the irreducible representation with highest weight λ. Then for all regular $g \in G$:*

$$\chi_\lambda(g) = \frac{A_{\lambda+\rho}(g)}{A_\rho(g)}.$$

Proof (Sketch based on Simon [188] Theorem IX.5.1, pp. 217–218). Let $\mathcal{W}(\pi)$ be the set of all weights of π. First, note that for all $X \in \mathfrak{t}$,

$$\chi_\lambda(X) = \mathrm{tr}(\exp(X)) = \sum_{\nu \in \mathcal{W}(\pi)} m_\nu e_\nu(X),$$

where $m_\nu \in \mathbb{N}$. Define $L_\lambda = \{\nu + s(\rho); \nu \in \mathcal{W}(\pi), s \in W\}$. Then by (2.5.24),

$$A_\rho \chi_\lambda = \sum_{\eta \in L_\lambda} n_\eta e_\eta,$$

where $n_\eta \in \mathbb{Z}$. Now since $\lambda + \rho$ can be written in only one way in the form $\nu + s(\rho)$, we can extract it from the last sum and rewrite:

$$A_\rho \chi_\lambda = n_\lambda A_{\rho+\lambda} + \sum_{\eta \in L_\lambda, \eta \neq s(\lambda+\rho)} n_\eta e_\eta.$$

By the usual Plancherel theorem on the r-torus and making use of (2.5.25), we obtain

$$\int_T |A_\rho(h)\chi_\lambda(h)|^2 dh = n_\lambda^2 \# W + \sum_{\eta \in L_\lambda, \eta \neq s(\lambda+\rho)} n_\eta^2. \tag{2.5.26}$$

But by the Weyl integral formula (Theorem 2.5.5) and orthonormality of characters (Theorem 2.4.3), we also have

$$1 = \int_G |\chi_\lambda(g)|^2 dg = \frac{1}{\# W} \int_T |A_\rho(h)\chi_\lambda(h)|^2 dh. \tag{2.5.27}$$

Comparing the Eqs. (2.5.26) and (2.5.27), we deduce that $n_\lambda = 1$ and $n_\eta = 0$ for all $\eta \in L_\lambda, \eta \neq s(\lambda + \rho)$. So $A_\rho \chi_\lambda = A_{\lambda+\rho}$, and the result follows. □

For $G = SU(2)$ (which has rank 1), the characters may be found by direct computation and we will present the calculations in Sect. 2.6.2. But we state the result now. As we have seen, in this case, there is a bijection between $\widehat{G}$ and $\mathbb{Z}_+$, and then for each $m \in \mathbb{Z}_+$,

$$\chi_m(\theta) = \frac{\sin((m+1)\theta)}{\sin(\theta)}$$

for $\theta \in (0, \pi) \cup (\pi, 2\pi)$.

The final key result in the section is Weyl's beautiful formula for the dimensions of the representation spaces. At first glance we might think that this is directly deducible from Theorem 2.5.6 as $d_\pi = \chi_\pi(e)$, but of course e is not a regular point, indeed it is a point of intersection of all the maximal tori in G. In fact there is a singularity of order $\#\mathcal{P}_+$ in A_ρ, and we need to use differentiation to remove this.

Theorem 2.5.7 (Weyl's Dimension Theorem) *For an irreducible representation π of G with highest weight λ,*

$$d_\lambda = \frac{\prod\limits_{\alpha \in \mathcal{P}_+} (\alpha, \lambda + \rho)}{\prod\limits_{\alpha \in \mathcal{P}_+} (\alpha, \rho)}. \tag{2.5.28}$$

Proof (Sketch based on Simon [188] Theorem IX.6.1, pp. 219–21). For each $\alpha \in \mathfrak{t}^*$, let Y_α be the unique vector in $\mathfrak{t}$ for which $(\alpha, \beta) = \beta(Y_\alpha)$ for all $\beta \in \mathfrak{t}^*$. For a smooth function f on $\mathcal{T}$, define its directional derivative D_α in the direction α as

$$D_\alpha f(X) = \lim_{t \to 0} \frac{1}{t}(f(X + tY_\alpha) - f(X)),$$

for all $X \in \mathcal{T}$. It is straightforward to verify that

$$\begin{aligned} D_\alpha e_\beta(X) &= D_\alpha e^{2\pi i \beta(X)} \\ &= 2\pi i \beta(Y_\alpha) e_\beta(X) \\ &= 2\pi i (\alpha, \beta) e_\beta(X). \end{aligned}$$

Now define $D := \prod_{\alpha \in \mathcal{P}_+} D_\alpha$. Then we have

$$d_\lambda = \chi_\lambda(e) = \frac{DA_{\lambda+\rho}(e)}{DA_\rho(e)}.$$

Repeated differentiation within (2.5.24) yields

$$TA_\beta = \#W(2\pi i)^{\#\mathcal{P}_+} \prod_{\alpha \in \mathcal{P}_+} (\alpha, \beta),$$

and the required result follows. □

The main use that we will make of Weyl's wonderful formulae within the main part of the book is an estimate that is obtained as a straightforward corollary of Weyl's dimensional formula.

Corollary 2.5.2 *There exists $C > 0$ so that for any irreducible representation π of G with highest weight λ*

$$d_\lambda \leq C|\lambda|^m,$$

where $m = \#\mathcal{P}_+$.

Proof Apply the Cauchy-Schwarz inequality within the numerator of the Weyl dimension formula and use Lemma 2.5.3 to deduce that

$$|\lambda + \rho| \leq |\lambda + \alpha_1 + \alpha_2 + \cdots + \alpha_m| \leq |\lambda|.$$

Here we have used the fact that $|\lambda + 2\rho| \geq |\lambda + \rho|$ which follows from $(\lambda, \rho) > 0$. We then obtain the required result with $C = \dfrac{\prod_{\alpha \in \mathcal{P}_+} |\alpha|}{\left|\prod_{\alpha \in \mathcal{P}_+} (\alpha, \rho)\right|}$. □

2.6 Irreducible Representations of $SU(2)$

In this section, we will compute all the irreducible representations of $SU(2)$. As we will see, these are explicitly related to the irreducible representations of its Lie algebra $\mathbf{su}(2)$, and that of the related complex Lie algebra $\mathbf{sl}(2, \mathbb{C})$ and our first task will be to understand these. Our approach is closely based on those of Faraut [63] pp. 136–142 and Želobenko [220] pp. 90–101. We have previously discussed representations of Lie algebras in the text, so we begin by recalling the definition. Let g be a Lie algebra over the complex numbers. We say that ξ is a *(finite dimensional) representation* of g in a finite-dimensional complex vector space V_ξ if ξ is a complex linear map from g to $\mathcal{L}(V_\xi)$ for which $\xi[X, Y] = [\xi(X), \xi(Y)]$ for all $X, Y \in g$, where the second bracket is understood in the sense of matrix multiplication. The notions of *subrepresentation* and *irreducible representation* are defined just as in the group theoretic case. Two representations ξ_1 and ξ_2 are *equivalent* if there is a complex-linear isomorphism $T : V_{\xi_1} \to V_{\xi_2}$ so that $T\xi_1(X) = \xi_2(X)T$ for all $X \in g$.

2.6.1 Irreducible Representations of $\mathbf{sl}(2, \mathbb{C})$

The real Lie algebra $\mathbf{su}(2)$ comprises all skew-hermitian 2×2 matrices having zero trace. A straightforward computation verifies that a generic element A of $\mathbf{su}(2)$ is of the form

$$A = \begin{pmatrix} i\alpha & \beta + i\gamma \\ -\beta + i\gamma & -i\alpha \end{pmatrix},$$

where $\alpha, \beta, \gamma \in \mathbb{R}$. Indeed, $\dim_{\mathbb{R}}(\mathbf{su}(2)) = 3$, and we easily see that a basis for $\mathbf{su}(2)$ is given by $\{L_0, L_1, L_2\}$, where

$$L_0 = \frac{i}{2}\begin{pmatrix} 1 & 0 \\ 0 & -1 \end{pmatrix}, \; L_1 = \frac{i}{2}\begin{pmatrix} 0 & 1 \\ 1 & 0 \end{pmatrix}, \; L_2 = \frac{i}{2}\begin{pmatrix} 0 & i \\ -i & 0 \end{pmatrix}.$$

Readers who have studied some quantum mechanics will recognise $\frac{2}{i}L_j (j = 0, 1, 2)$ as the *Pauli spin matrices*. Some routine algebra establishes the following commutation relations:

$$[L_0, L_1] = L_2, \; [L_2, L_0] = L_1, \; [L_1, L_2] = L_0. \tag{2.6.29}$$

We define the *raising operator*

$$L_+ = -iL_1 - L_2 = \begin{pmatrix} 0 & 1 \\ 0 & 0 \end{pmatrix}$$

and the *lowering operator*

$$L_- = -iL_1 + L_2 = \begin{pmatrix} 0 & 0 \\ 1 & 0 \end{pmatrix}.$$

We also find it convenient to make the minor change $L_3 := iL_0$ and then we have the commutation relations

$$[L_+, L_3] = L_+, \; [L_3, L_-] = L_-, \; [L_-, L_+] = 2L_3. \tag{2.6.30}$$

The matrices L_-, L_+ and L_3 are no longer elements of $\mathbf{su}(2)$, but they form a basis for the complex Lie algebra $\mathbf{sl}(2, \mathbb{C})$ of all 2×2 complex matrices having zero trace. Now $\mathbf{sl}(2, \mathbb{C})$ is the *complexification* of the real Lie algebra $\mathbf{su}(2)$, in that any matrix $X \in \mathbf{sl}(2, \mathbb{C})$ has the form $Y + iZ$ where $Y, Z \in \mathbf{su}(2)$. This suggests that representations of $\mathbf{su}(2)$ might be closely related to those of $\mathbf{sl}(2, \mathbb{C})$, and we will see that this is indeed the case. Another reason for introducing L_+ and L_- is that they enable us to compute the spectrum of L_3, and this is a key step in the analysis that follows.

Let ξ be a finite-dimensional representation of $\mathbf{sl}(2, \mathbb{C})$. To simplify notation we always write E_3, E_+ and E_- for $\xi(L_3)$, $\xi(L_+)$ and $\xi(L_-)$ (respectively).

Lemma 2.6.1 *Suppose that there exists $\psi \in V_\xi$ so that $E_3\psi = \lambda\psi$ for some $\lambda \in \mathbb{C}$. Then*

$$E_3(E_+\psi) = (\lambda - 1)E_+\psi, \; E_3(E_-\psi) = (\lambda + 1)E_-\psi.$$

Proof By the commutation relations (2.6.30),

$$E_3(E_+\psi) = E_+(E_3\psi) - E_+\psi = (\lambda - 1)E_+\psi,$$

and the other relation is derived similarly. □

Theorem 2.6.1 *1. Every non-trivial irreducible finite-dimensional representation of* $\mathbf{sl(2,\mathbb{C})}$ *acts in a complex vector space* V_m *of dimension* $m+1$ *where* $m \in \mathbb{N}$.

2. There exists a basis $\{v_0, v_1, \ldots, v_m\}$ *of* V_m *such that for all* $k = 0, 1, \ldots, m$,

$$E_3 v_k = \frac{1}{2}(m - 2k)v_k, \; E_+ v_k = v_{k+1}, \; E_- v_k = k(m - k + 1)v_{k-1} \quad (2.6.31)$$

Proof 1. We choose an arbitrary non-trivial irreducible finite-dimensional representation ξ of $\mathbf{sl(2,\mathbb{C})}$ acting in a vector space V_ξ. Let λ be that element of the spectrum of E_3 that has maximum real part and let v_0 be the corresponding eigenvector. Then by Lemma 2.6.1, $E_- v_0 = 0$. For each $k \in \mathbb{N}$, define $v_k = E_+^k v_0$. Then by Lemma 2.6.1 again, $E_3 v_k = (\lambda - k)v_k$ and by finite-dimensionality, $v_{m+1} = 0$ for some $m \in \mathbb{N}$. Since they are eigenvectors of the same matrix corresponding to distinct eigenvalues, $\{v_0, v_1, \ldots, v_m\}$ are linearly independent. Let W_m be the linear span of these $m+1$ vectors. Then W_m is invariant under the action of E_3 and E_+. We will show that it is also invariant for E_-. To do this we introduce the matrix[9]

$$C = E_3^2 + \frac{1}{2}(E_+E_- + E_-E_+).$$

Then by the commutation relations (2.6.30) we deduce that

$$C = E_+E_- + E_3(E_3 + 1) = E_-E_+ + E_3(E_3 - 1).$$

From the first of these formulae, we find that $Cv_0 = \lambda(\lambda + 1)v_0$. By straightforward algebra using (2.6.30), we see that C commutes with E_+ and so for all $k = 1, 2, \ldots, m$

$$Cv_k = CE_+^k v_0 = \lambda(\lambda + 1)v_k,$$

and hence by linearity $C = \lambda(\lambda + 1)I$ on W. Now we have

$$E_-E_+ = C - E_3(E_3 - 1)$$

and applying both sides of this identity to the vector v_{k-1} yields

[9] This is the negation of the Casimir operator for $\mathbf{su(2)}$, i.e. $-C = \xi(L_0)^2 + \xi(L_1)^2 + \xi(L_2)^2$.

$$\begin{aligned} E_- v_k &= [(\lambda(\lambda+1)I - E_3(E_3-1)]v_{k-1} \\ &= [(\lambda(\lambda+1) - (\lambda-k)(\lambda-k-1)]v_{k-1} \\ &= k(2\lambda-k+1)v_{k-1}. \end{aligned}$$

From this it follows that E_- preserves W, and then by (2.6.30) we deduce that W is an invariant subspace for the action of ξ. By irreducibility it follows that $W = V_\xi$.

2. The proof will follow from the actions we have found of E_+, E_- and E_3 on the given basis elements, if we can show that $\lambda = 2m$ where $m \in \mathbb{N}$. To see this, we first observe that $E_3 v_m = (\lambda - m)v_m$ and $E_3 v_{m+1} = 0$, and so $C v_m = E_3(E_3-1)v_m$. From this we find that $\lambda(\lambda+1) = (\lambda-m)(\lambda-m-1)$. This yields the quadratic equation $m^2 + m(1-2\lambda) - 2\lambda = 0$ and so $m = 2\lambda$ as required. □

It follows from the proof of Theorem 2.6.1 that the equivalence classes of irreducible finite-dimensional representations of $\mathbf{sl}(2,\mathbb{C})$ are labelled by the non-negative half-integer or integer $l = \frac{m}{2}$ to which there corresponds a representation having dimension $2l+1$.

2.6.2 Construction of Irreducible Representations of $SU(2)$

The (non-compact) Lie group $SL(2,\mathbb{C})$ consists of all 2×2 complex matrices $g = \begin{pmatrix} \alpha & \beta \\ \gamma & \delta \end{pmatrix}$ for which $\alpha\delta - \beta\gamma = 1$ and $SU(2)$ is the compact subgroup obtained when we put $\gamma = \overline{\beta}$ and $\delta = \overline{\alpha}$. Let $m \in \mathbb{N}$ and define $\mathcal{P}_m$ to be the linear space of all homogeneous polynomials of degree m in the two indeterminates x and y. A basis for $\mathcal{P}_m$ is the set of all monomials $x^j y^{m-j}$ where $j = 0, 1, \ldots, m$. We also define $\mathcal{P}_0 = \mathbb{C}$. Then $\dim(\mathcal{P}_m) = m+1$ for all $m \in \mathbb{Z}_+$. We easily check that we obtain a linear representation π_m of $SL(2,\mathbb{C})$ on $\mathcal{P}_m$ for $m \in \mathbb{N}$ by the prescription

$$(\pi_m(g)f)(x,y) = f(\alpha x + \gamma y, \beta x + \delta y), \tag{2.6.32}$$

where $f \in \mathcal{P}_m$, and we take π_0 to be the trivial representation. The representation π_m restricts to a representation of $SU(2)$, which we also denote by π_m for each $m \in \mathbb{Z}_+$. By compactness, we can and will choose an inner product in $\mathcal{P}_m$ so that the restricted representation of $SU(2)$ is unitary (see Proposition 2.2.1).

For now we work with $SL(2,\mathbb{C})$ and consider the one-parameter subgroups given for each $t \in \mathbb{R}$ by

$$g_+(t) := \exp(tL_+) = \begin{pmatrix} 1 & t \\ 0 & 1 \end{pmatrix},$$

$$g_-(t) := \exp(tL_-) = \begin{pmatrix} 1 & 0 \\ t & 1 \end{pmatrix},$$

$$g_3(t) := \exp(tL_+) = \begin{pmatrix} e^{-\frac{t}{2}} & 0 \\ 0 & e^{\frac{t}{2}} \end{pmatrix}.$$

We compute $d\pi_m(L_3)$, $d\pi_m(L_+)$ and $d\pi_m(L_-)$ on $\mathcal{P}_m$. For the first of these we have for each $f \in \mathcal{P}_m$

$$d\pi_m(L_3)f(x, y) = \frac{d}{dt}\pi_m(\exp(tL_3)f(x, y)\Big|_{t=0} = \frac{d}{dt}f(e^{-\frac{t}{2}}x, e^{\frac{t}{2}}y)\Big|_{t=0},$$

and hence

$$d\pi_m(L_3)f = \frac{1}{2}\left(y\frac{\partial f}{\partial y} - x\frac{\partial f}{\partial x}\right). \tag{2.6.33}$$

Similarly we obtain

$$d\pi_m(L_-)f(x, y) = \frac{d}{dt}f(x + ty, y)\Big|_{t=0},$$

and

$$d\pi_m(L_+)f(x, y) = \frac{d}{dt}f(x, tx + y)\Big|_{t=0},$$

which yields

$$d\pi_m(L_-)f = y\frac{\partial f}{\partial x}, \tag{2.6.34}$$

$$d\pi_m(L_+)f = x\frac{\partial f}{\partial y}. \tag{2.6.35}$$

Now apply (2.6.33), (2.6.34) and (2.6.35) to each of the basis monomials $f_j = x^j y^{m-j}$ for $j = 0, 1, \dots, m$ to obtain

$$d\pi_m(L_3)f_j = \frac{1}{2}(m - 2j)f_j, d\pi_m(L_-)f_j = jf_{j-1}, d\pi_m(L_+)f_j = (m - j)f_{j+1}. \tag{2.6.36}$$

The reader will find it instructive to compare (2.6.36) with (2.6.31).

Theorem 2.6.2 *Every finite-dimensional irreducible representation of* $\mathbf{sl}(2, \mathbb{C})$ *is equivalent to* $d\pi_m$ *for some* $m \in \mathbb{Z}_+$.

Proof We recall the notation and result of Theorem 2.6.1. The case $m = 0$ is clear as both representations are then trivial. For $m \in \mathbb{N}$ we follow Faraut [63] p. 140 to find a linear invertible map A_m from V_m to $\mathcal{P}_m$ so that $A_m \xi_m(X) = d\pi_m(X) A_m$ for all $X \in \mathbf{sl}(\mathbf{2}, \mathbb{C})$. In fact we define A_m by complex linear extension of the prescription $A_m v_k = \gamma_k^{(m)} f_k$ for $k = 0, 1, \ldots, m$, where $\gamma_0^{(m)} := 1$ and $\gamma_k^{(m)} := m(m-1)\cdots(m-k+1)$. It is sufficient by linearity to check that the required result holds, using (2.6.36) and (2.6.31), in each of the three cases $X = L_+, L_-, L_3$. □

Now we are ready for the main result of this section

Theorem 2.6.3 *Every irreducible representation of $SU(2)$ is equivalent to π_m for some $m \in \mathbb{Z}_+$.*

Proof This is based on Faraut [63] p. 141. Let ρ be an arbitrary non-trivial irreducible representation of $SU(2)$ acting in some space V. Then $d\rho$ is a finite-dimensional representation of the Lie algebra $\mathbf{su(2)}$, and so extends $\mathbb{C}$-linearly to a finite-dimensional representation of $\mathbf{sl}(\mathbf{2}, \mathbb{C})$. We will show that $d\rho$ is irreducible. To obtain a contradiction, suppose that this is not the case, and let W be a non-trivial invariant subspace of $\mathbf{sl}(\mathbf{2}, \mathbb{C})$. Then it is also invariant for $\mathbf{su(2)}$. Hence for all $X \in \mathbf{su(2)}$ the matrix exponential $e^{d\rho(X)} = \sum_{n=0}^{\infty} \frac{d\rho(X)^n}{n!}$ leaves W invariant. In other words, $\rho(\exp(X)) = e^{d\rho(X)}$ leaves W invariant for all $X \in \mathbf{su(2)}$, But $SU(2)$ is compact and connected, and so exp is surjective. But then $\rho(g)W \subseteq W$ for all $g \in SU(2)$, which yields our desired contradiction. We may now appeal to Theorem 2.6.2 to deduce that there exists an invertible linear mapping T_m from V to $\mathcal{P}_m$ for some $m \in Z_+$ so that $T_m d\rho(X) T_m^{-1} = d\pi_m(X)$ for all $X \in \mathbf{sl}(\mathbf{2}, \mathbb{C})$. Now restrict X to $\mathbf{su(2)}$ and take matrix exponentials of both sides to deduce that $T_m \rho(\exp(X)) = \pi_m(\exp(X)) T_m$ for all $X \in \mathbf{su(2)}$. The result then follows by surjectivity of exp as above. □

As an application of this result, we can directly calculate all the characters of the irreducible representations of $SU(2)$. For each $\theta \in [0, 2\pi)$, consider

$$e(\theta) := \exp(2\theta L_0) = \exp(-2i\theta L_3) = \begin{pmatrix} e^{i\theta} & 0 \\ 0 & e^{-i\theta} \end{pmatrix}.$$

It follows from (2.6.36) that for all $\theta \in [0, 2\pi)$, $m \in \mathbb{Z}_+$,

$$\pi_m(e(\theta)) f_j = e^{i(2j-m)\theta} f_j$$

for each $j = 0, 1, \ldots, m$. Now $\{e(\theta), \theta \in [0, 2\pi)\}$ is a maximal torus in $SU(2)$ and so we have all the necessary ingredients to compute the characters. Indeed, by conjugation, for each $g \in SU(2)$ there exists $\theta \in [0, 2\pi)$ so that for each $m \in \mathbb{Z}_+$,

$$\begin{aligned}\chi_m(g) &= \mathrm{tr}(\pi_m(g)) \\ &= \mathrm{tr}(\pi_m(e(\theta))) \\ &= \sum_{j=0}^{m} e^{i(2j-m)\theta} \\ &= e^{-im\theta}\sum_{j=0}^{m} e^{2ij\theta} \\ &= \frac{e^{-im\theta} - e^{i(m+2)\theta}}{1 - e^{2i\theta}} \\ &= \begin{cases} \frac{\sin((m+1)\theta)}{\sin(\theta)} & \text{if } \theta \in (0, \pi) \cup (\pi, 2\pi), \\ m & \text{if } \theta = 0, \\ (-1)^m m & \text{if } \theta = \pi. \end{cases}\end{aligned}$$

We also note that the weights of the representation π_m are parametrised by $k - \frac{m}{2}$ for $k = 0, 1, 2, \ldots, m$ with highest weight $\frac{m}{2}$ corresponding to the weight vector v_m constructed in Theorem 2.6.1.

Using the fact that $SU(2)$ is a double covering of $SO(3)$ we can obtain all irreducible representations of the latter group from those of the former (in fact from those π_m for which m is even). We can also realise these representations of $SO(3)$ via actions on homogeneous harmonic polynomials in three indeterminates. For details see e.g. Faraut (pp. 142–149).

2.6.3 Matrix Elements of Irreducible Representations of $SU(2)$

This section is based closely on Želobenko [220] pp. 96–98 (see also Ruzhansky and Turunen [172] pp. 616–620). As a consequence of Theorem 2.6.2 we know that all the non-trivial irreducible representations of $SU(2)$ are equivalent to π_m acting on $\mathcal{P}_m$ for some $m \in \mathbb{N}$ as in (2.6.32). Note that as we observed in the previous section, π_m is also a linear representation of $SL(2, \mathbb{C})$ and we retain this level of generality for now. Observe that by homogeneity we may write each $f \in \mathcal{P}_m$ as $f(x, y) = y^m F(z)$ where $F(z) := f(z, 1)$ and $z = \frac{x}{y}$. Then (2.6.32) takes the form

$$\pi_m(g)F(z) = (\beta z + \delta)^m F\left(\frac{\alpha z + \gamma}{\beta z + \delta}\right). \tag{2.6.37}$$

Remark Let $\alpha, \beta, \gamma, \delta \in \mathbb{Z}$, so that we restrict g to lie in the closed subgroup $SL(2, \mathbb{Z})$. Then any suitable holomorphic function F which satisfies $\pi_m(g)F = F$

for all $g \in SL(2,\mathbb{Z})$ is a *modular form of weight m*. These functions play an important role in the study of elliptic curves (see e.g. Lozano-Robledo [137]).

From now on we find it convenient to to use the fact that $m = 2l$ and employ the basis vectors $z_\mu := x^{l-\mu}$ where $\mu = -l, -l+1, \ldots, l$. So the matrix elements of the representation are given by

$$\pi_m(g)z_\nu = \sum_{\mu=-l}^{l} \pi_{\mu\nu}^{(m)}(g) z_\mu,$$

where $g \in SU(2)$. Then by Taylor's theorem

$$\pi_{\mu\nu}^{(m)}(g) = \frac{1}{(l-\mu)!}\frac{d^{l-\mu}}{dz^{l-\mu}}[(\beta z+\delta)^{l+\nu}(\alpha z+\gamma)^{l-\nu}]\bigg|_{z=0}. \tag{2.6.38}$$

Observe that if we let $u = \beta z + \delta$ and $v = \alpha z + \gamma$, then $\alpha u - \beta v = \alpha\delta - \beta\gamma = 1$. Hence if we substitute $t = \alpha u$, we have $v = \frac{1}{\beta}(t-1)$, and using the fact that $\frac{d}{dt} = \alpha\beta\frac{d}{dz}$, we obtain

$$\begin{aligned}\pi_{\mu\nu}^{(m)}(g) &= \frac{(-1)^{l-\nu}}{(l-\mu)!}\alpha^{-(\mu+\nu)}\beta^{\nu-\mu}\frac{d^{l-\mu}}{dt^{l-\mu}}[t^{l+\nu}(1-t)^{l-\nu}]\bigg|_{t=\alpha\delta} \qquad (2.6.39)\\ &= \frac{(-1)^{l-\nu}}{(l-\mu)!}\alpha^{-(\mu+\nu)}\beta^{\nu-\mu}P_{\mu\nu}^{l}(\alpha\delta),\end{aligned}$$

where $P_{\mu\nu}^{l}(t) := \frac{d^{l-\mu}}{dz^{l-\mu}}[t^{l+\nu}(1-t)^{l-\nu}]$ are the *Jacobi polynomials*.

Finally, when we restrict $g \in SU(2)$ we let $s = \alpha\delta = |\alpha|^2$. Then $|\beta|^2 = 1-s$, and if we write $\theta := \arg(\alpha)$ and $\phi = \arg(\beta)$, then we obtain

$$\pi_{\mu\nu}^{(m)}(g) = e^{-i[(\mu+\nu)\theta+(\nu-\mu)\phi]}s^{-\frac{\mu+\nu}{2}}(1-s)^{-\frac{\mu-\nu}{2}}P_{\mu\nu}^{l}(s). \tag{2.6.40}$$

Notes 1. Slightly different formulae can be found in the literature, where the phase factor in (2.6.40) is expressed in terms of *Euler angles* instead of θ and ϕ.

2. The Jacobi polynomials are classical orthogonal polynomials on the space $L^2((-1,1),\rho)$, where $\rho(dx) = (1-x)^\mu(1+x)^\nu dx$ for $\mu,\nu > -1$. For a modern treatment see Beals and Wong [22] pp. 116–120. It is worth pointing out that these may be expressed as hypergeometric functions (up to a multiplicative factor). The Legendre and Chebyshev polynomials arise as Jacobi polynomials for which $\mu = \nu$ (and in the Legendre case, $\mu = \nu = 0$).
3. The reader is directed to Chap. 11 of Ruzhansky and Turunen [172], where many more explicit calculations are carried out in $SU(2)$.

2.7 Notes and Further Reading

Representations of finite nonabelian groups began with the work of Ferdinand Frobenius (1849–1917). Other pioneers in this area were William Burnside (1852–1927) and Issai Schur (1875–1941). The extension to continuous groups (including Lie groups) was initiated by Elie Cartan and Hermann Weyl (1885–1955). The important role of weights in studying representations seems to be due to Cartan. In particular, Weyl's contributions (including his character formula and dimension formula) were all developed during the period 1923–1938 and the famous Peter-Weyl theorem, which he established with his student F. Peter, was published in 1927.

The deep and beautiful relationship between group representations and harmonic analysis has began to emerge in this chapter. It is one of the main themes of this book. A wonderful historical survey of this interplay can be found in an essay of George Mackey (1916–2006) "Harmonic analysis as the exploitation of symmetry—a historical survey." Originally published in *Bull. Amer. Math. Soc.* **3**, 543–699 (1980) it is reprinted in the volume [141]. Another great book by Mackey which explores the impact of group representations in many areas of mathematics, including probability theory, is [140].

Mackey was himself a major contributor towards the theory of group representations through his work on *induced representations* of a locally compact group from a closed subgroup. This topic lies outside the scope of the present volume, but it plays an important role in obtaining the irreducible representations of semisimple Lie groups. An introduction to these ideas in the finite case can be found in Chapter V of Simon [188]. A thorough treatment can be found in the monograph of Knapp [119] from the semisimple perspective. For the general case, see the celebrated "Chicago lecture notes" of Mackey [139].

Many of the books that are mentioned at the end of Chap. 1 contain material on representation theory of compact groups, see e.g. Chapter V of Knapp [120], Chapters VII to IX of Simon [188] and Chaps. 7–9 of Fegan [66]. Bröcker and tom Dieck [36] is a graduate level textbook that is dedicated to this topic. Here the emphasis is very much on the structural point of view. The approach of this chapter has been heavily influenced by Faraut [63] (Chap. 6), which is written in a more analytic spirit. Žebolenko [220] is an older book that contains a great deal of interesting information about the compact case.

At the end of Sect. 2.6, we saw that the Jacobi polynomials are obtained as matrix elements of irreducible representations of $SU(2)$. The use of Lie groups to investigate special functions is a vast topic. Two classic references are Miller [149] and Vilenkin [210]. The latter has now expanded to a three-volume work [211] (co-authored with Klimyk).

Chapter 3
Analysis on Compact Lie Groups

Abstract We study the Laplacian from an analytic viewpoint as a self-adjoint operator with discrete eigenvalues given by the Casimir spectrum. This leads naturally to a study of Sobolev spaces, which are also characterised from a Fourier analytic viewpoint. We introduce Sugiura's zeta function as a tool to study regularity of Fourier series on groups. In particular, we find conditions for absolute and uniform convergence, and for smoothness. Smoothness is characterised by means of the Sugiura space of rapidly decreasing functions defined on the space of highest weights, and we will utilise this in the next chapter to study probability measures on groups that have smooth densities.

Throughout this chapter, G is a compact connected Lie group with Lie algebra $\mathfrak{g}$. We equip $\mathfrak{g}$ with an Ad-invariant inner product and fix an orthonormal basis $\{X_1, \ldots, X_d\}$ in $\mathfrak{g}$.

3.1 The Laplacian in $L^2(G)$

3.1.1 *Self-Adjointness and Spectrum*

We consider each element of $\mathcal{U}(\mathfrak{g})$ as a densely defined linear operator having domain $C^\infty(G)$ in $L^2(G)$. In particular we focus attention on the elements X of $\mathfrak{g}$ and the Laplacian $\Delta = \sum_{i=1}^d X_i^2$, which was introduced in Sect. 1.3.5.

Proposition 3.1.1 *1. Each $X \in \mathfrak{g}$ is skew-symmetric in $L^2(G)$ with domain $C^\infty(G)$.*

2. The operator Δ is symmetric in $L^2(G)$ with domain $C^\infty(G)$.

3. The operator $-\Delta$ is positive in $L^2(G)$ with domain $C^\infty(G)$.

Proof 1. Using a straightforward dominated convergence argument, for each $f \in C^\infty(G)$ we have

D. Applebaum, *Probability on Compact Lie Groups*, Probability Theory and Stochastic Modelling 70, DOI: 10.1007/978-3-319-07842-7_3,

$$\int_G Xf(g)dg = \frac{d}{dt}\int_G f(g\exp(tX))dg\Big|_{t=0} = \frac{d}{dt}\int_G f(g)dg = 0.$$

Hence for each $f_1, f_2 \in C^\infty(G)$, using the derivation property,

$$X(f_1\overline{f_2}) = Xf_1.\overline{f_2} + f_1.\overline{Xf_2},$$

we find that

$$\begin{aligned} 0 &= \int_G X(f_1(g)\overline{f_2(g)})dg \\ &= \int_G Xf_1(g)\overline{f_2(g)}dg + \int_G f_1(g)\overline{Xf_2(g)}dg \\ &= \langle Xf_1, f_2\rangle + \langle f_1, Xf_2\rangle. \end{aligned}$$

2. This follows easily from (1).
3. For each $f \in C^\infty(G)$, $-\langle \Delta f, f\rangle = \sum_{i=1}^d ||X_i f||^2 \geq 0$. □

Now let $\widehat{G}$ be the unitary dual of G and recall from Peter-Weyl theory (Corollary 2.2.4) that $\{d_\pi^{\frac{1}{2}}\pi_{ij}; 1 \leq i, j \leq d_\pi, \pi \in \widehat{G}\}$ is a complete orthonormal basis for $L^2(G)$.

Theorem 3.1.1 *For each* $\pi \in \widehat{G}, 1 \leq i, j \leq d_\pi, \pi_{ij} \in \mathrm{Dom}(\Delta)$ *and*

$$\Delta\pi_{ij} = -\kappa_\pi\pi_{ij}, \tag{3.1.1}$$

where $\{\kappa_\pi, \pi \in \widehat{G}\}$ *is the Casimir spectrum for* G.

Proof For each $1 \leq i, j \leq d_\pi$, $\pi_{ij} \in C^\infty(G) \subseteq \mathrm{Dom}(\Delta)$ by Theorem 2.5.1 and Proposition 3.1.1 (2). For each $g \in G$, using Theorem 2.5.2,

$$\begin{aligned} \Delta\pi_{ij}(g) &= \sum_{k=1}^d \frac{d^2}{dt^2}\pi_{ij}(g\exp(tX_k))\Big|_{t=0} \\ &= \sum_{l=1}^{d_\pi} \pi_{il}(g)\frac{d^2}{dt^2}\sum_{k=1}^d \pi_{lj}(\exp(tX_k))\Big|_{t=0} \\ &= \sum_{l=1}^{d_\pi} \pi_{il}(g)\sum_{k=1}^d d\pi_{lj}(X_k)^2 \\ &= \sum_{l=1}^{d_\pi} \pi_{il}(g)(\Omega_\pi)_{lj} \end{aligned}$$

$$= -\kappa_\pi \sum_{l=1}^{d_\pi} \pi_{il}(g)\delta_{lj}$$
$$= -\kappa_\pi \pi_{ij}(g).$$
□

As the symmetric linear operator Δ has a complete orthonormal basis of eigenvectors, it is not difficult to see that it is essentially self-adjoint (i.e. it has a unique self-adjoint extension). As this result is so important, we give a formal proof.

Proposition 3.1.2 *The operator Δ is essentially self-adjoint in $L^2(G)$.*

Proof Δ is essentially self-adjoint if $\mathrm{Ran}(\Delta \pm iI)$ is dense in $L^2(G)$ (see e.g. Reed and Simon [166], the corollary to Theorem VIII.3 on p. 257). Assume that $\psi \in L^2(G)$ with $\psi \perp \mathrm{Ran}(\Delta \pm iI)$. Then for all $\pi \in \widehat{G}$, $1 \leq i, j \leq d_\pi$, $\langle(\Delta \pm iI)\pi_{ij}, \psi\rangle = 0$ and so $(-\kappa_\pi \pm i)\langle\pi_{ij}, \psi\rangle = 0$, i.e. $\langle\pi_{ij}, \psi\rangle = 0$ hence $\psi = 0$ and the required result follows. □

In the sequel we will continue to use Δ to denote the unique self-adjoint extension of the Laplacian. In particular, we may use spectral theory to construct self-adjoint operators that are functions of the Laplacian, e.g. if F is a continuous function from $\mathbb{R}$ to $\mathbb{R}$, then by Theorem 3.1.1

$$F(\Delta) = \sum_{\pi \in \widehat{G}} F(-\kappa_\pi) P_\pi, \tag{3.1.2}$$

where P_π is the orthogonal projection from $L^2(G)$ onto $\mathcal{M}_\pi$ and the domain of $F(\Delta)$ is $\mathrm{Dom}(F(\Delta)) = \left\{\psi \in L^2(G); \sum_{\pi \in \widehat{G}} |F(-\kappa_\pi)|^2 ||P_\pi \psi||^2 < \infty\right\}$. Note that $F(\Delta)$ is a bounded operator if the restriction of F to $(-\infty, 0)$ is a bounded function. In particular, if we take $F(x) = e^{tx}$ for $t \geq 0, x \in \mathbb{R}$, we obtain the *heat semigroup* $(T_t, t \geq 0)$ where $T_t = e^{t\Delta}$. It follows from Theorem 1.3.3 (2) that if $C^\infty(G) \subseteq \mathrm{Dom}(F(\Delta))$, then for all $X \in \mathfrak{g}$, $f \in C^\infty(G)$,

$$F(\Delta)Xf = XF(\Delta)f. \tag{3.1.3}$$

The next result shows how to compute the Fourier transforms of the operators $F(\Delta)$:

Theorem 3.1.2 *If $f \in \mathrm{Dom}(\mathrm{F}(\Delta))$, then for all $\pi \in \widehat{G}$*

$$\widehat{F(\Delta)f}(\pi) = F(-\kappa_\pi)\widehat{f}(\pi). \tag{3.1.4}$$

Proof For all $\pi \in \widehat{G}$, $1 \leq i, j \leq d_\pi$, by (3.1.2)

$$\widehat{F(\Delta)f}(\pi)_{ij} = \int_G \sum_{\pi' \in \widehat{G}} F(-\kappa_{\pi'}) P_{\pi'} f(g) \pi_{ij}(g^{-1}) dg$$

$$
\begin{aligned}
&= \sum_{\pi' \in \widehat{G}} F(-\kappa_{\pi'}) \int_G P_{\pi'} f(g) \pi_{ij}(g^{-1}) dg \\
&= \sum_{\pi' \in \widehat{G}} F(-\kappa_{\pi'}) \langle P_{\pi'} f, \pi_{ji} \rangle \\
&= \sum_{\pi' \in \widehat{G}} F(-\kappa_{\pi'}) \langle f, P_{\pi'} \pi_{ji} \rangle \\
&= F(-\kappa_{\pi}) \langle f, \pi_{ji} \rangle \\
&= F(-\kappa_{\pi}) \widehat{f}(\pi)_{ij}.
\end{aligned}
$$

The interchange of integral and summation is justified by Fubini's theorem. Indeed, from the argument given above we see that it is sufficient to prove that

$$\int_G \sum_{\pi' \in \widehat{G}} |F(-\kappa_{\pi'}) P_{\pi'} f(g) P_{\pi'} \pi_{ij}(g^{-1})| dg < \infty.$$

However, by the Cauchy-Schwarz inequality

$$
\begin{aligned}
&\int_G \sum_{\pi' \in \widehat{G}} |F(-\kappa_{\pi'}) P_{\pi'} f(g) P_{\pi'} \pi_{ij}(g^{-1})| dg \\
&\leq \int_G \left(\sum_{\pi' \in \widehat{G}} |F(-\kappa_{\pi'}) P_{\pi'} f(g)|^2 \right)^{\frac{1}{2}} \left(\sum_{\pi' \in \widehat{G}} |P_{\pi'} \pi_{ij}(g^{-1})|^2 \right)^{\frac{1}{2}} dg \\
&\leq \left(\int_G \sum_{\pi' \in \widehat{G}} |F(-\kappa_{\pi'})|^2 |P_{\pi'} f(g)|^2 dg \right)^{\frac{1}{2}} \left(\int_G \sum_{\pi' \in \widehat{G}} |P_{\pi'} \pi_{ij}(g^{-1})|^2 dg \right)^{\frac{1}{2}} \\
&= ||F(\Delta) f||.||\pi_{ij}|| < \infty.
\end{aligned}
$$

□

Let $p \in \mathbb{N}$ and $f \in C^{2p}(G, \mathbb{C}) \subseteq \mathrm{Dom}(\Delta^p)$. Then by (3.1.4) we have for each $\pi \in \widehat{G}$, $\widehat{\Delta^p f}(\pi) = (-1)^p \kappa_\pi^p \widehat{f}(\pi)$. If π is non-trivial, we may invert this expression to find

$$\widehat{f}(\pi) = (-1)^p \kappa_\pi^{-p} \widehat{\Delta^p f}(\pi), \tag{3.1.5}$$

and we will find a use for this identity in Sect. 3.3 below.

The *heat equation* is the parabolic pde:

$$\frac{\partial u}{\partial t} = \Delta u. \tag{3.1.6}$$

It is shown in e.g. Rosenberg [169] that it has a fundamental solution $k \in C^\infty((0,\infty)\times G)$ called the *heat kernel*.[1] We write $k_t(g) := k(t,g)$ for all $t > 0, g \in G$, and we always normalise k_t so that $\int_G k_t(g)dg = 1$. In the (non-compact) case of $G = \mathbb{R}^d$, k_t is a Gaussian, i.e. $k_t(x) = \frac{1}{(4\pi t)^{\frac{d}{2}}} e^{-\frac{|x|^2}{4t}}$ while on the d-torus, k_t is a product of Jacobi theta functions (see e.g. Heyer [95] p. 375). In the general case of compact Lie groups there is typically no explicit known formula for k_t but we do have the following *heat kernel estimates* (see e.g. Varopoulos et al. [209]):

$$K_1 t^{-\frac{d}{2}} \exp\left\{-\frac{\rho(g)^2}{L_1 t}\right\} \leq k_t(g) \leq K_2 t^{-\frac{d}{2}} \exp\left\{-\frac{\rho(g)^2}{L_2 t}\right\}, \tag{3.1.7}$$

for each $g \in G$, where $K_i, L_i > 0$ $(i = 1, 2)$ and ρ is the metric induced on G by the inner product on $\mathfrak{g}$. The general solution of the initial value problem given by (3.1.6), together with the initial condition $u(0,\cdot) = f$, where $f \in C(G)$, is given by

$$u(t,\sigma) = \int_G f(\sigma\tau)k_t(\tau)d\tau, \tag{3.1.8}$$

and the initial condition is related to the general solution by the heat semigroup

$$T_t f(\sigma) = u(t,\sigma),$$

for all $t \geq 0, \sigma \in G$.

We will discuss the heat semigroup and its generalisations in greater detail later in Chap. 5.

3.1.2 Sobolev Spaces

For each $p \in \mathbb{N}$ we define a norm $||\cdot||_{p,2}$ on $C^\infty(G)$ by

$$||f||_{p,2} := ||(I-\Delta)^{\frac{p}{2}} f|| = \langle (I-\Delta)^p f, f\rangle^{\frac{1}{2}}. \tag{3.1.9}$$

The completion of $C^\infty(G)$ in the associated metric is called the *L^2-Sobolev space of order p* on G and denoted by $H_p(G)$. Clearly $H_q(G) \subseteq H_p(G)$ if $q > p$. An equivalent norm is found from the following estimate:

[1] Technically speaking, the heat kernel is the mapping $\tilde{k} \in C^\infty((0,\infty)\times G\times G, \mathbb{R})$ given by $\tilde{k}(t,g,h) = k(t, g^{-1}h)$.

Proposition 3.1.3 *For each $p \in \mathbb{N}$ there exists $0 < c_p < C_p < \infty$ such that for all $f \in C^\infty(G)$,*

$$c_p||f||_{p,2}^2 \le ||f||^2 + \sum_{k=1}^{p} \sum_{i_1,\ldots,i_k=1,\ldots,d} ||X_{i_1} \cdots X_{i_k} f||^2 \le C_p||f||_{p,2}^2. \quad (3.1.10)$$

Proof We proceed by induction. The case $p = 1$ is easy since by (3.1.9)

$$||f||_{1,2}^2 = ||f||^2 - \langle \Delta f, f \rangle = ||f||^2 + \sum_{i=1}^{d} ||X_i f||^2.$$

Now assume that the required result holds for some p. To establish the result for $p + 1$, replace f in (3.1.10) with $(I - \Delta)^{\frac{1}{2}} f$ and use the fact that, by (3.1.3), $(I - \Delta)^{\frac{1}{2}} X_{i_1} \cdots X_{i_k} f = X_{i_1} \cdots X_{i_k} (I - \Delta)^{\frac{1}{2}} f$ for each $i_1, \ldots, i_k = 1, \ldots, d, 1 \le k \le p$ to obtain

$$\begin{aligned} c_p||f||_{p+1,2}^2 \le ||(I - \Delta)^{\frac{1}{2}} f||^2 \sum_{k=1}^{p} \sum_{i_1,\ldots,i_k=1,\ldots,d} ||(I - \Delta)^{\frac{1}{2}} X_{i_1} \cdots X_{i_k} f||^2 \\ \le C_p||f||_{p+1,2}^2. \end{aligned}$$

But $||(I - \Delta)^{\frac{1}{2}} X_{i_1} \cdots X_{i_k} f||^2 = \langle (I - \Delta) X_{i_1} \cdots X_{i_k} f, X_{i_1} \cdots X_{i_k} f \rangle$, and the required result follows after some algebra. □

The Fourier transform gives us a very convenient expression for the Sobolev norm.

Proposition 3.1.4 *For each $p \in \mathbb{N}$ and for each $f \in C^\infty(G)$,*

$$||f||_{p,2}^2 = \sum_{\pi \in \widehat{G}} d_\pi (1 + \kappa_\pi)^p \text{tr}(\widehat{f}(\pi) \widehat{f}(\pi)^*).$$

Proof Using (3.1.9), the Parseval-Plancherel identity (2.3.9) and (3.1.4) we have

$$\begin{aligned} ||f||_{p,2}^2 &= \langle (I - \Delta)^p f, f \rangle \\ &= \sum_{\pi \in \widehat{G}} d_\pi \text{tr}(\widehat{(I - \Delta)^p f}(\pi) \widehat{f}(\pi)^*) \\ &= \sum_{\pi \in \widehat{G}} d_\pi (1 + \kappa_\pi)^p \text{tr}(\widehat{f}(\pi) \widehat{f}(\pi)^*). \end{aligned}$$

□

G is a Riemannian manifold under the given metric and the Sobolev spaces that we have introduced coincide with those that are known from that context. The following two important results are proved in e.g. Rosenberg [169] pp. 24–26, and we will be content to state them here.

Theorem 3.1.3 (Sobolev Embedding Theorem) *If $f \in H_p(G)$, then $f \in C^s(G)$ for all $s < p - \frac{d}{2}$.*

In particular, it follows that $\bigcap_{p\in\mathbb{N}} H_p(G) \subseteq C^\infty(G)$.

Theorem 3.1.4 (Rellich-Kondrachov Compactness Theorem) *If $p < q$, the inclusion of $H_q(G)$ into $H_p(G)$ is a compact operator.*

Theorem 3.1.4 may be applied to show that the heat semigroup operator T_t is compact in $L^2(G)$ for $t > 0$. We will give a different proof of this fact later on using a probabilistic approach, and we will show that these operators are also trace class.

3.2 Sugiura's Zeta Function

In this section, we introduce a function on the space D of highest weights of a compact group G which is in some sense, a generalisation of the famous Riemann zeta function. Our rationale for introducing this mapping is to use it as a tool to study smoothness, and we will not explore its number theoretic properties here.

Before we discuss Suguira's zeta function we establish two useful preparatory results.

Proposition 3.2.1 *Consider the formal Dirichlet series $S := \sum_{n=1}^{\infty} \frac{a_n}{n^s}$, where $s \in \mathbb{C}$ and $a_n \in \mathbb{N}$ for all $n \in \mathbb{N}$. Let $A_n = \sum_{i=1}^{n} a_i$. If $A_n = O(n^k)$ where $k > 0$, then S converges absolutely if $\Re(s) > k$.*

Proof Define $A_0 := 0$. Then

$$\begin{aligned} S &= \sum_{n=1}^{\infty} \frac{A_n - A_{n-1}}{n^s} \\ &= \sum_{n=1}^{\infty} A_n \left(\frac{1}{n^s} - \frac{1}{(n+1)^s} \right) \\ &= \sum_{n=1}^{\infty} A_n \left(\int_n^{n+1} \frac{s}{x^{s+1}} dx \right). \end{aligned}$$

Now there exists $N \in \mathbb{N}$ and $C > 0$ such that $|A_n| \leq Cn^k$ for all $n \geq N$ and so

$$\sum_{n=N}^{\infty} \left| \frac{a_n}{n^s} \right| \leq C \sum_{n=N}^{\infty} \int_n^{n+1} \frac{|s|}{x^{\Re(s)+1-k}} dx$$

$$= C|s| \int_{N}^{\infty} \frac{1}{x^{\Re(s)+1-k}} dx,$$

and the result follows. □

For the next result we consider r-tuples of non-negative integers $n = (n_1, \dots, n_r)$ as elements of $\mathbb{Z}_+^r$.

Proposition 3.2.2 *The series* $\sum_{n \in \mathbb{Z}_+^r - \{0\}} \frac{1}{(n_1^2 + \cdots + n_r^2)^s}$ *converges absolutely whenever* $\Re(s) > \frac{r}{2}$.

Proof We rewrite the series of interest as $\sum_{m=1}^{\infty} \frac{a_m}{m^s}$, where

$$a_m := \#\{(n_1, \dots, n_r); n_1^2 + \cdots + n_r^2 = m\}.$$

Then (see Chamizo and Iwaniec [42] and references therein[2])

$$\sum_{i=1}^{m} a_i \le V(r; m^{\frac{1}{2}}) + O(m^{\frac{\theta}{2}}),$$

where $\theta < r$ and

$$V(r; \rho) = \frac{\pi^{\frac{r}{2}}}{\Gamma\left(\frac{r}{2} + 1\right)} \rho^r$$

is the volume of a sphere in $\mathbb{R}^r$ having radius ρ. The result then follows from Proposition 3.2.1. □

Recall that D is the set of all highest weights of the group G. From now on let $D_0 = D - \{0\}$, and consider the formal power series, where $s \in \mathbb{C}$:

$$\zeta(s) = \sum_{\lambda \in D_0} \frac{1}{|\lambda|^{2s}}. \tag{3.2.11}$$

We call ζ the *Sugiura zeta function*. Our main result of this section is the following (where we recall that r is the rank of G, as defined in Sect. 2.5.1):

Theorem 3.2.1 *The series* $\sum_{\lambda \in D_0} \frac{1}{|\lambda|^{2s}}$ *converges absolutely if* $2\Re(s) > r$.

[2] This is a generalisation of the celebrated Gauss circle problem.

Proof We follow Sugiura [200] p. 38. Recall from Sect. 2.5.3 the lattice of weights $\mathbb{I}^*$, and let $\mathbb{I}_0^* := \mathbb{I}^* - \{0\}$. If we can establish the convergence of $\sum_{\lambda \in \mathbb{I}_0^*} \frac{1}{|\lambda|^{2s}}$ on the required range of s, then the result follows. Let $\{\lambda_1, \ldots, \lambda_r\}$ be a basis for $\mathbb{I}^*$, so that if $\lambda \in \mathbb{I}^*$, then $\lambda = \sum_{i=1}^r n_i \lambda_i$, where $n_i \in \mathbb{Z} (1 \leq i \leq r)$. Now define an inner product $\langle \cdot, \cdot \rangle$ on $\mathfrak{t}^*$ as follows: if $x, y \in \mathfrak{t}^*$ with $x = \sum_{i=1}^r x_i \lambda_i$, $y = \sum_{i=1}^r y_i \lambda_i$ with $x_i, y_i \in \mathbb{R}$ $(1 \leq i \leq r)$, then $\langle x, y \rangle = \sum_{i=1}^r x_i y_i$. Let $|| \cdot ||$ be the associated norm on $\mathfrak{t}^*$. By the equivalence of norms in finite-dimensional vector spaces, there exist $c_1, c_2 > 0$ so that for all $\lambda \in \mathbb{I}^*$ with $\lambda = \sum_{i=1}^r n_i \lambda i$,

$$c_1 \sum_{i=1}^r n_i^2 \leq |\lambda|^2 \leq c_2 \sum_{i=1}^r n_i^2,$$

and the required result may then be deduced from Proposition 3.2.2. □

The proof of the next result gives a nice interplay between Sobolev spaces and Sugiura's zeta-function. We will find an application for it in Chap. 6. For this proof and also some of the results that follow, we recall that $m = \#\mathcal{P}_+$ is the number of positive roots of G.

Theorem 3.2.2 *If $f \in H_s(G)$ for some $s > \frac{d}{2}$, then $f \in L^\infty(G)$. Moreover the inclusion map is continuous.*

Proof If $f \in H_s(G)$, then by the Fourier expansion (2.3.7), and the fact that convergence in L^2 implies almost everywhere convergence of a subsequence, it follows that there exists $G_0 \in \mathcal{B}(G)$ with $m(G_0) = 1$ so that for all $\sigma \in G_0$,

$$|f(\sigma)| \leq \sum_{\pi \in \widehat{G}} d_\pi |\mathrm{tr}(\widehat{f}(\pi)\pi(\sigma))|.$$

Now (see (3.3.12) in the next section)

$$\begin{aligned} |\mathrm{tr}(\widehat{f}(\pi)\pi(\sigma))|^2 &\leq \mathrm{tr}(\widehat{f}(\pi)\widehat{f}(\pi)^*)\mathrm{tr}(\pi(\sigma)\pi(\sigma)^*) \\ &= d_\pi \mathrm{tr}(\widehat{f}(\pi)\widehat{f}(\pi)^*). \end{aligned}$$

Hence by the Cauchy-Schwarz inequality

$$\begin{aligned} |f(\sigma)|^2 &\leq \left(\sum_{\pi \in \widehat{G}} d_\pi^{\frac{3}{2}} (1+\kappa_\pi)^{\frac{s}{2}} (1+\kappa_\pi)^{-\frac{s}{2}} \mathrm{tr}(\widehat{f}(\pi)\widehat{f}(\pi)^*)^{\frac{1}{2}} \right)^2 \\ &\leq \sum_{\pi \in \widehat{G}} d_\pi (1+\kappa_\pi)^s \mathrm{tr}(\widehat{f}(\pi)\widehat{f}(\pi)^*) \sum_{\pi \in \widehat{G}} d_\pi^2 (1+\kappa_\pi)^{-s} \\ &= ||f||_{H_s(G)}^2 \sum_{\pi \in \widehat{G}} d_\pi^2 (1+\kappa_\pi)^{-s}. \end{aligned}$$

Using Corollaries 2.5.1 and 2.5.2, there exists $N > 0$ such that

$$\sum_{\pi \in \widehat{G}} d_\pi^2 (1 + \kappa_\pi)^{-s} \leq N^2 \sum_{\pi \in \widehat{G}} \frac{1}{|\lambda_\pi|^{2s-2m}} = \zeta(s - m).$$

By Theorem 3.2.1, the right hand side of the last inequality converges absolutely if $2s - 2m > r$, i.e. $s > \frac{d}{2}$. Since the bound is independent of $\sigma \in G_0$, we conclude that if $s > \frac{d}{2}$, then $f \in L^\infty(G)$ as required. Continuity of the inclusion map is clear from the estimates. □

3.3 Uniform Convergence of Fourier Series

The material within this section and also the following one is based on Sugiura [200].

From Chap. 2, we know that any $f \in L^2(G)$ has a Fourier series expansion $\sum_{\pi \in \widehat{G}} d_\pi \mathrm{tr}(\widehat{f}(\pi)\pi)$. In this section we will seek to find some classes of functions for which convergence of the series holds absolutely and uniformly. In that case we have the pointwise equality $f(g) = \sum_{\pi \in \widehat{G}} d_\pi \mathrm{tr}(\widehat{f}(\pi)\pi(g))$ for all $g \in G$. Since the uniform limit of a sequence of continuous functions is continuous, we have

Proposition 3.3.1 *A necessary condition for the Fourier series of $f \in L^2(G)$ to converge uniformly is that f is continuous.*

For the remainder of this section we will be concerned with seeking sufficient conditions.

We introduce the inner product $\langle \cdot, \cdot \rangle_2$ on spaces $M_d(\mathbb{C})$ given by $\langle X, Y \rangle_2 = \mathrm{tr}(XY^*)$ for each $X, Y \in M_d(\mathbb{C})$. In this case the Cauchy-Schwarz inequality takes the form

$$|\langle X, Y \rangle_2| \leq ||X||_{HS} ||Y||_{HS}. \tag{3.3.12}$$

In particular, we will frequently use (3.3.12) in the case where Y is a unitary matrix, which we here denote as U. Since $||U||_{HS}^2 = \mathrm{tr}(UU^*) = \mathrm{tr}(I_d) = d$, we find that

$$|\mathrm{tr}(XU)| \leq \sqrt{d} ||X||_{HS}. \tag{3.3.13}$$

In the sequel we will identify elements of $\widehat{G}$ with dominant weights in D (as described in Sect. 2.5.7), and write the Fourier series $\sum_{\pi \in \widehat{G}} d_\pi \mathrm{tr}(\widehat{f}(\pi)\pi) = \sum_{\lambda \in D} d_\lambda \mathrm{tr}(\widehat{f}(\lambda)\pi_\lambda)$. The next result will be crucial for much of what follows in both this and the next section. Recall the notion of *compatible matrices* from Sect. 2.3.

Proposition 3.3.2 *Let $(M(\lambda), \lambda \in D)$ be a family of compatible matrices for which $||M(\lambda)||_{HS} = O(|\lambda|^{-s})$ as $|\lambda| \to \infty$, where $s > 0$. If $s > \alpha + r + \dfrac{3m}{2}$, then for $\alpha > 0$, the series $\sum_{\lambda \in D} d_\lambda |\lambda|^\alpha \mathrm{tr}(M(\lambda)\pi_\lambda)$ converges absolutely and uniformly.*

Proof Using (3.3.13) and the dimension estimate of Corollary 2.5.2, we obtain for all $g \in G$

$$\sum_{\lambda \in D} d_\lambda |\lambda|^\alpha |\mathrm{tr}(M(\lambda)\pi_\lambda(g))| \le \sum_{\lambda \in D} d_\lambda^{\frac{3}{2}} |\lambda|^\alpha ||M(\lambda)||_{HS}$$
$$\le C^{\frac{3}{2}} \sum_{\lambda \in D} |\lambda|^{\alpha + \frac{3m}{2}} ||M(\lambda)||_{HS}.$$

By assumption, given any $K > 0$ we can find $\lambda_0 \in D_0$ such that if $|\lambda| > |\lambda_0|$, then $||M(\lambda)||_{HS} < K|\lambda|^{-s}$, and so

$$\sum_{|\lambda| > |\lambda_0|} |\lambda|^{\alpha + \frac{3m}{2}} ||M(\lambda)||_{HS} < K \sum_{|\lambda| > |\lambda_0|} |\lambda|^{\alpha + \frac{3m}{2} - s}.$$

The required result follows from Theorem 3.2.1. □

Before we state and prove the next theorem, we apply (3.1.5) and Corollary 2.5.1 to deduce the useful estimate

$$||\widehat{f}(\lambda)||_{HS} \le |\lambda|^{-2p}.||\widehat{\Delta^p f}(\lambda)||_{HS} \tag{3.3.14}$$

for $f \in C^{2p}(G)$, where $p \in \mathbb{N}$ and $\lambda \in D_0$.

Theorem 3.3.1 *Let $f \in C(G)$. Its Fourier series $\sum_{\lambda \in D} d_\lambda \mathrm{tr}(\widehat{f}(\lambda)\pi_\lambda)$ converges absolutely and uniformly to f if either of the following conditions hold.*

(i) $||\widehat{f}(\lambda)||_{HS} = O(|\lambda|^{-s})$ *as* $|\lambda| \to \infty$ *with* $s > r + \dfrac{3m}{2}$.

(ii) $f \in C^{2p}(G, \mathbb{C})$ *where* $p \in \mathbb{N}$ *with* $4p > d$.

Proof (i) This follows immediately from Proposition 3.3.2.

(ii) We argue as at the beginning of the proof of Proposition 3.3.2, and use the estimate (3.3.14) and the Cauchy-Schwarz inequality to obtain for all $g \in G$

$$\sum_{\lambda \in D_0} d_\lambda |\mathrm{tr}(\widehat{f}(\lambda)\pi_\lambda(g))| \le \sum_{\lambda \in D_0} d_\lambda^{\frac{3}{2}} ||\widehat{f}(\lambda)||_{HS}$$
$$\le \sum_{\lambda \in D_0} d_\lambda^{\frac{3}{2}} |\lambda|^{-2p}.||\widehat{\Delta^p f}(\lambda)||_{HS}$$
$$\le \left(\sum_{\lambda \in D_0} d_\lambda^2 |\lambda|^{-4p} \right)^{\frac{1}{2}} \left(\sum_{\lambda \in D_0} d_\lambda ||\widehat{\Delta^p f}(\lambda)||_{HS}^2 \right)^{\frac{1}{2}}$$
$$= \left(\sum_{\lambda \in D_0} d_\lambda^2 |\lambda|^{-4p} \right)^{\frac{1}{2}} ||\Delta^p f||_2,$$

by the Parseval-Plancherel identity (2.3.8). Now by using the dimension estimate of Corollary 2.5.2 we find that

$$\sum_{\lambda \in D_0} d_\lambda^2 |\lambda|^{-4p} \leq C^2 \sum_{\lambda \in D_0} |\lambda|^{2m-4p}.$$

By Theorem 3.2.1 the last series converges if $4p > 2m + r = d$ and we have our required result. □

Note that by Theorem 3.3.1 (ii), every function in $C^\infty(G)$ has a uniformly convergent Fourier series.

3.4 Fourier Series and Smoothness

In this section we will show that there is a deep connection between the smoothness of a function and the rapid decay of its Fourier transform as the dominant weights move to infinity. First we establish a useful estimate for induced Lie algebra representations. In the following the norm $||\cdot||$ on $\mathfrak{g}$ is the one that is induced by the given Ad-invariant inner product.

Theorem 3.4.1 *For all $\lambda \in D_0$, $X \in \mathfrak{g}$,*

$$||d\pi_\lambda(X)||_{HS}^2 \leq C|\lambda|^{m+2}||X||^2, \tag{3.4.15}$$

where C is the constant of Corollary 2.5.2.

Proof Fix a maximal torus T in G having Lie algebra $\mathfrak{t}$. We first show that it is sufficient to prove the result for $X \in \mathfrak{t}$. Indeed (see Sect. 2.5.1), if $X \in \mathfrak{g}$, then there exists $Y \in \mathfrak{t}$ and $g \in G$ such that $g \exp(X) g^{-1} = \exp(Y)$, and so $\text{Ad}(g)X = Y$ by (1.3.5). It follows immediately that $||Y|| = ||X||$, and we also have that for all $\lambda \in D$,

$$||d\pi_\lambda(Y)||_{HS} = ||\pi_\lambda(g) d\pi_\lambda(X) \pi_\lambda(g^{-1})||_{HS} = ||d\pi_\lambda(X)||_{HS}.$$

So we may take $X \in \mathfrak{t}$ henceforth. With $\{\mu_1, \ldots, \mu_{d_\lambda}\}$ the set of weights of the representation π_λ, there exists a non-singular matrix Q such that $Q d\pi_\lambda(X) Q^{-1} = \text{diag}(i\mu_1(X), \ldots, i\mu_{d_\lambda}(X))$. By the Cauchy-Schwarz inequality and the fact that $\max_{1 \leq i \leq d_\lambda} |\mu_i| = |\lambda|$, we obtain

$$||d\pi_\lambda(X)||_{HS}^2 = \sum_{i=1}^{d_\lambda} \mu_i(X)^2 = \sum_{i=1}^{d_\lambda} (\mu_i, X)^2 \leq \sum_{i=1}^{d_\lambda} |\mu_i|.||X||^2 \leq d_\lambda |\lambda|^2 ||X||^2.$$

The result then follows from the dimension estimate of Corollary 2.5.2. □

Proposition 3.4.1 *If $f \in C^{2p}(G)$ where $p \in \mathbb{N}$, then $||\widehat{f}(\lambda)||_{HS} = o(|\lambda|^{-2p})$ as $|\lambda| \to \infty$.*

Proof Since $C^{2p}(G) \subseteq \mathrm{Dom}(\Delta^p)$ for all $f \in C^{2p}(G)$, by the Parseval-Plancherel identity:

$$||\Delta^p f||^2 = \sum_{\lambda \in D} d_\lambda ||\widehat{\Delta^p f}(\lambda)||_{HS}^2,$$

and as $\lim_{|\lambda| \to \infty} d_\lambda = \infty$, by convergence of the infinite series we must have $\lim_{|\lambda| \to \infty} ||\widehat{\Delta^p f}(\lambda)||_{HS} = 0$. The required result follows by use of the estimate (3.3.14). □

Theorem 3.4.2 *If $f \in C(G)$, then $f \in C^\infty(G)$ if and only if $||\widehat{f}(\lambda)||_{HS} = o(|\lambda|^{-p})$ for all $p \in \mathbb{N}$.*

Proof Necessity is an immediate consequence of Proposition 3.4.1. For sufficiency, we first show that if f satisfies the given condition, then Xf exists and is continuous for all $X \in \mathfrak{g}$. It follows by Theorem 3.3.1(i) that for all $g \in G, t \in \mathbb{R}$

$$f(g \exp(tX)) = \sum_{\lambda \in D} d_\lambda \mathrm{tr}(\widehat{f}(\lambda)\pi_\lambda(g \exp(tX))).$$

The formal derivative of the right hand side is the series $\sum_{\lambda \in D} d_\lambda \mathrm{tr}(\widehat{f}(\lambda)\pi_\lambda(g \exp(tX))d\pi_\lambda(X))$. This series converges absolutely and uniformly in the variables g and t since by (3.4.15)

$$\begin{aligned}
&\sum_{\lambda \in D} d_\lambda |\mathrm{tr}(\widehat{f}(\lambda)\pi_\lambda(g \exp(tX))d\pi_\lambda(X))| \\
&\quad \leq \sum_{\lambda \in D} d_\lambda ||\widehat{f}(\lambda)\pi_\lambda(g \exp(tX))||_{HS} ||d\pi_\lambda(X)||_{HS} \\
&\quad = \sum_{\lambda \in D} d_\lambda ||\widehat{f}(\lambda)||_{HS} ||d\pi_\lambda(X)||_{HS} \\
&\quad \leq C^{\frac{1}{2}} ||X|| \sum_{\lambda \in D} d_\lambda |\lambda|^{\frac{m+2}{2}} ||\widehat{f}(\lambda)||_{HS},
\end{aligned}$$

and the final series in this display converges by using Corollary 2.5.2 and then taking p sufficiently large as to be able to apply Theorem 3.2.1. Then by the mean value theorem, for all $t \in \mathbb{R}$ we can find $0 < \theta < 1$ so that

$$\begin{aligned}
\frac{f(g \exp(tX)) - f(g)}{t} &= \sum_{\lambda \in D} d_\lambda \mathrm{tr}\left(\widehat{f}(\lambda)\pi_\lambda(g)\frac{\pi_\lambda(\exp(tX)) - I_{\pi_\lambda}}{t}\right) \\
&= \sum_{\lambda \in D} d_\lambda \mathrm{tr}(\widehat{f}(\lambda)\pi_\lambda(g \exp(\theta t X))d\pi_\lambda(X)).
\end{aligned}$$

But since the series on the right hand side of this display converges absolutely and uniformly we can use dominated convergence to deduce that

$$X(f)(g) = \frac{d}{dt} f(g \exp(tX))\Big|_{t=0} = \sum_{\lambda \in D} d_\lambda \mathrm{tr}(\widehat{f}(\lambda)\pi_\lambda(g \exp(tX)) d\pi_\lambda(X)).$$

A similar argument and use of induction shows that for all $p \in \mathbb{N}$ and $X_1, \ldots, X_p \in \mathfrak{g}$ we have that $X_1 \cdots X_p f$ is defined and continuous. The result then follows by Theorem 1.3.5. □

We introduce the *Sugiura space of rapidly decreasing functions* to be the set $\mathcal{S}(D)$ of all compatible matrix-valued functions F on D for which

$$\lim_{|\lambda| \to \infty} |\lambda|^p ||F(\lambda)||_{HS} = 0$$

for all $p \in \mathbb{N}$. $\mathcal{S}(D)$ is a complex vector space with respect to pointwise addition and scalar multiplication. The main result of this section is the following.

Theorem 3.4.3 *The Fourier transformation $\mathcal{F} : f \to \widehat{f}$ is a linear isomorphism between $C^\infty(G)$ and $\mathcal{S}(D)$.*

Proof By Theorem 3.4.2 and properties of the Fourier transform we see that $\mathcal{F}$ maps $C^\infty(G)$ linearly to $\mathcal{S}(D)$. Since a continuous function is uniquely determined by its Fourier series, the mapping is injective. To see that it is surjective, let $F \in \mathcal{S}(D)$ and consider the formal series $\sum_{\lambda \in D} \mathrm{tr}(F(\lambda)\pi(\cdot))$. This converges absolutely and uniformly on G by Proposition 3.3.2, and so we can define a function $f \in C(G)$ by $f(g) = \sum_{\lambda \in D} \mathrm{tr}(F(\lambda)\pi(g))$ for all $g \in G$. Then $F = \widehat{f}$, by uniqueness of Fourier series expansions, and so $f \in C^\infty(G)$ by Theorem 3.4.2. The result is now established. □

Although we won't require it, Sugiura [200] proved a stronger result than Theorem 3.4.3. $C^\infty(G)$ is a locally convex topological vector space with respect to the seminorms $||f||_U = \sup_{\sigma \in G} |Uf(\sigma)|$ where $U \in \mathcal{U}(\mathfrak{g})$, and $\mathcal{S}(D)$ is a locally convex topological vector space with respect to the seminorms $||F||_s = \sup_{\lambda \in D} |\lambda|^s |||F(\lambda)|||$, where $s \geq 0$. When both spaces are equipped with these topologies we have:

Theorem 3.4.4 *The Fourier transformation $\mathcal{F} : f \to \widehat{f}$ is a topological linear isomorphism between $C^\infty(G)$ and $\mathcal{S}(D)$.*

3.5 Notes and Further Reading

There are many interesting ways in which analysis interacts with the theory of Lie groups. A key theme of the current chapter which will resonate throughout this book is harmonic analysis on Lie groups, which is essentially the study of Fourier series and Fourier transforms for classes of complex-valued functions on the group. There are four different levels at which there is a rich, fully developed theory, these being (i) locally compact abelian groups, (ii) compact groups, (iii) semisimple Lie groups (and the more general reductive groups) and (iv) nilpotent Lie groups, and we have mainly considered (ii) in the current chapter.

A comprehensive source for this theory at the level of locally compact abelian groups and also compact groups (without assuming any Lie structure) is Volume 2 of Hewitt and Ross [92]. A graduate-level introduction is given by Edwards [61]. In both of these sources the emphasis is on treating the unitary dual $\widehat{G}$ in terms of irreducible representations. When there is Lie structure and the group is compact and connected, we have seen the virtues of parametrising $\widehat{G}$ by using dominant weights. To learn more about this, we recommend Sugiura's paper [200] and also his book [201]. For a dedicated treatment of the abelian case, Rudin [171] is also strongly recommended. For L^p-convergence of Fourier series on compact Lie groups, see Stanton [190] and Stanton and Tomas [191].

The extension of harmonic analysis to (non-compact) semi-simple Lie groups is a deep and beautiful subject. The Indian mathematician Harish-Chandra (1923–1983) was one of the leading contributers to this theory in the twentieth century. He was strongly motivated by the desire to obtain a general Plancherel formula in this context. Some insight into these ideas can be found in the graduate text by Varadarajan [207], but see also the survey article Harish-Chandra [80]. Section A.5 of Folland [68] gives a nice introduction to the topic of Plancherel formulae in general locally compact groups. Harish-Chandra also generalised spherical harmonics on the sphere to general *spherical functions* on semi-simple Lie groups, and these beautiful objects have also found applications in probability. A thorough account of these can be found in Helgason [89]. Taylor [203] is also recommended as giving a valuable introduction to harmonic analysis in a range of different Lie groups, with a greater emphasis on the connection with pdes. In particular, this is a good place to begin to learn about harmonic analysis on nilpotent Lie groups, and the important example of the Heisenberg group is an excellent place to start. Another comprehensive introduction to representation theory and harmonic analysis on groups with an emphasis on the treatment of Laplacians is Gurarie [76].

Consideration of pdes leads us naturally to the heat equation on a Lie group and more general semigroups of linear operators. This will be a key theme of Chap. 5. We point out here that the Littlewood-Paley g-function for such (symmetric) semigroups on compact Lie groups is studied in Stein [192]. A key theme in modern analysis is the interaction between heat-kernel estimates and functional inequalities in metric measure spaces. A nice introduction to these ideas in a Lie group setting can be found in the lecture notes by Varapoulos et al. [209]. For example, if G is a unmodular Lie group of dimension n, under certain assumptions on the decay of the heat kernel in time, we have the *Sobolev inequality*

$$||f||_{\frac{n}{n-1}} \leq C||\nabla f||_1,$$

for all $f \in C_c^\infty(G)$ where $C > 0$. An informative survey of heat-kernel estimates and applications may be found in Saloff-Coste [176]. Grigor'yan [74] is an excellent primer for heat kernels on manifolds.

Pseudo-differential operators are another important area of modern analysis. We remind the reader that a linear operator A acting on (say) $C_0(\mathbb{R}^n)$ with $C_c^\infty(\mathbb{R}^n) \subseteq$ Dom(A) is a *pseudo-differential operator* if it has a suitably regular symbol $\eta : \mathbb{R}^n \times \mathbb{R}^n \to \mathbb{C}$ for which $(Af)(x) = \int_{\mathbb{R}^n} e^{iu\cdot x}\eta(x,u)\widehat{f}(u)du$ for all $f \in C_c^\infty(\mathbb{R}^n), x \in \mathbb{R}^n$. Ruzhansky and Turunen [172] have generalised these ideas to compact Lie groups, and the analogous operators may be written as $(Af)(\sigma) = \sum_{\pi \in \widehat{G}} d_\pi \mathrm{tr}(\pi(\sigma)\eta(\sigma,\pi)\widehat{f}(\pi))$ for all $f \in C_c^\infty(G), \sigma \in G$. In this case the symbol $\eta : G \times \widehat{G} \to \mathbb{C}$. These ideas are also finding applications to probability theory (see Jacob [160, 107] for the Euclidean space case and Applebaum [10, 11] for compact groups).

Chapter 4
Probability Measures on Compact Lie Groups

Abstract We introduce the space of Borel probability measures (and the important subspaces of central and symmetric measures) on a group, and topologise this space with the topology of weak convergence. A key tool for studying such measures is the (non-commutative) Fourier transform, which we extend from its action on functions that we described in Chap. 2. We discuss Lo-Ng positivity as a possible replacement for Bochner's theorem in this context. The theorems of Raikov-Williamson and Raikov are presented that give necessary and sufficient conditions for absolute continuity with respect to Haar measure. We then use the Fourier transform to find conditions for square-integrable densities, and the Sugiura space techniques of Chap. 3 to investigate smoothness of densities. Next we turn our attention to classifying idempotent measures and present the Kawada-Itô equidistribution theorem for the convergence of convolution powers of a measure to the uniform distribution. We introduce and establish key properties of convolution operators, including the notion of associated (sub/super-)harmonic functions. Finally we study some properties of recurrent measures on groups.

4.1 Classes of Probability Measures and Convolution

Let $\mathcal{P}(G)$ be the set of all Borel probability measures defined on an arbitrary Lie group G. As discussed in Appendix A.5, every $\mu \in \mathcal{P}(G)$ is both regular and Radon.[1] We equip $\mathcal{P}(G)$ with the topology of *weak convergence*, so if $(\mu_n, n \in \mathbb{N})$ is a sequence of measures in $\mathcal{P}(G)$ and $\mu \in \mathcal{P}(G)$, we say that the sequence converges to μ weakly as $n \to \infty$ if $\lim_{n\to\infty} \int_G f(x)\mu_n(dx) = \int_G f(x)\mu(dx)$ for all $f \in C_b(G, \mathbb{R})$. In this case we sometimes write $\mu_n \underset{w}{\to} \mu$ as $n \to \infty$. In the Chap. 5

[1] But if we drop the condition that G be a Lie group, we should work instead with *regular* Borel probability measures on the topological group G.

D. Applebaum, *Probability on Compact Lie Groups*, Probability Theory and Stochastic Modelling 70, DOI: 10.1007/978-3-319-07842-7_4,

we will also need *vague convergence* of probability measures and this is defined in exactly the same manner as weak convergence, except that the part of $C_b(G, \mathbb{R})$ is played by $C_c(G, \mathbb{R})$.

If $\mu \in \mathcal{P}(G)$, its *reversed measure* is $\widetilde{\mu} \in \mathcal{P}(G)$ where $\widetilde{\mu}(A) := \mu(A^{-1})$ for all $A \in \mathcal{P}(G)$. We say that $\mu \in \mathcal{P}(G)$ is *symmetric* if $\mu = \widetilde{\mu}$ and *central* (or *conjugate invariant*) if $\mu(gAg^{-1}) = \mu(A)$ for all $g \in G$ and all $A \in \mathcal{B}(G)$. We write $\mathcal{P}_s(G)$ and $\mathcal{P}_c(G)$ to denote the spaces of symmetric and central Borel probability measures defined on G (respectively), and we define $\mathcal{P}_{sc}(G) = \mathcal{P}_c(G) \cap \mathcal{P}_s(G)$.

If we are given a (left or right) Haar measure on G (which is always, as usual, assumed to be normalised when G is compact), we define $\mathcal{P}_{ac}(G)$ to be the corresponding subset of $\mathcal{P}(G)$ comprising absolutely continuous measures, so $\mu \in \mathcal{P}_{ac}(G)$ if there exists $f_\mu \in L^1(G)$ so that $\mu(A) = \int_A f_\mu(g)dg$ for all $A \in \mathcal{B}(G)$. The Radon-Nikodym derivative f_μ is called the *density* of the measure μ (with respect to the given Haar measure). If G is compact and $\mu \in \mathcal{P}_{ac}$, then $\mu \in \mathcal{P}_s$ if and only if $f_\mu(g) = f_\mu(g^{-1})$ for almost all $g \in G$, and $\mu \in \mathcal{P}_c$ if and only if for all $h \in G$, $f_\mu(hgh^{-1}) = f_\mu(g)$ for almost all $g \in G$.

To see that $\mathcal{P}(G) \neq \emptyset$, consider the *Dirac mass* δ_g at the point $g \in G$ which is defined for each $A \in \mathcal{B}(G)$ by

$$\delta_g(A) = \begin{cases} 1 & \text{if } g \in A \\ 0 & \text{if } g \notin A \end{cases}.$$

Clearly $\delta_g \in \mathcal{P}(G)$ and $\widetilde{\delta_g} = \delta_{g^{-1}}$. We may also form measures in $\mathcal{P}(G)$ by taking convex combinations of distinct Dirac masses. We will consider many more interesting examples as this and the subsequent chapter unfold.

Let $\mu_1, \mu_2 \in \mathcal{P}(G)$. Using the Riesz representation theorem we may assert the existence in $\mathcal{P}(G)$ of the left and right *convolution* products $\mu_1 *_L \mu_2$ and $\mu_1 *_R \mu_2$, which are defined (respectively) for all $f \in C_c(G)$ by

$$\int_G f(g)(\mu_1 *_L \mu_2)(dg) = \int_G \int_G f(gh)\mu_1(dg)\mu_2(dh),$$

$$\int_G f(g)(\mu_1 *_R \mu_2)(dg) = \int_G \int_G f(hg)\mu_1(dg)\mu_2(dh).$$

It is easily verified that $\mu_1 *_R \mu_2 = \mu_2 *_L \mu_1$. From now on we will only deal with left convolution, and we will write $\mu_1 * \mu_2 := \mu_1 *_L \mu_2$. It can be shown (see e.g. Stromberg [197]) that for all $B \in \mathcal{B}(G)$

$$\begin{aligned}(\mu_1 * \mu_2)(B) &= \int_G \int_G 1_B(gh)\mu_1(dg)\mu_2(dh) \\ &= \int_G \mu_1(Bh^{-1})\mu_2(dh) = \int_G \mu_2(g^{-1}B)\mu_1(dg).\end{aligned} \tag{4.1.1}$$

Convolution is associative, i.e. if $\mu_1, \mu_2, \mu_3 \in \mathcal{P}(G)$, then $(\mu_1 * \mu_2) * \mu_3 = \mu_1 * (\mu_2 * \mu_3)$, and so $(\mathcal{P}(G), *)$ is a semigroup . But note that if G is not abelian, we cannot expect commutativity to hold. Indeed, you can easily check that for $g, h \in G$, $\delta_g * \delta_h = \delta_{gh}$, and so $\delta_g * \delta_h = \delta_h * \delta_g$ if and only if $gh = hg$. In the general case $(\mathcal{P}(G), *)$ is a monoid (i.e. a semigroup with an identity element), since $\mu * \delta_e = \delta_e * \mu$ for all $\mu \in \mathcal{P}(G)$.

If $\mu \in \mathcal{P}_{ac}(G)$ and $\nu \in \mathcal{P}(G)$, we write $f_\mu * \nu := \mu * \nu$ and $\nu * f_\mu := \nu * \mu$. By using Fubini's theorem we easily verify that if we employ a right-invariant Haar measure, then $f_\mu * \nu \in \mathcal{P}_{ac}(G)$ with density $\int_G f_\mu(gh^{-1})\nu(dh)$ and if we choose a left-invariant Haar measure, then $\nu * f_\mu \in \mathcal{P}_{ac}(G)$ with density $\int_G f_\mu(g^{-1}h)\nu(dh)$.

The operation $\tilde{\cdot}$ acts as an involution on $(\mathcal{P}(G), *)$. Indeed, we have $\tilde{\tilde{\mu}} = \mu$ for all $\mu \in \mathcal{P}(G)$, $\widetilde{\mu_1 * \mu_2} = \widetilde{\mu_2} * \widetilde{\mu_1}$ for all $\mu_1, \mu_2 \in \mathcal{P}(G)$ and $\widetilde{\delta_e} = \delta_e$.

The *support* of $\mu \in \mathcal{P}(G)$, which we denote by supp(μ), is the set of all $g \in G$ for which every Borel neighbourhood of g has strictly positive μ-measure. It is clear that supp(μ) is a closed subset of G. It is shown in Wendel [217] (pp. 925–926) that if μ_1, μ_2 are regular probability measures on G, then

$$\text{supp}(\mu_1 * \mu_2) = \text{supp}(\mu_1)\text{supp}(\mu_2), \tag{4.1.2}$$

where if $A, B \in \mathcal{B}(G)$, $AB := \{gh, g \in A, h \in B\}$ (and for later usage $A^2 := AA$).

Although we won't use it in the sequel, the next result may be of interest.

Proposition 4.1.1 *If G is a compact group, then the space $\mathcal{P}(G)$, equipped with the weak topology, is compact.*

Proof By identifying each $\mu \in \mathcal{P}(G)$ with the linear functional I_μ on $C(G, \mathbb{R})$ defined by $I_\mu(f) = \int_G f(g)\mu(dg)$ for $f \in C(G, \mathbb{R})$, we embed $\mathcal{P}(G)$ into the topological dual space $C(G, \mathbb{R})^*$, and recognise that the weak topology on $\mathcal{P}(G)$ is in fact the restriction of the weak-$*$ topology on $C(G, \mathbb{R})^*$. By the Banach-Alaoglu theorem, the unit ball in $C(G, \mathbb{R})^*$ is weak-$*$ compact. However, $\mathcal{P}(G)$ is easily verified to be a closed subset of this ball, and the result follows. □

Note that the mapping $g \to \delta_g$ is a continuous embedding of G into a closed subspace of $\mathcal{P}(G)$.

We recall that a family of Borel probability measures $(\mu_\alpha \in \mathcal{I})$ defined on some locally compact space X (where $\mathcal{I}$ is some index set) is *tight* if given any $\epsilon > 0$ there exists a compact set K_ϵ such that $\mu_\alpha(K_\epsilon) > 1 - \epsilon$ for all $\alpha \in \mathcal{I}$. If X is itself compact, then it is clear that any family of probability measures is tight (just take $K_\epsilon = X$ for all ϵ). So on a compact group G, by Prohorov's theorem, (see e.g. Heyer [95] Theorem 1.1.11, p. 26), any family of Borel probability measures $(\mu_\alpha \in \mathcal{I})$ contains a convergent sequence.

Let $(\Omega, \mathcal{F}, P)$ be a probability space. A G-valued *random variable* is a measurable function from $(\Omega, \mathcal{F})$ to $(G, \mathcal{B}(G))$. If X is such a random variable, its *law* or *distribution* is the measure $\mu_X \in \mathcal{P}(G)$ defined by $\mu_X(B) = P(X^{-1}(B))$ for all

$B \in \mathcal{B}(G)$. The product of two random variables X and Y is the random variable XY whose value at $\omega \in \Omega$ is $X(\omega)Y(\omega)$.[2] If X and Y are independent then the law of XY is the convolution $\mu_X * \mu_Y$.

4.2 The Fourier Transform of a Probability Measure

Let $\text{Rep}(G)$ be the set of all unitary representations of G. So for each $\pi \in \text{Rep}(G)$, $g \in G$, $\pi(g)$ acts as a unitary operator on the complex separable Hilbert space V_π. For each $\mu \in \mathcal{P}(G)$, we define its *Fourier transform* or *characteristic function* $\widehat{\mu}(\pi)$ at $\pi \in \text{Rep}(G)$ to be the bounded linear operator on V_π defined as a Bochner integral (see e.g. Cohn [50] Appendix E, pp. 350–354) by:

$$\widehat{\mu}(\pi)\psi = \int_G \pi(g^{-1})\psi\mu(dg), \tag{4.2.3}$$

for each $\psi \in V_\pi$. Equivalently, it may be defined as a Pettis integral to be the unique bounded linear operator on V_π for which

$$\langle\widehat{\mu}(\pi)\phi, \psi\rangle = \int_G \langle\pi(g^{-1})\phi, \psi\rangle\mu(dg), \tag{4.2.4}$$

for all $\phi, \psi \in V_\pi$ (c.f. Heyer [93], Siebert [185] and Hewitt and Ross [92], pp. 77–87).

Note that if μ is absolutely continuous with respect to a given left Haar measure on μ and has density $f \in L^1(G)$, then our definition is such that $\widehat{\mu}(\pi) = \widehat{f}(\pi)$, where $\widehat{f}(\pi)$ is as defined in Chap. 2.[3]

*From now until Sect.*4.7, *we will take G to be a compact Lie group and restrict* π *to be an irreducible representation.*[4]

So (observing our usual convention of identifying equivalence classes with representative elements) we will from now on always take $\pi \in \widehat{G}$. Then $\widehat{\mu}(\pi)$ is a $d_\pi \times d_\pi$ matrix and both (4.2.3) and (4.2.4) are equivalent to defining the matrix elements

$$\widehat{\mu}(\pi)_{ij} = \int_G \pi_{ij}(g^{-1})\mu(dg), \tag{4.2.5}$$

---

[2] If G is abelian, then the binary operation in the group is usually written additively.

[3] It is common in the literature to see the alternative definition "$\widehat{\mu}(\pi) = \int_G \pi(g)\mu(dg)$" which is natural for probabilists but which clashes with the analysts' convention that we introduced in Chap. 2.

[4] Many theorems that we state hold under more general conditions on G. The reader who wants minimal assumptions may consult the original sources, or check what is really needed from the proof.

for $1 \leq i, j \leq d_\pi$.

It is often convenient to write (4.2.5) using the simplified notation:

$$\widehat{\mu}(\pi) := \int_G \pi(g^{-1})\mu(dg).$$

Example 1 Dirac Mass. If $\mu = \delta_g$ for some $g \in G$, then it is easily verified that for all $\pi \in \widehat{G}, \widehat{\mu}(\pi) = \pi(g^{-1})$. In particular, $\widehat{\delta_e} = I_\pi$.

Example 2 Normalised Haar measure. We again denote this measure by m. It is easy to see that $m \in \mathcal{P}_{sc}(G)$. We have

$$\widehat{m}(\pi) = \begin{cases} 0 & \text{if } \pi \neq \pi_0 \\ 1 & \text{if } \pi = \pi_0 \end{cases}.$$

To see this, it is sufficient to observe that for all $\pi \in \widehat{G}, 1 \leq i, j \leq d_\pi$,

$$\widehat{m}(\pi)_{ij} = \int_G \pi_{ij}(g^{-1})dg = \langle 1, \pi_{ij} \rangle_{L^2(G)},$$

and the result then follows by Peter-Weyl theory (Theorem 2.2.4).

Example 3 Standard Gaussian Measures. We recall the discussion of the heat kernel in Sect. 3.1.1. Now fix a parameter $\sigma > 0$ and consider the heat equation:

$$\frac{\partial u}{\partial t} = \sigma \Delta u. \tag{4.2.6}$$

We write the corresponding heat kernel as $k_\sigma \in C^\infty((0, \infty) \times G, \mathbb{R})$, and for fixed $t > 0$ we write $k_{t,\sigma}(\cdot) := k_\sigma(t, \cdot) \in C^\infty(G, \mathbb{R})$. Taking $f = 1$ in (3.1.8), we see immediately from (4.2.6) that $\int_G k_{t,\sigma}(g)dg = 1$, and so $k_{t,\sigma}$ is the density of a measure $\gamma_{t,\sigma} \in \mathcal{P}(G)$ which we call a *standard Gaussian measure* with parameter σ.[5] We now compute the Fourier transform. Using the smoothness of $t \to k_{t,\sigma}$ and dominated convergence, we deduce that for all $\pi \in \widehat{G}$,

$$\int_G \pi(g^{-1})\frac{\partial k_{t,\sigma}(g)}{\partial t}dg = \frac{\partial}{\partial t}\int_G \pi(g^{-1})k_{t,\sigma}(g)dg,$$

and so the mapping $t \to \widehat{k_{t,\sigma}}(\pi)$ is differentiable. Taking Fourier transforms of both sides of (4.2.6) then yields that

[5] If we were to take a strict analogy with the well-known theory in Euclidean space, we would only use the terminology "standard" Gaussian measure for the case where $\sigma t = \frac{1}{2}$.

$$\frac{\partial \widehat{k_{t,\sigma}}(\pi)}{\partial t} = \sigma \widehat{\Delta k_{t,\sigma}}(\pi)$$
$$= -\sigma \kappa_\pi \widehat{k_{t,\sigma}}(\pi).$$

Since $\widehat{k_{0,\sigma}}(\pi) = \widehat{\delta_e}(\pi) = I_\pi$, we deduce that

$$\widehat{k_{t,\sigma}}(\pi) = e^{-t\sigma\kappa_\pi} I_\pi. \tag{4.2.7}$$

The next theorem summarises some key properties of the Fourier transform (see also Heyer [93]):

Theorem 4.2.1 *For all* $\mu, \mu_1, \mu_2 \in \mathcal{P}(G), \pi \in \widehat{G}$,

1. $\widehat{\mu}(\pi_0) = 1$,
2. $\widehat{\mu_1 * \mu_2}(\pi) = \widehat{\mu_2}(\pi)\widehat{\mu_1}(\pi)$,
3. $||\widehat{\mu}(\pi)||_{op} \leq 1$,
4. $\widehat{\widetilde{\mu}}(\pi) = \widehat{\mu}(\pi)^*$.

Proof 1. is obvious.
2. For all $1 \leq i, j \leq d_\pi$

$$\widehat{\mu_1 * \mu_2}(\pi)_{ij} = \int_G \pi_{ij}(h^{-1}g^{-1})\mu_1(dg)\mu_2(dh)$$
$$= \sum_{k=1}^{d_\pi} \left(\int_G \pi_{ik}(h^{-1})\mu_2(dh)\right)\left(\int_G \pi_{kj}(g^{-1})\mu_1(dg)\right)$$
$$= [\widehat{\mu_2}(\pi)\widehat{\mu_1}(\pi)]_{ij}.$$

3. For all $\phi \in V_\pi$,

$$||\widehat{\mu}(\pi)\phi|| = \left|\left|\int_G \pi(g^{-1})\phi\mu(dg)\right|\right|$$
$$\leq \int_G ||\pi(g^{-1})\phi||\mu(dg)$$
$$= ||\phi||.$$

4.

$$\widehat{\widetilde{\mu}}(\pi) = \int_G \pi(g^{-1})\widetilde{\mu}(dg)$$
$$= \int_G \pi(g)\mu(dg)$$
$$= \left(\int_G \pi(g^{-1})\mu(dg)\right)^* = \widehat{\mu}(\pi)^*.$$

□

Corollary 4.2.1 *The measure $\mu \in \mathcal{P}_s(G)$ if and only if the matrix $\widehat{\mu}(\pi)$ is self-adjoint*[6] *for all $\pi \in \widehat{G}$.*

Proof Necessity is immediate from Theorem 4.2.1 (4). For sufficiency it is enough to observe that if the self-adjointness condition holds, then for all $\pi \in \widehat{G}$, $\phi, \psi \in V_\pi$,

$$\int_G \langle \pi(g)\phi, \psi\rangle \mu(dg) = \int_G \langle \pi(g)\phi, \psi\rangle \widetilde{\mu}(dg).$$

By linearity we find that $\int_G f(g)\mu(dg) = \int_G f(g)\widetilde{\mu}(dg)$ for all $f \in \mathcal{E}(G)$ which is norm dense in $C(G)$ by the Peter-Weyl theorem (Theorem 2.2.4). By extension of bounded linear functionals, we then see that $\int_G f(g)\mu(dg) = \int_G f(g)\widetilde{\mu}(dg)$ for all $f \in C(G)$, and the result follows from the Riesz representation theorem. □

The next theorem generalises Theorem 2.4.1 (see also Hewitt and Ross [92] Theorem 28.48, pp. 84–85).

Theorem 4.2.2 *The measure $\mu \in \mathcal{P}_c(G)$ if and only if $\widehat{\mu}(\pi) = c_\pi I_\pi$, where $c_\pi \in \mathbb{C}$, for all $\pi \in \widehat{G}$.*

Proof Necessity is established by Schur's lemma just as in Theorem 2.4.1. For sufficiency, for each $h \in G$ define $\mu^h \in \mathcal{P}(G)$ by $\mu^h(A) = \mu(hAh^{-1})$ for $A \in \mathcal{B}(G)$. Then arguing as in the proof of Theorem 2.4.1 we obtain for all $\pi \in \widehat{G}$, $\int_G \pi(g^{-1})\mu(dg) = \int_G \pi(g^{-1})\mu^h(dg)$, and so for all $\phi, \psi \in V_\pi$,

$$\int_G \langle \pi(g)\phi, \psi\rangle \mu(dg) = \int_G \langle \pi(g)\phi, \psi\rangle \mu^h(dg).$$

We can now reach our desired conclusion by proceeding as in the proof of Corollary 4.2.1 □

Corollary 4.2.2 *The measure $\mu \in \mathcal{P}_{sc}(G)$ if and only if $\widehat{\mu}(\pi) = c_\pi I_\pi$, where $c_\pi \in \mathbb{R}$, for all $\pi \in \widehat{G}$.*

---

[6] i.e. hermitian, if you prefer that terminology.

Proof This follows immediately from Corollary 4.2.1 and Theorem 4.2.2. □

For example we find by (4.2.7) that standard Gaussian measure is both central and symmetric.

The remaining results in this section were originally due to Kawada and Itô [114]. The first of these establishes the injectivity of the Fourier transform:

Theorem 4.2.3 *Let $\mu_1, \mu_2 \in \mathcal{P}(G)$. Then $\widehat{\mu_1}(\pi) = \widehat{\mu_2}(\pi)$ for all $\pi \in \widehat{G}$ if and only if $\mu_1 = \mu_2$.*

Proof Sufficiency is immediate. For necessity let $f \in C(G)$, and $\epsilon > 0$ be arbitrary. By the Peter-Weyl theorem (Theorem 2.2.4) there exists $\widehat{G}_0 \subset \widehat{G}$ with $\#\widehat{G}_0 \in \mathbb{N}$ such that

$$\sup_{g \in G} \left| f(g) - \sum_{\pi \in \widehat{G}_0} \sum_{i,j=1}^{d_\pi} \alpha_{ij}^{(\pi)} \pi_{ij}(g) \right| < \frac{\epsilon}{2},$$

where $\alpha_{ij}^{(\pi)} \in \mathbb{C}$ $(1 \leq i, j \leq d_\pi)$. Then for $k = 1, 2$ we find that

$$\left| \int_G f(g) \mu_k(dg) - \sum_{\pi \in \widehat{G}_0} \sum_{i,j=1}^{d_\pi} \alpha_{ij}^{(\pi)} \widehat{\mu_k}(\pi)_{ij} \right| < \frac{\epsilon}{2}.$$

But since $\widehat{\mu_1}(\pi)_{ij} = \widehat{\mu_2}(\pi)_{ij}$ for all $1 \leq i, j \leq d_\pi$, we deduce that

$$\left| \int_G f(g) \mu_1(dg) - \int_G f(g) \mu_2(dg) \right| < \epsilon,$$

and the result follows by the fact that ϵ is arbitrary and by use of the Riesz representation theorem. □

Theorem 4.2.4 *Let $\mu_1, \mu_2 \in \mathcal{P}(G)$. Then $\mu_1 * \mu_2 = \mu_2 * \mu_1$ if and only if $\widehat{\mu_1}(\pi)\widehat{\mu_2}(\pi) = \widehat{\mu_2}(\pi)\widehat{\mu_1}(\pi)$ for all $\pi \in \widehat{G}$.*

Proof Necessity follows immediately from Theorem 4.2.1(2). For sufficiency, observe that by Theorem 4.2.1(2) again

$$\widehat{\mu_1 * \mu_2}(\pi) = \widehat{\mu_2}(\pi)\widehat{\mu_1}(\pi) = \widehat{\mu_1}(\pi)\widehat{\mu_2}(\pi) = \widehat{\mu_2 * \mu_1}(\pi),$$

and then apply Theorem 4.2.3. □

Theorem 4.2.5 *Let $(\mu_n, n \in \mathbb{N})$ be a sequence of measures in $\mathcal{P}(G)$. Then $\mu_n \underset{w}{\rightarrow} \mu$ as $n \to \infty$ if and only if $\widehat{\mu_n}(\pi)_{ij} \to \widehat{\mu}(\pi)_{ij}$ as $n \to \infty$ for all $1 \leq i, j \leq d_\pi, \pi \in \widehat{G}$.*

Proof If $\mu_n \underset{w}{\to} \mu$ as $n \to \infty$, then

$$\lim_{n\to\infty} \widehat{\mu_n}(\pi)_{ij} = \lim_{n\to\infty} \int_G \pi_{ij}(g^{-1})\mu_n(dg) = \int_G \pi_{ij}(g^{-1})\mu(dg) = \widehat{\mu}(\pi)_{ij}.$$

Conversely, if $\widehat{\mu_n}(\pi)_{ij} \to \widehat{\mu}(\pi)_{ij}$ as $n \to \infty$ for all $1 \leq i, j \leq d_\pi, \pi \in \widehat{G}$, then using the same notation, and a similar argument to that given in the proof of Theorem 4.2.3, we first observe that for any $f \in C(G)$, $\epsilon > 0$ there exists $\widehat{G}_0 \subset \widehat{G}$ with $\#\widehat{G}_0 \in \mathbb{N}$ so that for all $n \in \mathbb{N}$

$$\left| \int_G f(g)\mu_n(dg) - \sum_{\pi\in\widehat{G}_0} \sum_{i,j=1}^{d_\pi} \alpha_{ij}^{(\pi)} \widehat{\mu_n}(\pi)_{ij} \right| < \frac{\epsilon}{3},$$

and also

$$\left| \int_G f(g)\mu(dg) - \sum_{\pi\in\widehat{G}_0} \sum_{i,j=1}^{d_\pi} \alpha_{ij}^{(\pi)} \widehat{\mu}(\pi)_{ij} \right| < \frac{\epsilon}{3}.$$

But we can also find $n_1 \in \mathbb{N}$ so that if $n > n_1$ we have

$$|\widehat{\mu_n}(\pi)_{ij} - \widehat{\mu}(\pi)_{ij}| < \frac{\epsilon}{3C}$$

for all $1 \leq i, j \leq d_\pi, \pi \in \widehat{G}_0$ where $C := \sum_{\pi\in\widehat{G}_0} \sum_{i,j=1}^{d_\pi} |\alpha_{ij}^{(\pi)}|$. From these estimates we deduce that for all $n > n_1$,

$$\left| \int_G f(g)\mu_n(dg) - \int_G f(g)\mu(dg) \right| < \epsilon,$$

and this gives the desired weak convergence. □

Then final result of this section gives a compact Lie group version of the celebrated *Lévy convergence theorem* for sequences of probability measures in Euclidean space.

Theorem 4.2.6 (Kawada,Itô,Lévy convergence theorem) *Suppose that* $(\mu_n, n \in \mathbb{N})$ *is a sequences of measures in* $\mathcal{P}(G)$ *and that there exists a family of compatible matrices* $(Y(\pi), \pi \in \widehat{G})$ *so that* $\widehat{\mu_n}(\pi)_{ij} \to Y(\pi)_{ij}$ *as* $n \to \infty$ *for all* $1 \leq i, j \leq d_\pi, \pi \in \widehat{G}$. *Then there exists* $\mu \in \mathcal{P}(G)$ *for which* $\mu_n \underset{w}{\to} \mu$ *as* $n \to \infty$ *and* $\widehat{\mu}(\pi) = Y(\pi)$ *for all* $\pi \in \widehat{G}$.

Proof Let $f \in C(G)$. Once again using (a straightforward variation of) the same notation to that used in the proof of Theorem 4.2.3, we can assert that given any $m \in \mathbb{N}$ there exists $\widehat{G}_0 \subset \widehat{G}$ with $\#\widehat{G}_0 \in \mathbb{N}$ so that

$$\sup_{g \in G} \left| f(g) - \sum_{\pi \in \widehat{G}_0} \sum_{i,j=1}^{d_\pi} \alpha_{ij}^{(\pi,m)}(f)\pi_{ij}(g) \right| < \frac{1}{2^m},$$

and so for all $n \in \mathbb{N}$

$$\left| \int_G f(g)\mu_n(dg) - \sum_{\pi \in \widehat{G}_0} \sum_{i,j=1}^{d_\pi} \alpha_{ij}^{(\pi,m)}(f)\widehat{\mu_n}(\pi)_{ij} \right| < \frac{1}{2^m}.$$

Now given any $\epsilon > 0$ and choosing n sufficiently large, we obtain for such n and arbitrary m that:

$$\left| \int_G f(g)\mu_n(dg) - \sum_{\pi \in \widehat{G}_0} \sum_{i,j=1}^{d_\pi} \alpha_{ij}^{(\pi,m)}(f)Y(\pi)_{ij} \right| < \frac{1}{2^m} + \epsilon.$$

Define $\Gamma_m(f) := \sum_{\pi \in \widehat{G}_0} \sum_{i,j=1}^{d_\pi} \alpha_{ij}^{(\pi,m)}(f)Y(\pi)_{ij}$. Then from the last inequality we deduce that $(\Gamma_m(f), m \in \mathbb{N})$ is a Cauchy sequence, and hence convergent to $\Gamma(f) \in \mathbb{C}$. Again from the last inequality, we deduce that

$$\Gamma(f) = \lim_{n \to \infty} \int_G f(g)\mu_n(dg),$$

from which it follows that $f \to \Gamma(f)$ is a positive linear functional on $C(G)$ for which $\Gamma(1) = 1$. Hence by the Riesz representation theorem, there exists a probability measure $\mu \in \mathcal{P}(G)$ for which

$$\Gamma(f) = \int_G f(g)\mu(dg),$$

for all $f \in C(G)$ and this gives the required weak convergence. The fact that $\widehat{\mu}(\pi) = Y(\pi)$ for all $\pi \in \widehat{G}$ then follows from Theorem 4.2.3. □

4.3 Lo-Ng Positivity

Let μ be a Borel probability measure defined on a locally compact abelian group G (with group composition written additively). Let $\widehat{G}$ be the (abelian) dual group of characters (see Sect. 2.2.2) and let the neutral element in $\widehat{G}$ be $\widehat{e}$. In this case we have $\widehat{\mu}(\chi) = \int_G \overline{\chi(g)}\mu(dg)$ for all $\chi \in \widehat{G}$. Let $F : \widehat{G} \to \mathbb{C}$. The celebrated *Bochner theorem* gives a necessary and sufficient condition for $F = \widehat{\mu}$, for some $\mu \in \mathcal{P}(G)$, and this is precisely that $F(\widehat{e}) = 1$, F is continuous at $\widehat{e}$ and F is positive definite, i.e.

$$\sum_{i,j=1}^{n} c_i \overline{c_j} F(x_i - x_j) \geq 0$$

for all $n \in \mathbb{N}, c_1, \ldots, c_n \in \mathbb{C}$ and all $x_1, \ldots, x_n \in \widehat{G}$ (see e.g. Heyer [96], pp. 162–184 or Rudin [171], pp. 19–21). There is an analogue of this result if G is a compact Lie group which we now describe. Further details and proofs are in Heyer [95], pp. 57–59.[7]

We recall the coefficient algebra $\mathcal{E}(G)$ of G from Chap. 2. We say that a linear functional $\phi : \mathcal{E}(G) \to \mathbb{C}$ is continuous if given any sequence $(f_n, n \in \mathbb{N})$ converging uniformly to $f \in \mathcal{E}(G)$, we have that $(\phi(f_n), n \in \mathbb{N})$ converges to $\phi(f)$. We say that ϕ is positive if $\phi(\overline{f} f) \geq 0$ for all $f \in \mathcal{E}(G)$.

Theorem 4.3.1 *If G is a compact Lie group, then for any positive continuous linear functional ϕ on $\mathcal{E}(G)$ for which $\phi(1) = 1$, there exists $\mu \in \mathcal{P}(G)$ so that*

$$\langle \widehat{\mu}(\pi)x, y \rangle = \phi(\langle \pi(\cdot)x, y \rangle),$$

for all $\pi \in \widehat{G}, x, y \in V_\pi$.

As an alternative to Bochner's theorem, we can find an interesting necessary and sufficient condition for a family of compatible matrices to be the Fourier transform of a finite measure if we introduce a new notion of positivity due to Lo and Ng [136], as we will now demonstrate. To this end let $C : \widehat{G} \to \mathcal{M}(\widehat{G})$ be a compatible mapping. We say that it is *Lo-Ng positive* if the following holds: Whenever $B : \widehat{G} \to \mathcal{M}(\widehat{G})$ is any other compatible mapping for which

$$\sum_{\pi \in S} d_\pi \mathrm{tr}(\pi(g)B(\pi)) \geq 0$$

for all $g \in G$ for some finite subset[8] S of $\widehat{G}$, then

$$\sum_{\pi \in S} d_\pi \mathrm{tr}(\pi(g)C(\pi)B(\pi)) \geq 0$$

for all $g \in G$. It is immediate that if C is Lo-Ng positive and $a \geq 0$, then aC is also Lo-Ng positive. The following gives a useful alternative criterion for Lo-Ng positivity:

Lemma 4.3.1 *The compatible mapping C is Lo-Ng positive if and only for all compatible mappings $B : \widehat{G} \to \mathcal{M}(\widehat{G})$, $\sum_{\pi \in S} d_\pi \mathrm{tr}(\pi(g)B(\pi)) \geq 0$ for all $g \in G$ for some finite subset S of $\widehat{G}$ implies that $\sum_{\pi \in S} d_\pi \mathrm{tr}(B(\pi)C(\pi)) \geq 0$.*

---

[7] As a result of reading an early version of this manuscript, Herbert Heyer [97] was inspired to prove a new Bochner-type theorem for *central* probability measures on compact groups.

[8] Our definition is slightly different from that of Lo and Ng, who introduce an ordering of the countable set $\widehat{G}$ and instead of taking arbitrary finite subsets of $\widehat{G}$ as we do, choose sets of the form $\{1, 2, \ldots, n\}$, with respect to their given ordering.

Proof First suppose that C is indeed Lo-Ng positive. Then the required result follows by taking $g = e$ in the definition. Conversely suppose the given condition holds on some finite subset S of $\widehat{G}$. By the assumption on B we have

$$\sum_{\pi \in S} d_\pi \mathrm{tr}(\pi(gh)B(\pi)) \geq 0$$

for all $g, h \in G$. It follows that

$$\sum_{\pi \in S} d_\pi \mathrm{tr}(\pi(g)(B(\pi)\pi(h))) \geq 0$$

for all $g \in G$. Then by the given condition, for all $h \in G$,

$$\sum_{\pi \in S} d_\pi \mathrm{tr}(\pi(h)C(\pi)B(\pi)) = \sum_{\pi \in S} d_\pi \mathrm{tr}(C(\pi)(B(\pi)\pi(h))) \geq 0,$$

and Lo-Ng positivity is established. □

Lemma 4.3.1 equips us with the tool to show that the set of all Lo-Ng positive compatible mappings is closed under taking adjoints. To be precise, let $C : \widehat{G} \to \mathcal{M}(\widehat{G})$ be a compatible mapping and define its adjoint $C^* : \widehat{G} \to \mathcal{M}(\widehat{G})$ by the prescription $C^*(\pi) := C(\pi)^*$ for all $\pi \in \widehat{G}$.

Lemma 4.3.2 *If C is a Lo-Ng positive compatible mapping, then so is C^*.*

Proof Let $B : \widehat{G} \to \mathcal{M}(\widehat{G})$ be a compatible mapping for which

$$\sum_{\pi \in S} d_\pi \mathrm{tr}(\pi(g)B(\pi)) \geq 0$$

for all $g \in G$ for some finite subset S of $\widehat{G}$. Then

$$\begin{aligned}\sum_{\pi \in S} d_\pi \mathrm{tr}(\pi(g)B(\pi)^*) &= \sum_{\pi \in S} d_\pi \mathrm{tr}(B(\pi)^*\pi(g)) \\ &= \sum_{\pi \in S} d_\pi \overline{\mathrm{tr}(\pi(g^{-1})B(\pi))} \\ &= \sum_{\pi \in S} d_\pi \mathrm{tr}(\pi(g^{-1})B(\pi)) \geq 0.\end{aligned}$$

So by Lemma 4.3.1,

$$\sum_{\pi \in S} d_\pi \mathrm{tr}(C(\pi)^*B(\pi)) = \sum_{\pi \in S} d_\pi \overline{\mathrm{tr}(B(\pi)^*C(\pi))} = \sum_{\pi \in S} d_\pi \mathrm{tr}(B(\pi)^*C(\pi)) \geq 0,$$

and the result follows. □

Before we proceed further we state a useful technical lemma

Lemma 4.3.3 *Let $B, C : \widehat{G} \to \mathcal{M}(\widehat{G})$ be compatible mappings and let S, S' be finite subsets of $\widehat{G}$ with $S' \subseteq S$. Then*

$$\int_G \left(\sum_{\pi' \in S'} d_{\pi'} \mathrm{tr}(\pi'(g^{-1})B(\pi'))\right)\left(\sum_{\pi \in S} d_\pi \mathrm{tr}(\pi(g)C(\pi))\right) dg \tag{4.3.8}$$
$$= \sum_{\pi \in S'} d_\pi \mathrm{tr}(B(\pi)C(\pi)).$$

Proof Write both traces on the left hand side of (4.3.8) as finite sums and then use the Schur orthogonality relations (Corollary 2.2.3). □

The next result begins to establish the link between Lo-Ng positivity and the Fourier transform. Let S be a finite subset of $\widehat{G}$ and $C : S \to \mathcal{M}(\widehat{G})$ be compatible (we may consider C as extended to the whole of $\widehat{G}$ by defining it to be the zero matrix on $\widehat{G} - S$). Note that $f_{S,C} \in C(G)$, where for each $g \in G$, $f_{S,C}(g) := \sum_{\pi \in S} d_\pi \mathrm{tr}(C(\pi)\pi(g))$.

Proposition 4.3.1 *Let S, C and $f_{S,C}$ be as above.*

1. *For all $\pi \in \widehat{G}$, $C(\pi) = \widehat{f_{S,C}}(\pi)$.*
2. *If $f_{S,C} \geq 0$, then C is Lo-Ng positive.*

Proof 1. This follows by uniqueness of Fourier coefficients in the Fourier expansion (2.3.7) of $f_{S,C}$.
2. Suppose that $B : \widehat{G} \to \mathcal{M}(\widehat{G})$ is a compatible mapping for which

$$\sum_{\pi \in S'} d_\pi \mathrm{tr}(\pi(g)B(\pi)) \geq 0$$

for all $g \in G$ and some finite subset S' of S. By the hypothesis on $f_{S,C}$ and (4.3.8), it follows that

$$\sum_{\pi \in S'} d_\pi \mathrm{tr}(B(\pi)C(\pi)) \geq 0$$

and so C is Lo-Ng positive by Lemma 4.3.1. □

Next we state another technical lemma:

Lemma 4.3.4 *There exists a sequence $(\psi_n, n \in \mathbb{N})$ of continuous non-negative functions on G, with each $\psi_n(g) = \sum_{\pi \in S_n} d_\pi z_\pi^{(n)} \chi_\pi(g)$ where S_n is a finite subset of $\widehat{G}$ and $z_\pi^{(n)} \in \mathbb{C}$ for all $\pi \in S_n, n \in \mathbb{N}$ which has the following properties:*

(i) $\int_G \psi_n(g)dg = 1$ for all $n \in \mathbb{N}$;
(ii) Given any neighbourhood U of e and any $\epsilon > 0$ there exists $n_0 \in \mathbb{N}$ such that $\psi_n(g) < \epsilon$ for all $g \in U^c$ and all $n \geq n_0$,

(iii) $\lim_{n\to\infty} z_\pi^{(n)} = 1$ *for all* $\pi \in \widehat{G}$.

Proof We follow Talman [202] Theorem A.7.1, pp. 96–98 for (i) and (ii) and Lo-Ng [136] for (iii).

(i) First note that if π is a finite-dimensional representation of G, then by (3.3.13) we easily deduce that

$$\sup_{g\in G} |\chi_\pi(g)| \leq d_\pi.$$

Next observe that since G is a compact Lie group, it has a faithful finite-dimensional representation π (see e.g. Theorem 4.1 in Bröcker and tom Dieck [36], pp. 136–137) and for all $g, h \in G$ with $g \neq h$,

$$\sum_{i,j=1}^{d_\pi} |\pi_{ij}(g) - \pi_{ij}(h)|^2 > 0,$$

indeed, if there were equality, π would not be injective. Now

$$\begin{aligned}
&\sum_{i,j=1}^{d_\pi} |\pi_{ij}(g) - \pi_{ij}(h)|^2 \\
&= \sum_{i,j=1}^{d_\pi} (\pi_{ij}(g) - \pi_{ij}(h))\overline{(\pi_{ij}(g) - \pi_{ij}(h))} \\
&= \sum_{i,j=1}^{d_\pi} (\pi_{ij}(g) - \pi_{ij}(h))(\pi_{ji}(g^{-1}) - \pi_{ji}(h^{-1})) \\
&= 2\chi_\pi(e) - \chi_\pi(gh^{-1}) - \overline{\chi_\pi(gh^{-1})}.
\end{aligned}$$

Let $\pi' := \pi \oplus \overline{\pi}$. Then for all $g \in G$, $\chi_{\pi'}(g) = \chi_\pi(g) + \overline{\chi_\pi(g)}$, and we deduce from the last display that for all $g \in G \setminus \{e\}$

$$\chi_{\pi'}(g) < \chi_{\pi'}(e) = d_{\pi'}.$$

Incorporating this with our earlier estimate, we see that for all $g \in G \setminus \{e\}$

$$-d_{\pi'} \leq \chi_{\pi'}(g) < d_{\pi'}.$$

Now define a new representation π'' of G to be the direct sum of π' and $d_{\pi'}$ copies of the trivial representation. Then for all $g \in G$,

$$\chi_{\pi''}(g) = d_{\pi'} + \chi_{\pi'}(g),$$

and the estimate just established yields, for all $g \in G \setminus \{e\}$,

$$0 \leq \chi_{\pi''}(g) < 2d_{\pi'} = d_{\pi''}.$$

Now for each $n \in \mathbb{N}, g \in G$ define $\psi_n(g) := c_n \chi_{\pi''}(g)^n$, where $c_n := \left(\int_G \chi_{\pi''}(g)^n dg\right)^{-1}$. Then by construction ψ_n is continuous, non-negative and $\int_G \psi_n(g)dg = 1$. By Theorem 2.4.2 (ii), $\chi_{\pi''}(g)^n$ is the value at g of the character of the n-fold tensor product of π'', and so by Theorem 2.4.2 (iv), $\chi_{\pi''}(g)^n = \sum_{\pi \in S_n} m_\pi^{(n)} \chi_\pi$ (where $m_\pi^{(n)}$ is a non-negative integer). Hence the complex numbers $z_\pi^{(n)}$ appearing in the statement of the lemma are given by $z_\pi^{(n)} = \dfrac{c_n m_\pi^{(n)}}{d_\pi}$.

(ii) For simplicity we write $\chi := \chi_{\pi''}$ and $d := d_{\pi''}$ for the remainder of this proof. Let U be an open neighbourhood of e. Then $G \setminus U$ is compact, and so there exists $g_0 \in G \setminus U$ for which $\chi(g_0) = \sup_{g \in G \setminus U} \chi(g)$ and we have $\chi(g_0) < d$. By continuity of $g \to \chi(g)$ at $g = e$, given any $\varepsilon > 0$ there exists an open neighbourhood V of e so that if $g \in V$, then $d - \varepsilon < \chi(g) < d + \varepsilon$. Now choose $\varepsilon = \dfrac{d - \chi(g_0)}{2}$ and we see that for all $g \in G, \chi(g) > \dfrac{d + \chi(g_0)}{2}$. Consequently, for each $n \in \mathbb{N}$,

$$\int_V \chi(g)^n dg > m(V) \left(\frac{d + \chi(g_0)}{2}\right)^n.$$

Now

$$\begin{aligned} c_n &< \left(\int_V \chi(g)^n dg\right)^{-1} \\ &< \frac{1}{m(V)} \left(\frac{2}{d + \chi(g_0)}\right)^n. \end{aligned}$$

Then for all $g \in G \setminus U$ we have

$$\begin{aligned} \psi_n(g) &< \frac{1}{m(V)} \left(\frac{2\chi(g)}{d + \chi(g_0)}\right)^n \\ &\leq \frac{1}{m(V)} \left(\frac{2\chi(g_0)}{d + \chi(g_0)}\right)^n, \end{aligned}$$

and we can make the quantity on the right hand side arbitrarily small by taking n to be sufficiently large.

(iii) If we take the inner product in $L^2(G)$ of ψ_n with the character of an arbitrary representation in $\widehat{G}$ and use Theorem 2.4.3, we can easily deduce that for each $n \in \mathbb{N}, \pi \in S_n, z_\pi^{(n)} = \frac{1}{d_\pi} \int_G \psi_n(g) \chi_\pi(g^{-1}) dg$. Then we find that

$$|z_\pi^{(n)} - 1| = \frac{1}{d_\pi}\left|\int_G \psi_n(g)\chi_\pi(g^{-1})dg - d_\pi \int_G \psi_n(g)dg\right|$$
$$\leq \frac{1}{d_\pi}\int_{U^c} \psi_n(g)|\chi_\pi(g^{-1}) - d_\pi|dg + \frac{1}{d_\pi}\int_U \psi_n(g)|\chi_\pi(g^{-1}) - d_\pi|dg,$$

and the required result follows by taking U sufficiently small, n sufficiently large, and using the result of (ii) and the fact that $g \to \chi_\pi(g^{-1})$ is continuous, and takes the value d_π at e. □

The next result is the main one of this section.

Theorem 4.3.2 (The Lo-Ng Criterion) *Let $C : \widehat{G} \to \mathcal{M}(\widehat{G})$ be compatible. Then $C(\pi) = \widehat{\mu}(\pi)$ for all $\pi \in \widehat{G}$ where $\mu \in \mathcal{P}(G)$ if and only if C is Lo-Ng positive with $C(\pi_0) = 1$. Furthermore, μ is the weak limit of a sequence $(\mu_n, n \in \mathbb{N})$, where for each $n \in \mathbb{N}$, $\mu_n \in \mathcal{P}(G)$ is absolutely continuous with respect to Haar measure and has Radon-Nikodym derivative*

$$h_n(g) = \sum_{\pi \in S_n} z_\pi^{(n)} d_\pi \mathrm{tr}(\pi(g)C(\pi))$$

for all $g \in G$, where $\#S_m < \#S_n < \infty$ if $m < n$ and $\pi_0 \in S_n$ for all $n \in \mathbb{N}$.

Proof Assume that $\mu \in \mathcal{P}(G)$ and $\sum_{\pi \in S} d_\pi \mathrm{tr}(\pi(g)B_\pi) \geq 0$ for all $g \in G$ and some finite subset S of $\widehat{G}$. Then

$$\sum_{\pi \in S} d_\pi \mathrm{tr}(\widehat{\mu}(\pi)B_\pi) = \int_G \sum_{\pi \in S} d_\pi \mathrm{tr}(\pi(g^{-1})B_\pi)\mu(dg) \geq 0,$$

and so $\widehat{\mu} : \widehat{G} \to \mathcal{M}(\widehat{G})$ is Lo-Ng positive.

Conversely (and using the notation of Lemma 4.3.4), we have that for all $n \in \mathbb{N}$, $\psi_n(g) = \sum_{\pi \in S_n} d_\pi \mathrm{tr}(\pi(g)[z_\pi^{(n)} I_{d_\pi}]) \geq 0$ by Lemma 4.3.4, and so since C is assumed to be Lo-Ng positive, it follows that

$$\sum_{\pi \in S_n} d_\pi \mathrm{tr}(\pi(g) z_\pi^{(n)} C(\pi)) \geq 0.$$

By Lemma 4.3.1 we deduce that the compatible mapping whose value at $\pi \in S_n$ is $z_\pi^{(n)} C(\pi)$ (and whose value at $\pi \notin S_n$ is the zero matrix) is also Lo-Ng positive. By Proposition 4.3.1 (1),

$$z_\pi^{(n)} C(\pi) = \int_G \pi(g^{-1}) h_n(g) dg.$$

Since h_n is continuous it is integrable, and as h_n is non-negative, we can define a Borel measure μ_n on G whose Radon-Nikodym derivative is h_n. Using Peter-Weyl theory (Corollary 2.2.4), we have

$$\begin{aligned}\mu_n(G) &= \int_G h_n(g)dg \\ &= \int_G h_n(g)\pi_0(g)dg \\ &= z_{\pi_0}^{(n)} C(\pi_0) = 1.\end{aligned}$$

The fact that $z_{\pi_0}^{(n)} = 1$ follows from Lemma 4.3.4 (i) and the formula $z_\pi^{(n)} = \frac{1}{d_\pi}\int_G \psi_n(g)\chi_\pi(g^{-1})dg$ that is established within the proof of that same lemma. By Prohorov's theorem, we can now assert that there is a subsequence $(\mu_{n_k}, k \in \mathbb{N})$ that converges weakly to a probability measure μ. By Theorem 4.2.5, we have $\lim_{k\to\infty}\widehat{\mu_{n_k}}(\pi) = \widehat{\mu}(\pi)$ for all $\pi \in \widehat{G}$. But by construction $\lim_{k\to\infty}\widehat{\mu_{n_k}}(\pi) = \lim_{k\to\infty} z_\pi^{(n_k)} C(\pi) = C(\pi)$ by Lemma 4.3.4 (iii). Hence the converse is established.

To prove the last part of the theorem let $h \in C(G)$. Then by the Peter-Weyl theorem (Theorem 2.2.4), there exists a sequence of matrices $(H_n, n \in \mathbb{N})$ where each H_n acts in a finite-dimensional complex Hilbert space of dimension d_n such that $h(g) = \lim_{n\to\infty}\sum_{i=1}^n d_i \mathrm{tr}(\pi_i(g)^* H_i)$, and the convergence is uniform in $g \in G$. Using Schur orthogonality and (4.3.8), we find that

$$\begin{aligned}&\int_G h(g)\left(\sum_{\pi\in S_n} d_\pi z_\pi^{(n)}\mathrm{tr}(\pi(g)C(\pi))\right)dg \\ &= \lim_{m\to\infty}\int_G\left(\sum_{i=1}^m d_i \mathrm{tr}(\pi_i(g)^* H_i)\right)\left(\sum_{\pi\in S_n} d_\pi z_\pi^{(n)}\mathrm{tr}(\pi(g)C(\pi))\right)dg \\ &= \sum_{\pi\in S_n} d_\pi \mathrm{tr}(H(\pi) z_\pi^{(n)} C(\pi)) \\ &= \int_G\left(\sum_{\pi\in S_n} d_\pi \mathrm{tr}(\pi(g)^* H(\pi))\right)h_n(g)dg \\ &\to \int_G h(g)\mu(dg),\end{aligned}$$

as $n \to \infty$, using the dominated convergence theorem.

Remark

1. Although Lo-Ng positivity is an interesting theoretical result, it seems very difficult to use in practice to determine whether a given family of compatible matrices really is the Fourier transform of a finite measure.
2. As positive-definiteness (in the usual sense) is a key component of Bochner's theorem on locally compact abelian groups, it is worth pointing out that there is a general notion of positive definiteness for functions on a more general locally compact group G. Indeed, a continuous function $f : G \to \mathbb{C}$ is positive definite if and only if $\sum_{i,j=1}^{n} c_i\overline{c_j} f(g_i g_j^{-1}) \geq 0$ for all $g_1, \ldots, g_n \in G, c_1, \ldots, c_n \in \mathbb{C}, n \in \mathbb{N}$. You can learn about these functions in e.g. Sect. 2.8 of Edwards [61] or section 32 of Hewitt and Ross [92]. Note that there is even a *Bochner theorem* which describes the structure of such functions as linear combinations of certain elementary ones, but readers should be warned that it is not related to the Bochner theorem that we discussed at the beginning of this section (i.e. it does not give information about Fourier transforms of finite measures).

4.4 Absolute Continuity

We investigate absolute continuity of probability measures on G with respect to normalised Haar measure m. We follow the account in Wehn [216].

Theorem 4.4.1 (Raikov-Williamson) *Let $\mu \in \mathcal{P}(G)$. Then $\mu \in \mathcal{P}_{ac}(G)$ if and only if either $\mu(Eg) \to \mu(E)$ or $\mu(gE) \to \mu(E)$ as $g \to e$ for all $E \in \mathcal{B}(G)$.*

Proof We only deal here with the case $\mu(Eg) \to \mu(E)$ as $g \to e$. The other limit is dealt with by a similar argument.

First assume that $\mu \ll m$ and let $f_\mu := \dfrac{d\mu}{dm}$. Then for all $E \in \mathcal{B}(G)$,

$$\begin{aligned} |\mu(Eg) - \mu(E)| &\leq \int_E |f_\mu(hg^{-1}) - f_\mu(h)| dh \\ &\leq ||R_{g^{-1}} f_\mu - f_\mu||_1 \to 0 \text{ as } g \to e, \end{aligned}$$

by Proposition 1.2.1. Conversely, suppose that $\mu(Eg^{-1}) \to \mu(E)$ as $g \to e$ and suppose that $E \in \mathcal{B}(G)$ exists with $m(E) = 0$ and $\mu(E) > 0$. We seek a contradiction. Let $\rho \in L^1(G)$ be such that $\rho \geq 0$ and $\int_G \rho(g)dg = 1$. Then we may define a measure $\nu_\rho \in \mathcal{P}_{ac}(G)$ by $\nu_\rho(A) = \int_A \rho(g)dg$ for all $A \in \mathcal{B}(G)$. For all $g \in G$,

$$\begin{aligned} \nu_\rho(g^{-1}E) &= \int_G \rho(h) 1_E(gh) dh \\ &= \int_G \rho(g^{-1}h) 1_E(h) dh = 0, \end{aligned}$$

since $m(E) = 0$. Hence by (4.1.1)

$$(\mu * \nu_\rho)(E) = \int_G \nu_\rho(g^{-1}E)\mu(dg) = 0.$$

But again by (4.1.1), we have

$$(\mu * \nu_\rho)(E) = \int_G \mu(Eg^{-1})\nu_\rho(dg) > 0,$$

and this yields the required contradiction. □

For each $\mu \in \mathcal{P}(G)$ we define the associated *convolution operator* $T_\mu : B_b(G) \to B_b(G)$ by

$$(T_\mu f)(\sigma) := (f * \mu)(\sigma) = \int_G f(\sigma\tau)\mu(d\tau),$$

for all $f \in B_b(G), \sigma \in G$. It is easy to see that T_μ is linear and a contraction. Furthermore, if $\mu, \nu \in \mathcal{P}(G)$ we have

$$T_{\mu * \nu} = T_\mu T_\nu. \tag{4.4.9}$$

It is an important fact that $T_\mu : C(G) \to C(G)$. To see this, let $\sigma_1, \sigma_2 \in G$ and observe that for all $f \in C(G)$,

$$|T_\mu f(\sigma_1) - T_\mu f(\sigma_2)| \leq \int_G |f(\sigma_1\tau) - f(\sigma_2\tau)|\mu(d\tau) \leq ||L_{\sigma_1^{-1}} f - L_{\sigma_2^{-1}} f||_\infty,$$

and the result follows by left uniform continuity of f (see Theorem A.2.1 in Appendix A.2).

Before we state and prove the next result we recall that a subset S of $C(G)$ is *equicontinuous* if given any $\epsilon > 0$, each $g \in G$ has an open neighbourhood U_g so that if $h \in U_g$, then $|f(g) - f(h)| < \epsilon$ for all $f \in S$. The next theorem was originally established by Raikov [164], and we follow the account of Wehn [216].

Theorem 4.4.2 (Raikov) *Let* $\mu \in \mathcal{P}(G)$. *Then* $\mu \in \mathcal{P}_{ac}(G)$ *if and only if* $T_\mu : C(G) \to C(G)$ *is compact.*

Proof First suppose that $\mu \ll m$ and write $\rho_\mu := \dfrac{d\mu}{dm}$. Let $(f_n, n \in \mathbb{N})$ be a bounded sequence in $C(G)$. Then for all $g, h \in G, n \in \mathbb{N}$,

$$T_\mu f_n(g) - T_\mu f_n(h) = \int_G f_n(g\tau)\rho_\mu(\tau)d\tau - \int_G f_n(h\tau)\rho_\mu(\tau)d\tau$$
$$= \int_G f_n(\tau)(\rho_\mu(g^{-1}\tau) - \rho_\mu(h^{-1}\tau))d\tau,$$

from which we easily deduce that

$$|T_\mu f_n(g) - T_\mu f_n(h)| \leq \sup_{n\in\mathbb{N}} ||f_n||_\infty ||L_g\rho_\mu - L_h\rho_\mu||_1.$$

Then equicontinuity of $\{T_\mu f_n, n \in \mathbb{N}\}$ follows from Proposition 1.2.1. Uniform boundedness of $\{T_\mu f_n, n \in \mathbb{N}\}$ is easily verified. We can now appeal to the Arzelà-Ascoli theorem to deduce that $\{T_\mu f_n, n \in \mathbb{N}\}$ is relatively compact, and so contains a convergent subsequence. It follows that T_μ is compact.

Conversely, suppose that T_μ is compact and let E be an open set in G. Then since 1_E is lower semi-continuous, we can find a sequence $(f_n, n \in \mathbb{N})$ in $C(G)$ which increases monotonically to 1_E (see e.g. Nagami [152]). So in particular this sequence is bounded. Hence by assumption, its image contains a convergent subsequence $(T_\mu f_{n_k}, k \in \mathbb{N})$, and (uniformly in) $g \in G$,

$$\lim_{k\to\infty} T_\mu f_{n_k}(g) = T_\mu 1_E(g) = \mu(g^{-1}E).$$

It follows that the mapping $g \to \mu(g^{-1}E)$ is continuous, and so $\mu(g^{-1}E) \to \mu(E)$ as $g \to e$. A similar argument holds for the case where E is compact. By regularity of μ, the same limiting behaviour holds for $E \in \mathcal{B}(G)$. So by Theorem 4.4.1 we deduce that $\mu \ll m$, as required. □

In the case $G = \Pi^1$, the celebrated theorem of F. and M. Riesz gives a sufficient condition for a probability measure μ to be absolutely continuous. In that case $\widehat{G} = \mathbb{Z}$ and

$$\widehat{\mu}(n) = \frac{1}{2\pi}\int_0^{2\pi} e^{-inx}\mu(dx)$$

for each $n \in \mathbb{Z}$. Their sufficient condition for absolute continuity is that $\widehat{\mu}(n) = 0$ for all $n < 0$ (see e.g. Katznelson [113] p. 113). This result has been extended to compact Lie groups by Brummelhuis [38] (see also [37]). For ease of exposition, we state it here in the case where G is also connected and semisimple. Let $\pi \in \widehat{G}$ and recall that $V_\pi = \bigoplus_{\mu\in\mathcal{W}(\pi)} V_\mu$ where $\mathcal{W}(\pi)$ is the set of weights of π. Let λ be the highest weight and define $V_\pi^0 := V_\pi \ominus V_\lambda$.

Theorem 4.4.3 (Brummelhuis) *Let G be a compact, connected, semisimple Lie group. If $\mu \in \mathcal{P}(G)$ is such that $\widehat{\mu}(\pi)v = 0$ for all $v \in V_\pi^0$ and for all $\pi \in \widehat{G}$, then $\mu \ll m$.*

4.5 Regularity of Densities

In this section, we will investigate conditions for a probability measure on a compact group to have a square-integrable, continuous and smooth density of various orders. Although this topic is closely related to that of the Sect. 4.4, we will make no use of the results that we obtained there.

In this section we first examine the case where $\mu \in \mathcal{P}(G)$ has a square-integrable density. The following result is established in Applebaum [8].

Theorem 4.5.1 *Let G be a compact Lie group. Then $\mu \in \mathcal{P}(G)$ has an L^2-density f_μ if and only if*

$$\sum_{\pi \in \widehat{G}} d_\pi ||\widehat{\mu}(\pi)||_{HS}^2 < \infty. \tag{4.5.10}$$

In this case

$$f_\mu = \sum_{\pi \in \widehat{G}} d_\pi \mathrm{tr}(\widehat{\mu}(\pi)\pi(\cdot)) \tag{4.5.11}$$

Proof For necessity, suppose that $f_\mu \in L^2(G)$ is the density of μ. Then $\widehat{f_\mu}(\pi) = \widehat{\mu}(\pi)$ for all $\pi \in \widehat{G}$ and (4.5.10) follows from the Parseval-Plancherel identity (2.3.8).

For sufficiency define

$$f_\mu := \sum_{\pi \in \widehat{G}} d_\pi \mathrm{tr}(\widehat{\mu}(\pi)\pi).$$

Then $f_\mu \in L^2(G)$ since by (2.3.8),

$$||f_\mu||_2^2 = \sum_{\pi \in \widehat{G}} d_\pi ||\widehat{\mu}(\pi)||_{HS}^2 < \infty,$$

and by uniqueness of Fourier coefficients (in the Hilbert space sense) $\widehat{f_\mu}(\pi) = \widehat{\mu}(\pi)$ for all $\pi \in \widehat{G}$.[9]

Since Haar measure is finite, $L^2(G) \subseteq L^1(G)$, and so $f_\mu \in L^1(G)$. Recall that by Theorem 2.2.4 $\mathcal{E}(G)$, which is the algebra of all continuous functions on G that have only finitely many non-zero Fourier coefficients, is norm dense in $C(G)$. Let $h \in \mathcal{E}(G)$. Then there exists a finite subset S of $\widehat{G}$ so that

$$h(\sigma) = \sum_{\pi \in S} d_\pi \mathrm{tr}(\widehat{h}(\pi)\pi(\sigma))$$

for all $\sigma \in G$. Furthermore, by the Schur orthogonality relations, $\widehat{h}(\pi) = 0$ if $\pi \in S^c$. Using the Parseval-Plancherel identity (2.3.9), for each $h \in \mathcal{E}(G)$:

[9] To verify this directly, compute $\langle f_\mu, \pi'_{ij} \rangle$ for each $\pi' \in \widehat{G}$, $1 \le i, j \le d'$; it is then a straightforward application of the Peter-Weyl theorem (Corollary 2.2.4) to deduce that $\widehat{f_\mu}(\pi')_{ij} = \widehat{\mu}(\pi')_{ij}$.

$$\int_G h(\sigma)\overline{f_\mu(\sigma)}d\sigma = \sum_{\pi\in\widehat{G}} d_\pi \mathrm{tr}(\widehat{h}(\pi)\widehat{\mu}(\pi)^*)$$
$$= \sum_{\pi\in S} d_\pi \mathrm{tr}(\widehat{h}(\pi)\widehat{\mu}(\pi)^*)$$
$$= \int_G \sum_{\pi\in S} d_\pi \mathrm{tr}(\widehat{h}(\pi)\pi(\sigma))\mu(d\sigma)$$
$$= \int_G h(\sigma)\mu(d\sigma).$$

By a standard density argument, it then follows that

$$\int_G h(\sigma)\overline{f_\mu(\sigma)}d\sigma = \int_G h(\sigma)\mu(d\sigma),$$

for all $h \in C(G)$. The Riesz representation theorem implies that f_μ is real valued and $f_\mu(\sigma)d\sigma = \mu(d\sigma)$. The fact that f_μ is non-negative a.e. then follows from the Jordan decomposition for signed measures (see Appendix A.5). □

Note that we can also write (4.5.11) as

$$f_\mu = 1 + \sum_{\pi\in\widehat{G}\setminus\{\pi_0\}} d_\pi \mathrm{tr}(\widehat{\mu}(\pi)\pi(\cdot)),$$

and that such a representation is often found in the literature.

It is easily seen that if μ is central, so that $\widehat{\mu}(\pi) = c_\pi I_\pi$ for all $\pi \in \widehat{G}$ (where $c_\pi \in \mathbb{C}$), and has L^2-density f_μ, then f_μ is central (a.e.). Then from (4.5.11) we have

$$f_\mu = \sum_{\pi\in\widehat{G}} d_\pi c_\pi \chi_\pi, \tag{4.5.12}$$

in the L^2 sense.

Next we examine continuity of densities:

Proposition 4.5.1 *Let $\mu \in \mathcal{P}(G)$. A sufficient condition for μ to have a continuous density f_μ is that the infinite series $\sum_{\pi\in\widehat{G}} d_\pi \mathrm{tr}(\widehat{\mu}(\pi)\pi(\sigma))$ converges uniformly in $\sigma \in G$.*

Proof Define $f_\mu(\sigma) = \sum_{\pi\in\widehat{G}} d_\pi \mathrm{tr}(\widehat{\mu}(\pi)\pi(\sigma))$ for all $\sigma \in G$. Then $f_\mu \in C(G)$, and by uniqueness of Fourier coefficients, $\widehat{f_\mu}(\pi) = \widehat{\mu}(\pi)$ for all $\pi \in \widehat{G}$. Now argue as in the proof of Theorem 4.5.1. □

More concrete sufficient conditions for μ to have a continuous density are as follows. In the second of these, for each $\mu \in \mathcal{P}(G)$ we employ the notation $\widehat{\mu}(\lambda) := \widehat{\mu}(\pi_\lambda)$, where $\lambda \in D$ is the highest weight corresponding to $\pi_\lambda \in \widehat{G}$:

- $\sum_{\pi \in \widehat{G}} d_\pi^{\frac{3}{2}} ||\widehat{\mu}(\pi)||_{HS} < \infty$,
- $||\widehat{\mu}(\pi_\lambda)||_{HS} = O(|\lambda|^{-s})$ as $|\lambda| \to \infty$, where $s > r + \dfrac{m}{2}$.

The first of these is implicit in the first part of the proof of Proposition 3.3.2 (see also Faraut [63], pp. 117–119). and the second is a direct consequence of the statement of that same proposition.

Next we investigate differentiability of densities. Recall that $\{\kappa_\pi, \pi \in \widehat{G}\}$ is the Casimir spectrum of G.

Theorem 4.5.2 *If $\mu \in \mathcal{P}(G)$ and there exists $p \in \mathbb{N}$ so that*

$$\sum_{\pi \in \widehat{G}} d_\pi (1 + \kappa_\pi)^p ||\widehat{\mu}(\pi)||_{HS}^2 < \infty,$$

then μ has a C^k density for all $k < p - \dfrac{d}{2}$.

Proof Since $\kappa_\pi \geq 0$ for all $\pi \in \widehat{G}$, we have $\sum_{\pi \in \widehat{G}} d_\pi ||\widehat{\mu}(\pi)||_{HS}^2 < \infty$, and so by Theorem 4.5.1, μ has a L^2-density f_μ and $\widehat{f_\mu}(\pi) = \widehat{\mu}(\pi)$ for all $\pi \in \widehat{G}$. The result then follows by Proposition 3.1.4, and the Sobolev embedding theorem (Theorem 3.1.3). □

The next result establishes necessary and sufficient conditions for densities to exist and be C^∞. It was first established in Applebaum [12]. Recall that $\mathcal{S}(D)$ is the Sugiura space that was introduced in Sect. 3.4.

Theorem 4.5.3 *For G a compact connected Lie group, $\mu \in \mathcal{P}(G)$ has a C^∞ density if and only if $\widehat{\mu} \in \mathcal{S}(D)$.*

Proof Necessity is obvious. For sufficiency it is enough by Theorem 3.4.3 to show that μ has an L^2-density. Choose $s > r$, so that Suguira's zeta function (see below) converges (c.f. Theorem 3.2.1). Then using Theorem 4.5.1 we have

$$\begin{aligned}\sum_{\lambda \in D - \{0\}} d_\lambda ||\widehat{\mu}_\lambda||_{HS}^2 &\leq N \sum_{\lambda \in D - \{0\}} |\lambda|^m ||\widehat{\mu}_\lambda||_{HS}^2 \\ &\leq N \sup_{\lambda \in D - \{0\}} |\lambda|^{m+s} ||\widehat{\mu}_\lambda||_{HS}^2 \sum_{\lambda \in D - \{0\}} \frac{1}{|\lambda|^s} \\ &< \infty.\end{aligned}$$

□

The following result gives an application of Theorem 4.5.3. First we note a useful and easily verified inequality for matrices. If $A, B \in M_n(\mathbb{C})$, then

$$||AB||_{HS} \leq ||A||_{\mathrm{op}} ||B||_{HS} \text{ and } ||AB||_{HS} \leq ||B||_{\mathrm{op}} ||A||_{HS} \tag{4.5.13}$$

Corollary 4.5.1 *Let G be a compact connected Lie group. Let $\mu \in \mathcal{P}(G)$ be arbitrary and $\gamma_{t,\sigma}$ be a standard Gaussian measure with parameters $t, \sigma > 0$. Then the measures $\mu * \gamma_{t,\sigma}$ and $\gamma_{t,\sigma} * \mu$ have smooth densities.*

Proof It suffices to establish the result for $\mu * \gamma_{t,\sigma}$. First note that by Theorem 4.2.1, (4.5.13) and (4.2.7) for all $\lambda \in D$,

$$||\widehat{\mu * \gamma_{t,\sigma}}(\lambda)||_{HS} = ||\widehat{\gamma_{t,\sigma}}(\lambda)\widehat{\mu}(\lambda)||_{HS} \leq ||\widehat{\mu}(\lambda)||_{\text{op}}||\widehat{\gamma_{t,\sigma}}(\lambda)||_{HS} \leq d_\lambda e^{-t\sigma\kappa_\lambda}.$$

But then using the dimension estimate of Corollary 2.5.2 and (2.5.23) we obtain for all $p \in \mathbb{N}$,

$$\limsup_{|\lambda|\to\infty} |\lambda|^p ||\widehat{\mu * \gamma_{t,\sigma}}(\lambda)||_{HS} \leq \lim_{|\lambda|\to\infty} |\lambda|^p d_\lambda e^{-t\sigma\kappa_\lambda} \leq C \lim_{|\lambda|\to\infty} |\lambda|^{p+m} e^{-t\sigma|\lambda|^2} = 0,$$

and the result follows from Theorem 4.5.3. □

4.6 Idempotents and Convolution Powers

We say that $\mu \in \mathcal{P}(G)$ is *idempotent* if $\mu * \mu = \mu$. Equivalently by Theorems 4.2.1 and 4.2.3, μ is idempotent if and only if $\widehat{\mu}(\pi)^2 = \widehat{\mu}(\pi)$ for all $\pi \in \widehat{G}$. It is easy to see that normalised Haar measure m on G is idempotent. More generally, let H be a closed subgroup of G and let $m_H^{(0)}$ denote its normalised Haar measure. We extend $m_H^{(0)}$ to a measure $m_H \in \mathcal{P}(G)$ that has support H by the prescription

$$m_H(B) = m_H^{(0)}(B \cap H)$$

for all $B \in \mathcal{B}(G)$. For example if $H = \{e\}$, then $m_H = \delta_e$. It is again easy to see that m_H is always idempotent. The following result is due to Wendel [217] for compact groups, but note that it also holds in general locally compact groups (see Heyer [95] Theorem 1.2.10, p. 34).[10]

Theorem 4.6.1 *If $\mu \in \mathcal{P}(G)$ is idempotent, then $\mu = m_H$ for some closed subgroup H of G. Moreover, $H = supp(\mu)$.*

Proof For $H :=$ supp(μ), by (4.1.2) we have $H = H^2$, and so H is a semigroup under the group law. It is also closed, and hence compact. It is known that any subset of G that has these properties is a subgroup (see e.g. Lemma 2 in Gelbaum et. al. [71] and also Corollary 1.2.9. on p. 34 of Heyer [95]). Now let $f \in C(H, \mathbb{R})$, and define for each $h \in H$:

$$A_f(h) = \int_G f(gh)\mu(dg).$$

Using Proposition 1.2.1 it is easily verified that $A_f \in C(H, \mathbb{R})$. Now let h_0 be the point in G where A_f attains its maximum value. From now on we denote $f_1 := R_{h_o} f$.

[10] Our standing hypothesis remains that G is compact Lie, but observe that the proof of Theorem 4.6.1 requires no use of Lie structure.

Then A_{f_1} attains its maximum at e. Then since μ is idempotent:

$$\begin{aligned} A_{f_1}(e) &= \int_H f_1(g)\mu(dg) \\ &= \int_H f_1(g)(\mu * \mu)(dg) \\ &= \int_H \int_H f_1(g_1 g_2)\mu(dg_1)\mu(dg_2) \\ &= \int_H A_{f_1}(g_2)\mu(dg_2). \end{aligned}$$

Hence we see that $\int_H (A_{f_1}(e) - A_{f_1}(g_2))\mu(dg_2) = 0$, and so $A_{f_1}(e) - A_{f_1}(g)$ for all $g \in H$. It follows by uniqueness of Haar measure, and the fact that $\mu(H) = 1$, that $\mu = m_H$, as required. □

Let $\mu \in \mathcal{P}(G)$ and $n \in \mathbb{N}$. We define the nth convolution power of μ to be $\mu^{*(n)} = \mu * \cdots * \mu$ (n times). Note that we then have for all $\pi \in \widehat{G}$, $\widehat{\mu^{*(n)}}(\pi) = \widehat{\mu}(\pi)^n$. Let $(\Omega, \mathcal{F}, P)$ be a probability space and $(X_n, n \in \mathbb{N})$ be a sequence of independent, identically distributed (or i.i.d.) G–valued random variables. Let $(S_n, n \in \mathbb{N})$ be the associated G-valued *random walk*, so that for each $n \in \mathbb{N}$, $S_n = X_1 X_2 \ldots X_n$. Then the law of S_n is precisely $\mu^{*(n)}$. It is of interest to study the asymptotic behaviour of the random walk as for large n. In particular we might consider the weak limit of $\mu^{*(n)}$ as $n \to \infty$. It is clear that if the limit exists, it is an idempotent, and so by Theorem 4.6.1

$$\lim_{n \to \infty} \mu^{*(n)} = m_H,$$

for some closed subgroup H of G.

Necessary and sufficient conditions for the limit to exist were found by Stromberg [198]. We quote his result but omit the proof (see also Heyer [95] Theorem 2.1.4, pp. 91–92).

Theorem 4.6.2 *Let $\mu \in \mathcal{P}(G)$ and let K be the smallest closed subgroup of G containing supp(μ). Then $\lim_{n\to\infty} \mu^{*(n)}$ exists if and only if supp(μ) is not contained in any coset of a proper closed normal subgroup of K.*

Kawada and Itô [114] established an *equidistribution theorem* which gives conditions for the limit to exist and be normalised Haar measure m on the whole group. First we need a definition. We say that $\mu \in \mathcal{P}(G)$ is *aperiodic* if supp(μ) is not contained in a left or right coset of a proper closed normal subgroup of G. We then have the following:

Theorem 4.6.3 (Kawada-Itô equidistribution theorem) *If $\mu \in \mathcal{P}(G)$ is aperiodic, then $(\mu^{*(n)}, n \in \mathbb{N})$ converges weakly to normalised Haar measure.*

Proof This is based on the proof of Theorem 8 in [114]. By Theorem 4.2.6, it is sufficient to show that $\lim_{n\to\infty}\widehat{\mu}(\pi)^n = 0$ for all non-trivial $\pi \in \widehat{G}$. This is clearly equivalent to the requirement that all the eigenvalues of $\widehat{\mu}$ have modulus strictly less than 1. Note that since $\widehat{\mu}(\pi)$ is a contraction, its eigenvalues cannot have moduli that exceed 1. Now let λ be an eigenvalue of $\widehat{\mu}(\pi)$. Then we can find a unitary matrix U_π acting in V_π so that

$$U_\pi\widehat{\mu}(\pi)U_\pi^{-1} = \begin{pmatrix} \lambda & 0 & \cdot & \cdot & \cdot & 0 \\ 0 & & & & & \\ \cdot & & & & & \\ \cdot & & & D_\pi & & \\ \cdot & & & & & \\ 0 & & & & & \end{pmatrix},$$

where D_π is some $(d_\pi - 1) \times (d_\pi - 1)$ matrix. In particular, we have

$$\lambda = (U_\pi\widehat{\mu}(\pi)U_\pi^{-1})_{11} = \int_G (U_\pi\pi(g^{-1})U_\pi^{-1})_{11}\mu(dg),$$

and if $|\lambda| = 1$, $(U_\pi\pi(g^{-1})U_\pi^{-1})_{11} = \lambda$ for all $g \in G$ for which $g^{-1} \in \text{supp}(\mu)$.

Now suppose that $\lambda = 1$. Then we have

$$\text{supp}(\mu) \subseteq H := \left\{ g \in G, U_\pi\pi(g^{-1})U_\pi^{-1} = \begin{pmatrix} 1 & 0 & \cdot & \cdot & \cdot & 0 \\ 0 & & & & & \\ \cdot & & & & & \\ \cdot & & & E_\pi(g) & & \\ \cdot & & & & & \\ 0 & & & & & \end{pmatrix} \right\},$$

where $E_\pi(g)$ is some $(d_\pi - 1) \times (d_\pi - 1)$ matrix. But H is a proper closed subgroup of G and this contradicts aperiodicity of μ.

Now suppose that $\lambda = e^{i\theta}$ for some $\theta \in \mathbb{R}\backslash 2\pi\mathbb{Z}$. Then arguing as above we can find a unitary matrix U_π so that $e^{i\theta} = (U_\pi\widehat{\mu}(\pi)U_\pi^{-1})_{11} = \int_G (U_\pi\pi(g^{-1})U_\pi^{-1})_{11}\mu(dg)$ and

$$\text{supp}(\mu) \subseteq \Gamma := \left\{ g \in G, U_\pi\pi(g^{-1})U_\pi^{-1} = \begin{pmatrix} e^{i\theta} & 0 & \cdot & \cdot & \cdot & 0 \\ 0 & & & & & \\ \cdot & & & & & \\ \cdot & & & F_\pi(g) & & \\ \cdot & & & & & \\ 0 & & & & & \end{pmatrix} \right\},$$

where $F_\pi(g)$ is some $(d_\pi - 1) \times (d_\pi - 1)$ matrix. Now let $g_0 \in G$ be such that $(V_\pi \pi(g_0^{-1}) V_\pi^{-1})_{11} = e^{i\theta}$. Then it is easily verified that $\Gamma = g_0 H$, and this again contradicts aperiodicity. The required result follows. □

We briefly draw the reader's attention to more recent work in this area. Shlosman and Major [142] and Shlosman [179, 180] were able to extend the Kawada-Itô theorem to the case where μ has a density and there is uniform convergence of its convolution powers to the uniform density. Johnson and Suhov [109] used the Kullback-Liebler distance to obtain exponential rates of convergence and Harremoës [81] examined this from the perspective of uniform convergence of the rate distortion function.

4.7 Convolution Operators

In this and the Sect. 4.8 we will drop the condition that G be a compact Lie group. We work more generally and assume (unless otherwise stated) that G is a locally compact, Hausdorff and second countable topological group. Convolution operators were already introduced in Sect. 4.4 for compact Lie groups. Now we study them more systematically. Let $\mu \in \mathcal{P}(G)$. The associated *right convolution operator* $P_\mu^{(R)}$ is defined on $B_b(G)$ by the prescription $P_\mu^{(R)} f = f * \mu$ for $f \in B_b(G)$, so that

$$(P_\mu^{(R)} f)(g) = \int_G f(gh)\mu(dh)$$

for all $g \in G$. Similarly the *left convolution operator* $P_\mu^{(L)}$ is defined by

$$(P_\mu^{(L)} f)(g) = \int_G f(hg)\mu(dh).$$

The reader may check that these are related by the identity

$$P_\mu^{(L)} f = \widetilde{P_{\widetilde{\mu}}^{(R)} \widetilde{f}}.$$

Furthermore, if $\mu_1, \mu_2 \in \mathcal{P}(G)$, then $P_{\mu_1}^{(R)} P_{\mu_2}^{(L)} = P_{\mu_2}^{(L)} P_{\mu_1}^{(R)}$. We also have that for all $\mu \in \mathcal{P}(G), g \in G, L_g P_\mu^{(R)} = P_\mu^{(R)} L_g$ and $R_g P_\mu^{(L)} = P_\mu^{(L)} R_g$.

From now on we will almost always work with $P_\mu^{(R)}$, which we will denote simply as P_μ.

Proposition 4.7.1 *For each $\mu \in \mathcal{P}(G)$,*

1. *P_μ is linear,*

2. $P_\mu 1 = 1$,
3. *If* $f \geq 0$, *then* $P_\mu f \geq 0$,
4. P_μ *is a contraction,*
5. $P_\mu : C_0(G) \to C_0(G)$,
6. *If* $\mu_1, \mu_2 \in \mathcal{P}(G)$, *then*

$$P_{\mu_1 * \mu_2} = P_{\mu_1} P_{\mu_2}.$$

Proof (1)–(4) are all easy exercises. For (5) first note that continuity is straightforward to establish by using a dominated convergence argument (or see the discussion before Theorem 4.4.2). To see that if f vanishes at infinity then so does $P_\mu f$, let $G_\infty := G \cup \{\infty\}$ be the one-point compactification of G (see Appendix A.1). We extend μ to a probability measure on $(G_\infty, \mathcal{B}(G_\infty))$ by the requirement that $\mu(\{\infty\}) = 0$. Then P_μ extends to a linear operator from $C(G_\infty)$ to $C(G_\infty)$. If $F \in C(G_\infty)$ then $F \in C_0(G)$ if and only if $\lim_{g\to\infty} F(g) = 0$. Since $R_h : C_0(G) \to C_0(G)$ for all $h \in G$, the required result follows by using the dominated convergence theorem. To show (6), let $f \in B_b(G)$ and $g \in G$. By Fubini's theorem,

$$\begin{aligned}(P_{\mu_1 * \mu_2} f)(g) &= \int_G f(gh)(\mu_1 * \mu_2)(dh) \\ &= \int_G \int_G f(gh_1h_2)\mu_1(dh_1)\mu_2(dh_2) \\ &= \int_G \left(\int_G f(gh_1h_2)\mu_2(dh_2) \right) \mu_1(dh_1) \\ &= \int_G (P_{\mu_2} f)(gh_1)\mu_1(dh_1) \\ &= (P_{\mu_1} P_{\mu_2} f)(g). \qquad \square\end{aligned}$$

Note that if G is compact, then we use $C(G)$ in place of $C_0(G)$ in Proposition 4.7.1(5).

There is a useful link between the convolution operator and Fourier transform of the measure on compact groups.

Proposition 4.7.2 *If* μ *is a Borel probability measure on a compact Lie group* G, *then for all* $\pi \in \widehat{G}$, $1 \leq i, j \leq d_\pi$

$$\widehat{\mu}(\pi)_{ij} = P_\mu \pi_{ij}(e).$$

Proof

$$\begin{aligned} P_\mu \pi_{ij}(e) &= \int_G \pi_{ij}(\tau)\mu(d\tau) \\ &= \int_G \pi_{ij}(\tau^{-1})\widetilde{\mu}(d\tau) \\ &= \widehat{\widetilde{\mu}}(\pi)_{ij} \end{aligned}$$

□

Returning to the general case, we easily deduce from Proposition 4.7.1 (6) that for all $\mu \in \mathcal{P}(G), n \in \mathbb{N}$,

$$P_{\mu^{*(n)}} = P_\mu^n. \tag{4.7.14}$$

Note also that P_μ^n is the n-step transition operator for the random walk $(S(n), n \in \mathbb{N})$. Indeed, for all $f \in B_b(G), g \in G$,

$$(P_\mu^n f)(g) = \mathbb{E}(f(gS(n))).$$

In the following we fix a right Haar measure m_R on G and consider the p-norms $||\cdot||_p$ in the spaces $L^p(G, m_R)$ for $1 \leq p < \infty$.

Proposition 4.7.3 *For all* $\mu \in \mathcal{P}(G)$, $f \in C_c(G)$, $1 \leq p < \infty$,

$$||P_\mu f||_p \leq ||f||_p.$$

Proof By Jensen's inequality and Fubini's theorem,

$$\begin{aligned} ||P_\mu f||_p^p &= \int_G |P_\mu f(g)|^p m_R(dg) \\ &= \int_G \left| \int_G f(gh)\mu(dh) \right|^p m_R(dg) \\ &\leq \int_G \int_G |f(gh)|^p \mu(dh) m_R(dg) \\ &= \int_G \int_G |f(gh)|^p m_R(dg)\mu(dh) \\ &= \int_G |f(g)|^p m_R(dg) = ||f||_p^p. \end{aligned}$$

□

We have just shown that P_μ is a contraction from $C_c(G)$ to $L^p(G, m_R)$ and so it extends to a contraction on the whole of $L^p(G, m_R)$. We will continue to denote the operator by the same symbol P_μ whenever we consider it as acting in the L^p-spaces. For the case $p = 2$ the reader may verify the useful result

$$P_{\mu}^{*} = P_{\widetilde{\mu}}. \tag{4.7.15}$$

Let f be a non-negative measurable function defined on G. It is said to be *μ-harmonic* if $P_{\mu}f = f$, *μ-superharmonic* if $P_{\mu}f \leq f$ and *μ-subharmonic* if $P_{\mu}f \geq f$. Clearly non-negative constant functions are μ-harmonic. In the Sect. 4.8 we will investigate a condition under which they are the only ones. Note that if $(S_n, n \in \mathbb{N})$ is the random walk associated to μ and f is bounded and μ-superharmonic, then $(f(S_n), n \in \mathbb{N})$ is a supermartingale with respect to the filtration $(\mathcal{F}_n, n \in \mathbb{N})$ of $\mathcal{F}$ where for each $n \in \mathbb{N}$, $\mathcal{F}_n := \sigma\{X_1, \ldots, X_n\}$. To see this, observe that by the Markov property

$$\mathbb{E}(f(S_{n+1})|\mathcal{F}_n) = (P_{\mu}f)(S_n) \leq f(S_n).$$

Similarly we obtain a martingale (respectively, submartingale) if f is μ-harmonic (respectively, μ-subharmonic).

Now let $\mathcal{M}(G)$ be the space of all regular Borel measures on G. For each $\mu \in \mathcal{M}(G)$, we may consider the dual action P_{μ}^{*} of P_{μ} on $\mathcal{M}(G)$ defined by the prescription:

$$(P_{\mu}^{*}\rho)(f) = \rho(P_{\mu}f),$$

for all $\rho \in \mathcal{M}(G)$, $f \in C_c(G)$. We generalise the definitions we gave above for functions and say that $\rho \in \mathcal{M}(G)$ is μ-harmonic if $P_{\mu}^{*}\rho = \rho$, μ-superharmonic if $P_{\mu}^{*}\rho \leq \rho$ and μ-subharmonic if $P_{\mu}^{*}\rho \geq \rho$.[11] Now let $\rho \in \mathcal{M}(G)$ be absolutely continuous with respect to m_R with $f_{\rho} := \dfrac{d\rho}{dm_R}$. It is left as an exercise for the reader to check that f_{ρ} is μ-superharmonic if and only if ρ is $\widetilde{\mu}$-superharmonic.

We give some useful properties of μ-superharmonic functions.

Proposition 4.7.4 *Let $\mu \in \mathcal{P}(G)$*

1. *Suppose that $\lambda, \rho \in \mathcal{M}(G)$. If ρ is μ-superharmonic, then so is $\lambda * \rho$.*
2. *If f is μ-superharmonic, then so is $f \wedge c$ for any $c > 0$.*

Proof

1. For all $f \in C_c(G)_+$ we have

[11] If $\rho_1, \rho_2 \in \mathcal{M}(G)$ we write $\rho_1 \geq \rho_2$ if $\rho_1(f) \geq \rho_2(f)$ for all $f \in C_c(G)_+$.

$$\begin{aligned}P_\mu^*(\lambda * \rho)(f) &= \int_G \int_G (P_\mu^{(R)} f)(gh)\lambda(dg)\rho(dh) \\ &= \int_G (P_\lambda^{(L)} P_\mu^{(R)} f)(h)\rho(dh) \\ &= \int_G (P_\mu^{(R)} P_\lambda^{(L)} f)(h)\rho(dh) \\ &= (P_\mu^* \rho)(P_\lambda^{(L)} f) \\ &\leq \rho(P_\lambda^{(L)} f) = (\lambda * \rho)(f).\end{aligned}$$

2. This follows from the easily verified fact that $P_\mu(f \wedge c) \leq P_\mu f \wedge c$. □

Next we establish a connection between properties of convolution operators and existence and regularity of densities as discussed in Sect. 4.5. This result will be useful for us in the next chapter. Readers requiring background on Hilbert-Schmidt operators are referred to Appendix A.6. This result was obtained in Applebaum [9].

Theorem 4.7.1 *Let G be a compact Lie group and $\mu \in \mathcal{P}(G)$. The operator P_μ acting in $L^2(G)$ is Hilbert-Schmidt if and only if μ has a square-integrable density.*

Proof For sufficiency, assume that μ has density $f_\mu \in L^2(G, \mathbb{R})$. Then for all $g \in L^2(G), \sigma \in G$, $(P_\mu g)(\sigma) = \int_G g(\sigma\tau) f_\mu(\tau)d\tau = \int_G g(\tau) f_\mu(\sigma^{-1}\tau)d\tau$. Now define the mapping $k_\mu : G \times G \to \mathbb{R}$ by $k_\mu(\sigma, \tau) := f_\mu(\sigma^{-1}\tau)$. Then $k_\mu \in L^2(G \times G)$ since by left-invariance of (normalised) Haar measure, and Fubini's theorem

$$\int_G \int_G |k_\mu(\sigma, \tau)|^2 d\sigma d\tau = \int_G \int_G |f_\mu(\sigma^{-1}\tau)|^2 d\tau d\sigma = ||f_\mu||_2^2 < \infty,$$

and the result follows by Theorem A.6.4 in Appendix A.6. For necessity, suppose that P_μ is Hilbert-Schmidt. Then it has a kernel $k_\mu \in L^2(G \times G)$ and

$$(P_\mu f)(\sigma) = \int_G f(\tau) k_\mu(\sigma, \tau) d\tau.$$

In particular, for each $A \in \mathcal{B}(G)$,

$$\mu(A) = P_\mu 1_A(e) = \int_A k_\mu(e, \tau) d\tau.$$

Then for all $g \in C(G, \mathbb{R})$, $\int_G g(\sigma)\mu(d\sigma) = \int_G g(\sigma) k_\mu(e, \sigma) d\sigma$. It then follows by the argument used in the last part of the proof of Theorem 4.5.1 that μ is absolutely continuous with respect to m with density $f_\mu := k_\mu(e, \cdot)$, and we also have $f_\mu \in L^2(G, \mathbb{R})$. □

A linear operator $T : B_b(G) \to B_b(G)$ is said to be *strong Feller* if $\mathrm{Ran}(T) \subseteq C_b(G)$. The next theorem can essentially be found in Hewitt and Ross [91] (see (iv) on p. 298). The probabilistic interpretation was first observed by Hawkes [82] for the case $G = \mathbb{R}^n$, where a more detailed analysis appears to be possible.

Theorem 4.7.2 *If $\mu \in \mathcal{P}(G)$ has a continuous density g_μ with respect to left Haar measure, then the convolution operator P_μ is strong Feller.*

Proof We need only establish continuity. Let $\sigma \in G$ and $(\sigma_n, n \in \mathbb{N})$ be a sequence in G that converges to σ. Then for all $f \in B_b(G), n \in \mathbb{N}$,

$$\begin{aligned}|P_\mu f(\sigma) - P_\mu f(\sigma_n)| &= \left| \int_G f(\sigma\tau)\mu(d\tau) - \int_G f(\sigma_n\tau)\mu(d\tau) \right| \\ &\leq \int_G |f(\tau)||g_\mu(\sigma^{-1}\tau) - g_\mu(\sigma_n^{-1}\tau)|d\tau \\ &\leq \sup_{\tau\in G} |f(\tau)| \int_G |g_\mu(\sigma^{-1}\tau) - g_\mu(\sigma_n^{-1}\tau)|d\tau \\ &\leq \sup_{\tau\in G} |f(\tau)| \int_G (g_\mu(\sigma^{-1}\tau) + g_\mu(\sigma_n^{-1}\tau))d\tau \\ &\leq 2 \sup_{\tau\in G} |f(\tau)|.\end{aligned}$$

The result then follows by dominated convergence and the continuity of g_μ. □

Finally we establish a useful spectral property for the case where μ is a central measure and G is compact.

Theorem 4.7.3 *If G is compact and $\mu \in \mathcal{P}_c(G)$, then $\{\pi_{ij}, 1 \leq i, j \leq d_\pi, \pi \in \widehat{G}\}$ is a complete set of eigenfunctions for P_μ acting in $L^2(G)$. Moreover, we have*

$$P_\mu \pi_{ij} = \overline{c_\pi} \pi_{ij}$$

for all $1 \leq i, j \leq d_\pi, \pi \in \widehat{G}$, where $\widehat{\mu}(\pi) = c_\pi I_\pi$ (c.f. Corollary 4.2.2).

Proof For all $\sigma \in G$, using Theorem 4.2.1 (4) we have

$$\begin{aligned}P_\mu \pi_{ij}(\sigma) &= \int_G \pi_{ij}(\sigma\tau)\mu(d\tau) \\ &= \sum_{k=1}^{d_\pi} \pi_{ik}(\sigma) \int_G \pi_{kj}(\tau)\mu(d\tau)\end{aligned}$$

$$
\begin{aligned}
&= \sum_{k=1}^{d_\pi} \pi_{ik}(\sigma)\widehat{\widetilde{\mu}(\pi)}_{kj} \\
&= \sum_{k=1}^{d_\pi} \pi_{ik}(\sigma)\widehat{\mu(\pi)}^*_{kj} \\
&= \sum_{k=1}^{d_\pi} \pi_{ik}\overline{c_\pi}\delta_{kj} \\
&= \overline{c_\pi}\pi_{ij}(\sigma),
\end{aligned}
$$

and the result follows. □

4.8 Recurrence

If $\mu \in \mathcal{P}(G)$ we define $\mu^{*(0)} = \delta_e$. Throughout this section we will assume that $\mu \in \mathcal{P}(G)$ is regular and also *full*, i.e. the closed subgroup of G that is generated by $\text{supp}(\mu)$ is G. We define the *potential measure* V_μ of μ by the prescription

$$V_\mu := \sum_{n=0}^{\infty} \mu^{*(n)},$$

so that $V_\mu(f) = \sum_{n=0}^{\infty} \mu(P_\mu^n f)$ for each $f \in C_c(G)_+$. If V_μ is regular, then $V_\mu(f) < \infty$; otherwise it may take the value ∞.

We say that μ is *transient* if $V_\mu(A) < \infty$ for all open relatively compact subsets of G and *recurrent* if $V_\mu(A) = \infty$ for all non-empty open subsets of G. We say that the group G is *recurrent* if $\mathcal{P}(G)$ contains at least one full recurrent measure.

Theorem 4.8.1 (Recurrence-Transience Dichotomy) *Every full $\mu \in \mathcal{P}(G)$ is either recurrent or transient.*

We omit the proof, which can be obtained by combining the results of Theorem 22 (pp. 19–20) and Theorem 26 (pp. 23–24) in Guivarc'h et al. [75].

From the random-walk perspective, recurrence is equivalent to the requirement that for all $g \in G$, $P\left(\lim_{n\to\infty}(S_n \in V_g)|S(0) = e\right) = 1$ for every neighbourhood V_g of g. From the point of view of this monograph a key result is the following:

Proposition 4.8.1 *1. Any full probability measure on a compact group is recurrent.*
2. Every compact group G is recurrent.

Proof 1. Let $\mu \in \mathcal{P}(G)$ be full. By the recurrence-transience dichotomy (Theorem 4.8.1), if $V_\mu(A) = \infty$ for some open relatively compact subset A of G, then G cannot be transient and so must be recurrent. But we may take $A = G$ and then $V_\mu(G) = \sum_{n=0}^{\infty} \mu^{*(n)}(G) = \infty$.

2. Take μ to be normalised Haar measure on G. □

Lemma 4.8.1 *Let μ be a full measure in $\mathcal{P}(G)$ and f be continuous and μ-superharmonic. If μ is recurrent, then f is μ-harmonic.*

Proof We follow Guivarc'h et al [75] p. 43. Without loss of generality, we suppose that $f - P_\mu f > 0$ and seek a contradiction. By the recurrence assumption,

$$\infty = V_\mu(f - P_\mu f) = \sum_{n=0}^{\infty}(P_\mu^n(f - P_\mu f)),$$

and so

$$\infty = \lim_{n\to\infty}(f - P_\mu^n f) \leq f,$$

giving the required contradiction. □

The main result of this section is the following. Our proof closely mirrors that of Guivarc'h et al [75] Proposition 45, pp. 42–44.

Theorem 4.8.2 *Let $\mu \in \mathcal{P}(G)$ be full. The following are equivalent.*

(i) μ is recurrent.
(ii) Every μ-superharmonic continuous function on G is constant.
(iii) Every μ-superharmonic measure is a right Haar measure.
(iv) Every μ-superharmonic function on G is constant m_R-almost everywhere.

Proof (i) ⇒ (ii). Let f be a continuous μ-superharmonic function on G. By Lemma 4.8.1 it is μ-harmonic. Assume that f is bounded and suppose that it attains it maximum value, so there exists $g_0 \in G$ such that $f(g_0) = \sup_{g\in G} f(g)$. Then we have

$$f(g_0) = (P_\mu f)(g_0) = \int_G f(g_0 h)\mu(dh),$$

and so

$$\int_G (f(g_0) - f(g_0 h))\mu(dh) = 0.$$

It follows that $f(g_0) = f(g_0 h)$ for all $h \in g_0^{-1}\text{supp}(\mu)$. By repeatedly using the fact that f is harmonic we deduce that $f(g_0) = f(g_0 h_1 \ldots h_n)$ for all $h_1, \ldots, h_n, n \in \mathbb{N}$ such that $g_0 h_1 \ldots h_n \in \text{supp}(\mu)$. Repeatedly applying Proposition 4.7.4, wherein λ is taken to be $\widetilde{\mu}$, allows us to repeat the previous argument with any $h_i, i = 1, \ldots, n$ replaced by h_i^{-1}. Finally, using the continuity of f and the fact that μ is full, we deduce that f is constant. To extend this result to the general case, observe that if $f \neq 0$, we can replace f by $f \wedge c$ where $c > 0$ and appeal to Proposition 4.7.4(2).

(ii) ⇒ (i). We assume that μ is transient and seek a contradiction. Let $f \in C_c(G)_+$ be non-trivial and define $F_f = f * V_\mu$, so that for all $g \in G$, $F_f(g) =$

$\sum_{n=0}^{\infty} f(gh)\mu^{*(n)}(dh)$. By a standard use of dominated convergence, we can see that F_f is continuous. Then it is easily verified that

$$P_\mu F_f = F_f - f \leq F_f,$$

and so F_f is both continuous and μ-superharmonic. Hence it is constant, and so $f = P_\mu F_f - F_f = 0$, giving the desired contradiction.

(ii) $\Rightarrow$ (iii). Assume that $\rho \in \mathcal{M}(G)$ is $\widetilde{\mu}$-superharmonic. Choose an arbitrary $\psi \in C_c(G)$ and let α_ψ be the measure of compact support that has Radon-Nikodym derivative $\frac{d\alpha_\psi}{dm_R} = \psi$ with respect to right Haar measure. By Proposition 4.7.4(1) $\alpha_\psi * \rho$ is $\widetilde{\mu}$-superharmonic, and it is easily verified that this measure is absolutely continuous with respect to right Haar measure, and has Radon-Nikodym derivative F_ψ where $F_\psi(g) = \int_G \psi(gh^{-1})\rho(dh)$ for all $g \in G$. The mapping F_ψ is clearly non-negative, continuous and it is μ-superharmonic. Then it is constant by (ii). So $F_\psi(g) = F_\psi(e)$ for all $g \in G$. Then for all $\psi \in C_c(G)$, $g \in G$,

$$\int_G \psi(gh^{-1})\rho(dh) = \int_G \psi(h^{-1})\rho(dh).$$

It follows that $\widetilde{\rho}$ is a left Haar measure and so ρ is a right Haar measure.

(iii) $\Rightarrow$ (iv). Let $f \in B_b(G)_+$ and consider the measure $\rho_f \in \mathcal{M}(G)$ for which $\frac{d\rho_f}{dm_R} = f$. If f is μ-superharmonic, then ρ_f is $\widetilde{\mu}$-superharmonic, and so ρ_f is a right Haar measure. It follows that f is constant almost everywhere with respect to m_R. If f is not bounded, we can again replace f by $f \wedge c$ where $c > 0$ and use Proposition 4.7.4(2).

(iv) $\Rightarrow$ (ii) is obvious. □

Let M be a Hausdorff topological space and let $\mathcal{P}(M)$ be the space of all regular Borel probability measures on M. A continuous mapping $\alpha : G \times M \to M$ for which $\alpha(e, x) = x$ for all $x \in M$ is called an *action* of G on M. An action is *transitive* if for all $g_1, g_2 \in G, x \in M$, $\alpha(g_1, \alpha(g_2, x)) = \alpha(g_1 g_2, x)$. A locally compact Hausdorff group G is said to be *amenable in action*[12] if for every transitive action α on every compact space M, there exists a regular Borel probability measure μ_M on M such that for all $f \in C(M)$, $g \in G$

$$\int_M f(\alpha(g, x))\mu_M(dx) = \int_G f(x)\mu_M(dx).$$

[12] This term should not be confused with the notion of an *amenable group*, which is a group that possesses an invariant mean in the sense of the existence of a positive linear functional on $L^\infty(G)$ that is invariant with respect to left, or right translations. Such groups play an important role in ergodic theory, see e.g. Ornstein and Weiss [154].

The proof of the following theorem is based on that in Guivarc'h et al. [75] (pp. 45–47). We will need the notion of the convolution of $\mu \in \mathcal{P}(G)$ and $\lambda \in \mathcal{P}(M)$ relative to a given transitive action α. This is the measure $\mu *_\alpha \lambda \in \mathcal{P}(M)$ such that for all $A \in \mathcal{B}(M)$:

$$(\mu *_\alpha \lambda)(A) := \int_M \int_G 1_A(\alpha(g,x))\mu(dg)\lambda(dx).$$

Theorem 4.8.3 *Every recurrent group is amenable in action.*

Proof Let μ be a full recurrent probability measure on G and $\nu \in \mathcal{P}(M)$ be arbitrary. For each $n \in \mathbb{N}$ define $\mu_n := \frac{1}{n}\sum_{k=0}^{n-1}\mu^{*(k)}$. Then $(\mu_n *_\alpha \nu, n \in \mathbb{N})$ is weakly relatively compact and so has a subsequence $(\mu_{n_k} *_\alpha \nu, k \in \mathbb{N})$ that converges weakly to some $\lambda \in \mathcal{P}(M)$. If we can prove that λ is invariant, then we are done. We first show that $\mu *_\alpha \lambda = \lambda$. Indeed, we have for all $f \in C(M)$, using the transitivity of α,

$$\begin{aligned}
(\mu *_\alpha \lambda)(f) &= \int_M \int_G f(\alpha(g,x))\mu(dg)\lambda(dx) \\
&= \lim_{k\to\infty} \int_M \int_G \int_G f(\alpha(gh,x))\mu(dg)\mu_k(dh)\nu(dx) \\
&= \lim_{k\to\infty} \int_M f(x)[(\mu * \mu_k) *_\alpha \nu](dx) \\
&= \lim_{k\to\infty} \int_M f(x)(\mu_k *_\alpha \nu)(dx) = \lambda(f).
\end{aligned}$$

To see that λ is indeed invariant, let $f \in C(M)_+$ and define $\Phi_f \in C(G)_+$ by $\Phi_f(g) = \int_M f(\alpha(g,x))\lambda(dx)$ for $g \in G$. Then Φ_f is μ-harmonic, for by Fubini's theorem

$$\begin{aligned}
P_\mu \Phi_f(g) &= \int_M \int_G f(\alpha(gh,x))\mu(dh)\lambda(dx) \\
&= \int_M f(\alpha(g,x))(\mu *_\alpha \lambda)(dx) \\
&= \int_M f(\alpha(g,x))\lambda(dx) = \Phi_f(g).
\end{aligned}$$

So by Theorem 4.8.2, Φ_f is constant, and hence for all $g \in G$,

$$\int_M f(\alpha(g,x))\lambda(dx) = \Phi_f(g) = \Phi_f(e) = \int_M f(\alpha(e,x))\lambda(dx) = \int_M f(x)\lambda(dx),$$

and the result follows. □

For example, let H be a closed subgroup of a compact group G and $M = G/H$ be the (compact) homogeneous space of left cosets. Define the *natural action* of G on M by $\alpha(g, g'H) = gg'H$ for all $g, g' \in G$. This is clearly continuous and transitive. G is recurrent by Proposition 4.8.1, and so by Theorem 4.8.3 we can assert the existence of $\mu_M \in \mathcal{P}(M)$ so that for all $g \in G$

$$\int_M f(gx)\mu_M(dx) = \int_M f(x)\mu_M(dx).$$

In particular take $G = SO(n)$ and $H = SO(n-1)$, so that $M = S^{n-1}$. In this case μ_M is the normalised surface measure σ_{n-1}, for which we have the recursive formula

$$\int_{S^{n-1}} f(x)\sigma_{n-1}(dx)$$

$$= \frac{\Gamma\left(\frac{n}{2}\right)}{\sqrt{\pi}\Gamma\left(\frac{n-1}{2}\right)} \int_0^\pi \left(\int_{S^{n-2}} f(\sin(\theta)y) + \cos(\theta)e_n)\sigma_{n-2}(dy) \right) \sin^{n-2}(\theta)d\theta,$$

for all $f \in C(S^{n-1})$ where e_n is the "north pole" in S^{n-1} (see Faraut [63], pp. 186–190 for details).

Some detailed results on recurrent random walks on non-compact groups (at least in the abelian case) may be found in Chap. 9 of Revuz [167].

4.9 Notes and Further Reading

The interaction between probability and group theory covers a huge area and it is difficult to do this justice in such a short space. In particular this includes probability on discrete groups (which is not really the topic of this monograph) and has considerable overlap with "stochastic differential geometry", as random motion on a Lie group can be regarded as a special case of that on a more general manifold.

In the 1960s a considerable literature began to evolve on probability theory in such general mathematical structures as groups on the one hand and Banach spaces on the other. These two directions have now diverged considerably, but in 1963 Grenander [73] was able to justify including both themes within a single volume. From the continuous group point of view, he introduced the Fourier transform and gave some attention to limit theorems. He traces the historic roots of the subject back to work by

Perrin [159] in 1928 on Brownian motion in the rotation group. Hannan's survey paper [78] from 1965 is also of historical interest. He develops applications to second-order stationary processes, experimental design and ANOVA. Parthasarathy's book [155], which appeared in 1967, discusses probability on metric groups and gives an account of infinite divisibility, the Lévy-Khintchine formula and the central limit theorem on locally compact abelian groups. Ten years later, Heyer's highly influential treatise [95] appeared which gave a comprehensive and detailed state of the art account of probability on (general) locally compact groups. This monograph is now a classic and after 35 years is still a highly valuable resource for those doing research in this area. Highlights are the treatment of Hunt's classification of the generators of convolution semigroups and the central limit theorem. We will investigate both of these topics in the next chapter. Ten years later, Diaconis [56] published his beautiful lecture notes that demonstrate the fruitfulness of group actions in a variety of contexts within both probability and statistics, from card shuffling to ANOVA and spectral analysis of time series.

In more recent years there have been a number of books and monographs on more specific topics concerned with probability on groups. For example Hazod and Siebert [85] study stable laws on locally compact groups (where these make sense), Neuenschwander [153] investigates limit theorems and Brownian motion on the Heisenberg group, and Liao's monograph [132] is devoted to Lie group-valued Lévy processes. The treatise of Guivarc'h et al. [75] is a comprehensive account of random walks on groups. For a more recent survey, see Breuillard [35]. Random walks on the rotation groups $SO(n)$ were given an extensive treatment by Rosenthal [170]. For a study of random walks on spheres, see Bingham [29]. Central measures were introduced in this chapter and will feature prominently in the next one. These have been investigated by a number of analysts (see e.g. Ragozin [163] and Hare [79]) and probabilists (see e.g. Siebert [184]).

We can regard normalised Haar measure on $U(n)$ as the uniform distribution therein, and choosing unitary matrices according to this law plays an important role in *random matrix theory*; see e.g. Keating and Snaith [115] for intriguing connections to the Riemann hypothesis, and Diaconis and Shahshahani [57] for computations of the eigenvalues of random matrices in connection with a continuous generalisation of the classical matching problem. For a monograph treatment of this topic, see Anderson et al. [3].

It is worth pointing out that there is an interesting class of locally compact groups called *Moore groups*, whose defining property is that all of their irreducible representations are finite dimensional. So all compact groups are Moore groups and many probabilistic results that hold for Moore groups are automatically applicable to compact groups; see e.g. Sects. 1.3 and 1.4 of Heyer [95].

Chapter 5
Convolution Semigroups of Measures

Abstract Properties of convolution semigroups of probability measures and associated Hunt semigroups of operators on general Lie groups are described. We outline a proof of Hunt's theorem which gives a Lévy-Khintchine type representation for the infinitesimal generator of such semigroups. Then we study L^2-properties of these semigroups, specifically finding conditions which guarantee self-adjointness, injectivity, analyticity and the trace-class property. We then use the Fourier transform on a compact Lie group to find a direct analogue of the Lévy-Khintchine formula. We prove the central limit theorem for Gaussian semigroups of measures. Then we describe the technique of subordination, and discuss regularity of densities for some semigroups of measures that are obtained by subordinating the standard Gaussian. Finally we investigate the small-time asymptotic behaviour of some subordinated densities.

In this chapter, all functions spaces are real unless otherwise stated—so e.g. $C_0(G)$ means $C_0(G, \mathbb{R})$.

5.1 Definition and Examples

In this chapter we will continue to work primarily with general Lie groups (unless otherwise stated).

Let $(\mu_t, t \geq 0)$ be a family of probability measures on G. We say that it is a *weakly continuous convolution semigroup of probability measures* if it satisfies the following conditions:

(i) (The semigroup property) $\mu_{s+t} = \mu_s * \mu_t$ for all $s, t \geq 0$.
(ii) weak-$\lim_{s \to t} \mu_s = \mu_t$ for all $t \geq 0$.

By the semigroup property, μ_0 is an idempotent measure. For simplicity we always assume

D. Applebaum, *Probability on Compact Lie Groups*, Probability Theory and Stochastic Modelling 70, DOI: 10.1007/978-3-319-07842-7_5,

(iii) $\mu_0 = \delta_e$.

Indeed, once we assume (iii), we can replace (ii) with:

(ii)* weak-$\lim_{t \to 0} \mu_t = \delta_e$.

To see this, let $f \in C_b(G)$ and assume (ii)*. For all $s > 0$,

$$\begin{aligned}\lim_{t \downarrow s} \int_G f(g)\mu_t(dg) &= \lim_{r \to 0} \int_G f(g)\mu_{s+r}(dg) \\ &= \lim_{r \to 0} \int_G \left(\int_G f(gh)\mu_s(dg) \right) \mu_r(dh) \\ &= \int_G f(g)\mu_s(dg).\end{aligned}$$

The extension to the general case is left as an exercise for the reader.

From now on we will simply refer to any family of probability measures that satisfies (i), (ii) and (iii) (or equivalently (i), (ii)* and (iii)) as a *convolution semigroup.* It is left to the reader to verify that $(\mu_t, t \geq 0)$ is a convolution semigroup if and only if $(\widetilde{\mu_t}, t \geq 0)$ is.

Example 1 It is easy to see that $\mu_t = \delta_e$ for all $t \geq 0$ defines a convolution semigroup. We call it the *neutral semigroup.*

Theorem 5.1.1 *Let $(\mu_t, t \geq 0)$ be a convolution semigroup on a compact Lie group G and suppose that μ_s is idempotent for some $s > 0$. Then $(\mu_t, t \geq 0)$ is the neutral semigroup.*

Proof Let $\kappa = \mu_s$. Then for all $\pi \in \widehat{G}, n \in \mathbb{N}, \widehat{\mu_{\frac{s}{n}}}(\pi)^n = \widehat{\kappa}(\pi) = 0$ or I_π. It follows that $\mu_{\frac{s}{n}} = \kappa$, and by taking weak limits as $n \to \infty$ we deduce that $\kappa = \delta_e$. It then follows that $\mu_{\frac{ms}{n}} = \delta_e$ for all $m \in \mathbb{Z}_+, n \in \mathbb{N}$ and the required result follows by rational approximation (in the weak topology). □

Example 2 The Compound Poisson Semigroup. Fix $c > 0, \rho \in \mathcal{P}(G)$. The corresponding compound Poisson semigroup $(\pi_t^{(c,\rho)}, t \geq 0)$ is defined for all $A \in \mathcal{B}(G)$ by

$$\pi_t^{(c,\rho)}(A) := \begin{cases} \delta_e(A) & \text{if } t = 0, \\ e^{-ct} \displaystyle\sum_{n=0}^{\infty} \frac{(ct)^n}{n!} \rho^{*(n)}(A) & \text{if } t \neq 0. \end{cases}$$

In the sequel, if there is no ambiguity about c and ρ, we will always write $\pi_t := \pi_t^{(c,\rho)}$.

Proposition 5.1.1 $(\pi_t, t \geq 0)$ *is a convolution semigroup.*

Proof For all $s, t > 0$, $A \in \mathcal{B}(G)$,

$$
\begin{aligned}
(\pi_s * \pi_t)(A) &= e^{-c(s+t)} \sum_{m=0}^{\infty}\sum_{n=0}^{\infty} \frac{(cs)^m}{m!}\frac{(ct)^n}{n!}\rho^{*(m+n)}(A) \\
&= e^{-c(s+t)} \sum_{r=0}^{\infty}\sum_{m=0}^{r} \frac{1}{r!}\binom{r}{m}(cs)^m (ct)^{r-m}\rho^{*(r)}(A) \\
&= e^{-c(s+t)} \sum_{r=0}^{\infty} \frac{1}{r!}[c(s+t)]^r \rho^{*(r)}(A) \\
&= \pi_{s+t}(A).
\end{aligned}
$$

The proof of weak continuity is left as an exercise for the reader. □

A key role in our work so far has been played by the Fourier transform. So if $(\mu_t, t \geq 0)$ is a convolution semigroup we should be able to obtain some corresponding structure in the family of linear operators $(\widehat{\mu_t}(\pi), t \geq 0)$ acting in V_π for $\pi \in \widehat{G}$. For simplicity we restrict ourselves to the compact case.

Proposition 5.1.2 *Let G be a compact Lie group and $(\mu_t, t \geq 0)$ be a family of measures in $\mathcal{P}(G)$. Then $(\mu_t, t \geq 0)$ is a convolution semigroup if and only if for all $\pi \in \widehat{G}$:*

(i)′ $\widehat{\mu_{s+t}}(\pi) = \widehat{\mu_s}(\pi)\widehat{\mu_t}(\pi)$ *for all $s, t \geq 0$.*
(ii)′ $\lim_{t\to 0}\langle(\widehat{\mu_t}(\pi) - I_\pi)\phi, \psi\rangle = 0$ *for all $\phi, \psi \in V_\pi$.*
(iii)′ $\widehat{\mu_0}(\pi) = I_\pi$.

Proof That (i) is equivalent to (i)′ is a consequence of Theorem 4.2.1(ii) and the uniqueness of the Fourier transform, Theorem 4.2.3. The equivalence of (iii) and (iii)′ also follows from Theorem 4.2.3. Finally we see that (ii) is equivalent to (ii)′ by applying Theorem 4.2.5. □

By Proposition 5.1.2, we see that if $(\mu_t, t \geq 0)$ is a convolution semigroup on a compact Lie group G, then for every $\pi \in \widehat{G}$, $(\widehat{\mu_t}(\pi), t \geq 0)$ is a weakly continuous (and hence, since V_π is finite-dimensional) strongly continuous contraction semigroup of linear operators on V_π. Hence it has an infinitesimal generator A_π which is a $d_\pi \times d_\pi$ matrix and so for all $\pi \in \widehat{G}, t \geq 0$, we have

$$\widehat{\mu_t}(\pi) = e^{tA_\pi}, \tag{5.1.1}$$

with $A_{\pi_0} = 0$. The classification of all possible matrices A_π which can appear in (5.1.1) will concern us later in the chapter. Note that when $G = \mathbb{R}$ the solution to this problem is given by the classical Lévy-Khintchine formula (see e.g. Sato [177], pp. 37–45).

Returning to the case of compact G, by Corollary 4.2.1 $(\mu_t, t \geq 0)$ is a *symmetric convolution semigroup*, i.e. the measure μ_t is symmetric for all $t > 0$, if and only if A_π is a hermitian matrix for each $\pi \in \widehat{G}$. Using Theorem 4.2.2 we see that $(\mu_t, t \geq 0)$ is a *central convolution semigroup*, i.e. the measure μ_t is central for all $t > 0$, if and only if $A_\pi = a_\pi I_\pi$ for all $\pi \in \widehat{G}$ where $a_\pi \in \mathbb{C}$ with $\Re(a_\pi) \leq 0$. Finally it follows (see also Corollary 4.2.2) that a convolution semigroup $(\mu_t, t \geq 0)$ is both central and symmetric if and only if $A_\pi = -a_\pi I_\pi$ for all $\pi \in \widehat{G}$ where $a_\pi \geq 0$. This result is sufficiently important for later work to deserve summarising as a theorem:

Theorem 5.1.2 *$(\mu_t, t \geq 0)$ is a central convolution semigroup on a compact Lie group G if and only if for each $\pi \in \widehat{G}$, there exists $\alpha_\pi \in \mathbb{C}$ with $\Re(\alpha_\pi) \leq 0$ so that*

$$\widehat{\mu_t}(\pi) = e^{t\alpha_\pi} I_\pi. \tag{5.1.2}$$

$(\mu_t, t \geq 0)$ is a central symmetric convolution semigroup if and only $\alpha_\pi \in \mathbb{R}$ with $\alpha_\pi \leq 0$ in (5.1.2).

Example 3 Standard Gaussian semigroup on a compact Lie group.

By (4.2.7) for all $\sigma > 0$, we obtain a convolution semigroup $(k_t, t \geq 0)$ on the compact Lie group G with

$$\widehat{k_t}(\pi) = e^{-t\sigma^2 \kappa_\pi} I_\pi, \tag{5.1.3}$$

for all $\pi \in G, t \geq 0$. We call this a *standard Gaussian semigroup.* More general notions of Gaussianity in Lie groups will be explored later in the chapter. We have already pointed out that the measure k_t has a smooth density f_t for all $t > 0$. On $SU(2)$, using (4.5.12) it can be shown that

$$f_t(\theta) = \sum_{n=1}^{\infty} n \exp\left\{-\frac{\sigma^2(n^2-1)t}{32\pi^2}\right\} \frac{\sin(2\pi n\theta)}{\sin(2\pi\theta)},$$

where θ parametrizes a maximal torus (see Liao [131]). See pp. 100–102 of Liao [132] for a partial extension of this result to a wider class of convolution semigroups on $SU(2)$.

5.2 Infinite Divisibility and Lévy Processes

We say that $\mu \in \mathcal{P}(G)$ is *infinitely divisible* if it has a convolution nth root $\mu_n \in \mathcal{P}(G)$ for each $n \in \mathbb{N}$. Then $\mu = \mu_n^{*n}$ and we write $\mu_{\frac{1}{n}} := \mu_n$. When G is compact, it is easily verified that μ is infinitely divisible if and only if for all $n \in \mathbb{N}, \pi \in \widehat{G}$ there exists $\Gamma_n(\pi) \in \mathcal{L}(V_\pi)$ so that $\widehat{\mu}(\pi) = \Gamma_n(\pi)^n$ and that Γ_n is itself the Fourier transform of a probability measure on G. In fact we have $\Gamma_n = \widehat{\mu_{\frac{1}{n}}}$.

The relationship between infinite divisibility and convolution semigroups is rather easily established when G is abelian (see e.g. Applebaum [7] Corollary 1.4.6, p. 65 for the case $G = \mathbb{R}^d$). In the non-abelian case, it is a deep problem that has given rise to some highly non-trivial mathematics (see e.g. Dani and McCrudden [51, 52]). Of course it is easy to check that if $(\mu_t, t \geq 0)$ is a convolution semigroup, then

μ_t is infinitely divisible for each $t \geq 0$. It is the converse that is mathematically demanding. We state the relevant result only in the case where G is a compact Lie group, where the original work was due to Parthasarathy [156, 157]. We also draw readers' attention to the elegant result in the general locally compact case summarised in Theorem 3.5.8 on p. 220 of Heyer [95].

Theorem 5.2.1 (The Embedding Theorem) *Let G be a compact Lie group and $\rho \in \mathcal{P}(G)$ be infinitely divisible. Then there exists a convolution semigroup $(\mu_t, t \geq 0)$ for which $\mu_1 = \rho$.*

Let $\phi = (\phi(t), t \geq 0)$ be a stochastic process defined on some probability space $(\Omega, \mathcal{F}, P)$ and taking values in an arbitrary Lie group G. We call the random variable $\phi(s)^{-1}\phi(t)$, where $0 \leq s \leq t < \infty$, a *right increment* of the process ϕ. Similarly $\phi(t)\phi(s)^{-1}$ is a *left increment*. We say that ϕ has *stationary right increments* if for all $0 \leq s \leq t < \infty$, $\phi(s)^{-1}\phi(t)$ has the same law as $\phi(0)^{-1}\phi(t-s)$ and we say it has *independent right increments* if for all $n \in \mathbb{N}$ and all $0 < t_1 < t_2 < \cdots < t_n < \infty$, the random variables $\phi(0), \phi(0)^{-1}\phi(t_1), \phi(t_1)^{-1}\phi(t_2), \ldots, \phi(t_{n-1})^{-1}\phi(t_n)$ are independent. Finally we say that ϕ is a *left Lévy process* if

1. ϕ has stationary and independent right increments,
2. $\phi(0) = e$ (a.s.)
3. The process ϕ is stochastically continuous, i.e. $\lim_{t \to s} P(\phi(s)^{-1}\phi(t) \in A) = 0$ for all $A \in \mathcal{B}(G)$ with $e \neq \overline{A}$.

Right Lévy processes are defined similarly but with all instances of right increments replaced by left increments. From now on we will only deal with left Lévy processes and call these simply *Lévy processes*. It is not difficult to verify that if ϕ is a Lévy process and μ_t is its law at time t, i.e. $\mu_t(A) = P(\phi(t) \in A)$ for all $A \in \mathcal{B}(G)$, then $(\mu_t, t \geq 0)$ is a convolution semigroup. Conversely, given any weakly continuous convolution semigroup of probability measures $(\mu_t, t \geq 0)$ on G we can always construct a Lévy process $\phi = (\phi(t), t \geq 0)$ such that each $\phi(t)$ has law μ_t by taking Ω to be the space of all mappings from $\mathbb{R}^+$ to G and $\mathcal{F}$ to be the σ-algebra generated by cylinder sets. The existence of P then follows by Kolmogorov's construction and the time-ordered finite-dimensional distributions have the form

$$\begin{aligned}&P(\phi(t_1) \in A_1, \phi(t_2) \in A_2, \ldots, \phi(t_n) \in A_n)\\ &= \int_G \int_G \cdots \int_G 1_{A_1}(\sigma_1) 1_{A_2}(\sigma_1\sigma_2) \cdots 1_{A_n}(\sigma_1\sigma_2 \cdots \sigma_n) \mu_{t_1}(d\sigma_1)\mu_{t_2 - t_1}(d\sigma_2) \ldots\\ &\mu_{t_n - t_{n-1}}(d\sigma_n),\end{aligned}$$

for all $A_1, A_2, \ldots, A_n \in \mathcal{B}(G)$, $0 \leq t_1 \leq t_2 \leq \cdots \leq t_n < \infty$. For a proof in the case $G = \mathbb{R}^d$, see Theorem 10.4 on pp. 55–57 of Sato [177]. The extension to arbitrary G is straightforward.

5.3 The Hunt Semigroup and Its Generator

Let $(\mu_t, t \geq 0)$ be a convolution semigroup and for each $t \geq 0$, let $P_t := P_{\mu_t}$ be the associated convolution operator in $C_0(G)$. It follows from Proposition 4.7.1 that $(P_t, t \geq 0)$ is a contraction semigroup of linear operators on $C_0(G)$ with $P_0 = I$ which also satisfies $P_t f \geq 0$ whenever $f \geq 0$. Furthermore we have

$$L_\sigma P_t = P_t L_\sigma, \tag{5.3.4}$$

for all $t \geq 0, \sigma \in G$.

We will show that $(P_t, t \geq 0)$ is strongly continuous. First we need a lemma (due to Malliavin et al. [144] pp. 98–99).

Lemma 5.3.1 *Let O be an open set in G with $e \in O$. Then given $\epsilon > 0$ there exists $t_0 > 0$ so that $0 \leq t \leq t_0 \Rightarrow \mu_t(O^c) < \epsilon$.*

Proof Choose $f \in C_b(G)$ with $0 \leq f \leq 1$, $\mathrm{supp}(f) \subseteq O$ and $f(e) > 1 - \frac{\epsilon}{2}$. By weak continuity (ii)* there exists $t_0 > 0$ so that for all $0 \leq t \leq t_0$, $\left|\int_G (f(\tau) - f(e))\mu_t(d\tau)\right| < \dfrac{\epsilon}{2}$. Then

$$\begin{aligned}
\mu_t(O^c) &= 1 - \mu_t(O)\\
&\leq 1 - \int_O f(\tau)\mu_t(d\tau)\\
&= 1 - \int_G f(\tau)\mu_t(d\tau)\\
&= 1 - f(e) + \int_G (f(\tau) - f(e))\mu_t(d\tau)\\
&< \frac{\epsilon}{2} + \frac{\epsilon}{2} = \epsilon.
\end{aligned}$$

□

Proposition 5.3.1 *The semigroup $(P_t, t \geq 0)$ is strongly continuous.*

Proof We first establish the result when $f \in C_c(G)$ with $f \neq 0$. We need to show that $\lim_{t\to 0} \sup_{\sigma\in G} |P_t f(\sigma) - f(\sigma)| = 0$. By Theorem A.2.1 (in Appendix A.2) given $\epsilon > 0$ we can find an open set O in G with $e \in O$ so that for all $\tau \in O$, $\sup_{\sigma\in G} |f(\sigma\tau) - f(\sigma)| = \sup_{\sigma\in G} |R_\tau f(\sigma) - f(\sigma)| < \dfrac{\epsilon}{2}$. Moreover, by Lemma 5.3.1, we can find $t_0 > 0$ so that for $0 \leq t \leq t_0$, $\mu_t(O^c) < \dfrac{\epsilon}{2||f||_\infty}$. Then we have

$$\begin{aligned}
&\sup_{\sigma\in G}\left|\int_G (f(\sigma\tau) - f(\sigma))\mu_t(d\tau)\right|\\
&\leq \sup_{\sigma\in G}\left|\int_O (f(\sigma\tau) - f(\sigma))\mu_t(d\tau)\right| + \sup_{\sigma\in G}\left|\int_{O^c} (f(\sigma\tau) - f(\sigma))\mu_t(d\tau)\right|\\
&\leq \sup_{\tau\in O}\sup_{\sigma\in G} |f(\sigma\tau) - f(\sigma)| + 2||f||_\infty \mu_t(O^c)
\end{aligned}$$

$$< \frac{\epsilon}{2} + \frac{\epsilon}{2} = \epsilon.$$

The extension to $f \in C_0(G)$ is by a standard density argument. Indeed, given $\delta > 0$ we can find $g \in C_c(G)$ so that $||f - g||_\infty < \frac{\delta}{3}$ and we can find $t_0 > 0$ so that $0 \leq t < t_0 \Rightarrow ||P_t g - g||_\infty < \frac{\delta}{3}$. Then using the contraction property of P_t, we have for all $0 \leq t < t_0$

$$\begin{aligned} ||P_t f - f||_\infty &\leq ||P_t f - P_t g||_\infty + ||P_t g - g||_\infty + ||g - f||_\infty \\ &< 2\frac{\delta}{3} + \frac{\delta}{3} = \delta. \end{aligned}$$

□

We call $(P_t, t \geq 0)$ a *Hunt semigroup* in honour of the achievement of Hunt [103] in understanding its generic structure, and we will explore his results in this section. If $\phi = (\phi(t), t \geq 0)$ is a Lévy process on G, then

$$P_t f(\sigma) = \mathbb{E}(f(\sigma\phi(t)))$$

for all $t \geq 0$, $f \in C_0(G)$, $\sigma \in G$.

In the language of semigroup theory (see Appendix A.7 for background), $(P_t, t \geq 0)$ is a C_0-semigroup and so it has a densely defined generator $\mathcal{L}$ which is a closed linear operator on its domain $\mathcal{D}_\mathcal{L} :=Dom(\mathcal{L})$. Note that

$$\mathcal{D}_\mathcal{L} = \left\{ f \in C_0(G), \text{ there exists } h_f \in C_0(G) \text{ for which } \lim_{t \to 0} \left\| \frac{P_t f - f}{t} - h_f \right\|_\infty = 0 \right\}.$$

Moreover we have $\mathcal{L}f = h_f$. We call $\mathcal{L}$ a *Hunt generator* whenever it is the infinitesimal generator of a Hunt semigroup.

Example Hunt Generator of the Compound Poisson Semigroup.

In this case $\mathcal{D}_\mathcal{L} = C_0(G)$ and $\mathcal{L}$ is a bounded operator given by

$$\mathcal{L}f(\sigma) = \int_G (f(\sigma\tau) - f(\sigma))c\rho(d\tau), \tag{5.3.5}$$

for all $f \in C_0(G)$, $\sigma \in G$. We give a formal argument and leave it to the reader to make this fully rigorous. For all $t \geq 0$ we have

$$\frac{d}{dt} P_t f(\sigma) = e^{-ct} \sum_{n=0}^\infty \int_G f(\sigma\tau) \frac{(ct)^n}{n!} c\rho^{*(n+1)}(d\tau) - e^{-ct} \sum_{n=0}^\infty \int_G f(\sigma\tau) \frac{(ct)^n}{n!} c\rho^{*(n)}(d\tau),$$

and the result is obtained by putting $t = 0$ and rearranging the two surviving terms.

We now return to developing the general case.

Lemma 5.3.2 *For all $\sigma \in G$, $f \in \mathcal{D}_{\mathcal{L}}$ we have $L_\sigma \mathcal{D}_{\mathcal{L}} \subseteq \mathcal{D}_{\mathcal{L}}$ and $\mathcal{L}L_\sigma f = L_\sigma \mathcal{L} f$.*

Proof Since L_σ acts isometrically in $C_0(G)$, by (5.3.4) we have

$$\lim_{t\to 0}\left\| \frac{P_t L_\sigma f - L_\sigma f}{t} - L_\sigma \mathcal{L} f\right\|_\infty = \lim_{t\to 0}\left\| L_\sigma\left(\frac{P_t f - f}{t} - \mathcal{L} f\right)\right\|_\infty = 0,$$

and the result follows. □

We will find it convenient to define the bounded linear functionals $\mathcal{A}_t$ on $C_0(G)$ for all $t > 0$, $f \in C_0(G)$ by

$$\mathcal{A}_t f := \frac{P_t f(e) - f(e)}{t}.$$

We also need to consider a linear functional $\mathcal{A}$ which is no longer necessarily bounded and whose domain $\mathcal{D}_{\mathcal{A}}$ comprises precisely those $f \in C_0(G)$ for which $\lim_{t\to 0} \mathcal{A}_t$ exists. Then for all $f \in \mathcal{D}_{\mathcal{A}}$ we define $\mathcal{A} f = \lim_{t\to 0} \mathcal{A}_t f$. We call $\mathcal{A}$ the *generating functional* of the convolution semigroup $(\mu_t, t \geq 0)$. It will play an important role in the sequel. It is clear that $\mathcal{D}_{\mathcal{L}} \subseteq \mathcal{D}_{\mathcal{A}}$ and that $\mathcal{A} f = \mathcal{L} f(e)$ for all $f \in \mathcal{D}_{\mathcal{L}}$. Moreover for all $\sigma \in G$, $f \in \mathcal{D}_{\mathcal{L}}$,

$$\mathcal{L} f(\sigma) = \mathcal{A}(L_{\sigma^{-1}} f), \tag{5.3.6}$$

where Lemma 5.3.2 ensures that the right hand side of (5.3.6) is well-defined.

Let $\mathfrak{g}$ be the Lie algebra of G. We will find it convenient to distinguish between the left invariant vector field X^L and the right invariant vector field X^R which is associated to X. We fix once and for all a basis $\{X_1, \ldots, X_d\}$ of $\mathfrak{g}$ and define the linear subspace $C_0^{2,L}$ of $C_0(G)$ by

$$C_0^{2,L}(G) := \{f \in C_0(G);\ X_i^L f \in C_0(G) \text{ and } X_j^L X_k^L f \in C_0(G) \text{ for all } 1 \leq i, j, k \leq d\}.$$

Clearly $C_c^2(G) \subseteq C_0^{2,L}(G)$ and it follows that $C_0^{2,L}(G)$ is dense in $C_0(G)$. We will frequently regard $C_0^{2,L}(G)$ as a real Banach space with respect to the norm $||\cdot||_\infty^{2,L}$ defined on $f \in C_0^{2,L}(G)$ by

$$||f||_\infty^{2,L} := ||f||_\infty + \sum_{i=1}^d ||X_i^L f||_\infty + \sum_{j,k=1}^d ||X_j^L X_k^L f||_\infty.$$

We will also need to consider the analogous space $C_0^{2,R}(G)$, with associated norm $||\cdot||_\infty^{2,R}$, which is defined exactly as above except that each X^L is replaced by X^R.

It is easily verified that the mapping $f \to \tilde{f}$ is an isometric isomorphism between $C_0^{2,L}(G)$ and $C_0^{2,R}(G)$.

Proposition 5.3.2

1. *For all* $f \in C^1(G), X \in \mathfrak{g}, t \geq 0$

$$X^R P_t f = P_t X^R f.$$

2. *The restriction of* $(P_t, t \geq 0)$ *to the Banach space* $C_0^{2,R}(G)$ *is a* C_0*-contraction semigroup.*
3. $\mathcal{D}_{\mathcal{L}}^{2,R} \subseteq \mathcal{D}_{\mathcal{L}} \cap C_0^{2,R}(G)$ *where* $\mathcal{D}_{\mathcal{L}}^{2,R}$ *is the domain of the generator of* $(P_t, t \geq 0)$ *acting on* $C_0^{2,R}(G)$.

Proof

1. By (5.3.4) and continuity,

$$\begin{aligned} X^R P_t f &= \frac{d}{du} L_{\exp(-uX)} P_t f \bigg|_{u=0} \\ &= \frac{d}{du} P_t L_{\exp(-uX)} f \bigg|_{u=0} \\ &= P_t X^R f. \end{aligned}$$

2. This follows from (1).
3. This is straightforward. □

In order to state and prove Hunt's celebrated theorem which characterises the operator $\mathcal{L}$ we will need some technical tools.

5.3.1 Taylor's Theorem on Lie Groups

Let us fix a co-ordinate neighbourhood U of the neutral element in G and let $(x_1, \ldots, x_n)$ be canonical co-ordinate functions on U. We can and will always regard x_i as being extended to the whole of G so that $x_i \in C_c(G)$ for all $1 \leq i \leq d$. We have the following version of Taylor's theorem:

Theorem 5.3.1 *1. If* $f \in C_0^{2,L}(G), g \in G$ *for all* $h \in U$ *there exists* $h' \in U$ *so that*

$$f(gh) = f(g) + \sum_{i=1}^{d} x_i(h) X_i^L f(g) + \frac{1}{2} \sum_{j,k=1}^{d} x_j(h) x_k(h) X_j^L X_k^L f(gh').$$

2. *If* $f \in C_0^{2,R}(G)$, $g \in G$ *for all* $h \in U$ *there exists* $h'' \in U$ *so that*

$$f(hg) = f(g) + \sum_{i=1}^{d} x_i(h) X_i^R f(g) + \frac{1}{2} \sum_{j,k=1}^{d} x_j(h) x_k(h) X_j^R X_k^R f(h''g).$$

Proof These are obtained by applying the usual Taylor theorem to the C^2 functions from $\mathbb{R}$ to $\mathbb{R}$ given by $u \to f\left(g \exp\left(u \sum_{i=1}^d x_i(h) X_i\right)\right)$ and $u \to f\left(\exp\left(u \sum_{i=1}^d x_i(h) X_i\right) g\right)$, respectively. We then have $h' = \exp\left(\theta \sum_{i=1}^d x_i(h) X_i\right)$ and $h'' = \exp\left(\phi \sum_{i=1}^d x_i(h) X_i\right)$ for some $0 < \theta, \phi < 1$.

5.3.2 Lévy Measure

We say that a Borel measure ν on G is a *Lévy measure* if $\nu(\{e\}) = 0$ and for every co-ordinate neighbourhood U of the neutral element in G:

$$\int_G \left(\sum_{i=1}^{d} x_i(\tau)^2 \right) \nu(d\tau) < \infty \text{ and } \nu(U^c) < \infty, \tag{5.3.7}$$

where $(x_1, \ldots, x_n)$ are canonical co-ordinate functions on U as above.

The following technical lemma will play an important role later on in the study of Lévy measures, and so we include it here. Here and below we fix the co-ordinate neighbourhood U.

Lemma 5.3.3 *There exist* $y_1, \ldots, y_d, \psi \in \mathcal{D}_{\mathcal{L}}^{2,R} \subseteq \mathcal{D}_{\mathcal{L}} \cap C_0^{2,R}(G)$ *so that*

1. $y_i(e) = 0$ *and* $X_j^R y_k(e) = \delta_{jk}$ *for* $1 \leq i, j, k \leq d$.
2. $\psi(e) = X_i^R \psi(e) = 0$ *and* $X_j^R X_k^R \psi(e) = 2\delta_{jk}$, $(1 \leq i, j, k \leq d)$.
3. *There exists a neighbourhood* V *of* e *with* $V \subseteq U$ *and there exists* $k > 0$ *so that for all* $\tau \in V$,

$$\psi(\tau) \geq k \sum_{i=1}^{d} x_i(\tau)^2.$$

Proof We direct the reader to Liao [132] Lemma 3.4 p. 54 for the proof of existence of $y_1, \ldots, y_d$ and ψ satisfying (1) and (2). For the proof of (3), by Taylor's theorem (Theorem 5.3.1(2)), for all $\tau \in U$ there exists $\tau' \in U$ so that

$$\psi(\tau) = \frac{1}{2} \sum_{j,k=1}^{d} x_j(\tau) x_k(\tau) X_j^R X_k^R \psi(\tau').$$

By (2), given any $\epsilon > 0$ we can find a neighbourhood of e, $V \subseteq U$ so that for all $\tau \in V, 1 \leq j \neq k \leq d, 1 \leq i \leq d$:

$$-\epsilon < X_j^R X_k^R \psi(\tau) < \epsilon$$

and

$$2 - \epsilon < (X_i^R)^2 \psi(\tau) < 2 + \epsilon.$$

Then we have

$$\psi(\tau) > \frac{(2-\epsilon)}{2} \sum_{i=1}^{d} x_i(\tau)^2 - \frac{\epsilon}{2} \sum_{k=1}^{d} \sum_{j \neq k} x_j(\tau) x_k(\tau).$$

Since ϵ is arbitrary the result follows by taking the limit as $\epsilon \to 0$ in the last inequality. □

5.3.3 *Ramaswami's Lemmas*

We now present two key lemmas due to Ramaswami [165] (see also Liao [132] pp. 52–55 for some improvements). We need to introduce $C_u(G)$ which is the real Banach space (with respect to the supremum norm) of all bounded, left uniformly continuous real valued functions defined on G. By Corollary A.2.1 in Appendix A.2, $C_0(G) \subseteq C_u(G)$.

Lemma 5.3.4 (Ramaswami's First Lemma) *Given any Borel neighbourhood N of e,*

$$\sup_{t>0} \frac{1}{t} \mu_t(N^c) < \infty.$$

Proof We extend the semigroup $(P_t, t \geq 0)$ to a C_0-semigroup on $C_u(G)$ and let $\mathcal{D}'_{\mathcal{L}}$ be the domain of its generator in that space. We continue to use the notation $\mathcal{A}$ for the generating functional in the larger space. It suffices to prove that there exists a function $J \in \mathcal{D}'_{\mathcal{L}}$ and $c_J > 0$ so that $J \geq 0$, $J(e) = 0$ and $J \geq c_J$ on N^c. For if this is the case we then have

$$\begin{aligned} c_J \sup_{t>0} \frac{\mu_t(N^c)}{t} &\leq \frac{1}{t} \int_{N^c} J(\tau) \mu_t(d\tau) \\ &\leq \frac{1}{t} \int_G J(\tau) \mu_t(d\tau) < \infty, \end{aligned}$$

since $J \in \mathcal{D}'_{\mathcal{L}}$, and so $\lim_{t \to 0} \frac{1}{t} \int_G J(\tau) \mu_t(d\tau) = \mathcal{A}(J)$ is finite.

Now to conclude the proof of the lemma we must show that such a J exists. We begin by choosing a compact neighbourhood N_1 of e which has the property that $N_1 = N_1^{-1}$ and $N_1 N_1 \subset N$. Replacing N_1 by a smaller set if necessary, we have $N_1 N^c \subset N_1^c$. Let $f \in C_u(G)$ be such $f(e) = 0, 0 \leq f \leq 1$ and $f = 1$ on N_1^c. Choose $0 < \epsilon < \frac{1}{4}$. Since $\mathcal{D}'_{\mathcal{L}}$ is dense in $C_u(G)$, we can find $g \in \mathcal{D}'_{\mathcal{L}}$ so that

$$f(\tau) - \epsilon < g(\tau) < f(\tau) + \epsilon$$

for all $\tau \in G$. By compactness, we can find $\tau_1 \in N_1$ so that $f(\tau_1) = d := \inf_{\tau \in N_1} g(\tau)$. Then $d \leq g(e) < \epsilon$. For all $\tau \in N_1^c$,

$$g(\tau) > f(\tau) - \epsilon = 1 - \epsilon > \epsilon.$$

Hence $d = \inf_{\tau \in G} g(\tau)$. Define $h := g - d$. Then $h \in \mathcal{D}'_{\mathcal{L}}, h \geq 0$ and $h(\tau_1) = 0$. Finally we define $J := L_{\tau_1^{-1}} h$. Then $J \in \mathcal{D}'_{\mathcal{L}}$, by a slight extension of Lemma 5.3.2. Clearly $J \geq 0$ and $J(e) = 0$. Finally we show that we may take $c_J = \frac{1}{2}$. Indeed, for $\tau \in N^c$, since $\tau_1 \tau \in N_1^c$ we have

$$J(\tau) = g(\tau_1 \tau) - d > 1 - \epsilon - d \geq 1 - 2\epsilon > \frac{1}{2}.$$

□

Lemma 5.3.5 (Ramaswami's Second Lemma)

$$\sup_{t>0} \frac{1}{t} \int_U \left(\sum_{i=1}^d x_i(\tau)^2 \right) \mu_t(d\tau) < \infty.$$

Proof Recall the function ψ and the neighbourhood V of e of Lemma 5.3.3. Since $\psi \in \mathcal{D}_{\mathcal{L}}$ and $\psi(e) = 0$ we deduce that $\sup_{t>0} \frac{1}{t} \int_G \psi(\tau) \mu_t(d\tau)$ is finite. By Lemma 5.3.4, $\sup_{t>0} \dfrac{1}{t} \mu_t(V^c) < \infty$, hence $\sup_{t>0} \frac{1}{t} \int_V \psi(\tau) \mu_t(d\tau)$ is finite. By Lemma 5.3.3 (3) we then find that $\sup_{t>0} \frac{1}{t} \int_V \left(\sum_{i=1}^d x_i(\tau)^2 \right) \mu_t(d\tau) < \infty$. The result follows by Lemma 5.3.4 again, since

$$\sup_{t>0} \frac{1}{t} \int_{V^c \cap U} \left(\sum_{i=1}^d x_i(\tau)^2 \right) \mu_t(d\tau) \leq \sup_{\sigma \in U} \left(\sum_{i=1}^d x_i(\tau)^2 \right) \sup_{t>0} \frac{1}{t} \mu_t(V^c) < \infty.$$

□

Corollary 5.3.1

$$\sup_{t>0} \frac{1}{t} \int_G \left(\sum_{i=1}^d x_i(\tau)^2 \right) \mu_t(d\tau) < \infty.$$

Proof This follows easily from Lemmas 5.3.4 and 5.3.5, using the fact that x_i has compact support $(1 \leq i \leq d)$. □

5.3.4 The Domain of The Generating Function

We will find it convenient to introduce the notation $D_i f = X_i^R f(e)(= X_i^L f(e))$ and $D_{jk} f = X_j^R X_k^R f(e)$ for all $f \in C_0^{2,R}(G) \cup C_0^{2,L}(G)$, $1 \leq i, j, k \leq d$.

Theorem 5.3.2

1. *The linear functionals* $(\mathcal{A}_t, t \geq 0)$ *on the Banach space* $C_0^{2,R}(G)$ *are uniformly bounded in* $t > 0$.
2. $C_0^{2,R}(G) \subseteq \mathcal{D}_{\mathcal{A}}$.
3. $C_0^{2,L}(G) \subseteq \mathcal{D}_{\mathcal{A}}$.

Proof

1. For all $f \in C_0^{2,R}(G)$ define $g_f \in C_0^{2,R}(G)$ by

$$g_f(\tau) = f(\tau) - f(e) - \sum_{i=1}^{d} (D_i f) y_i(\tau).$$

By Theorem 5.3.1(2), we can find a neighbourhood W of e with $W \subset U$ with $\tau' \in W$ so that for all $\tau \in W$,

$$g_f(\tau) = \frac{1}{2} \sum_{j,k=1}^{d} x_j(\tau) x_k(\tau) X_j^R X_k^R g_f(\tau').$$

It follows easily that there exists $K > 0$ so that for all $\tau \in W$,

$$|g_f(\tau)| \leq K ||g_f||_{\infty}^{2,R} \sum_{i=1}^{d} x_i(\tau)^2.$$

Then by Lemma 5.3.5,

$$\sup_{t>0} \left| \frac{1}{t} \int_W g_f(\tau) \mu_t(d\tau) \right| < \infty.$$

But by Lemma 5.3.4,

$$\sup_{t>0} \left| \frac{1}{t} \int_{W^c} g_f(\tau) \mu_t(d\tau) \right| \leq ||g_f||_{\infty} \sup_{t>0} \frac{\mu_t(W^c)}{t} < \infty.$$

So we conclude that there exists $L > 0$ so that

$$\sup_{t>0} \frac{1}{t} \left| \int_G \left(f(\tau) - f(e) - \sum_{i=1}^{d} (D_i f) y_i(\tau) \right) \mu_t(d\tau) \right| < L,$$

i.e.

$$\sup_{t>0}\left|\mathcal{A}_t f - \frac{1}{t}\sum_{i=1}^{d}(D_i f)P_t y_i(e)\right| < L.$$

But $y_i \in \mathcal{D}_{\mathcal{L}}, 1 \leq i \leq d$ and so $\sup_{t>0}\left|\frac{1}{t}\sum_{i=1}^{d}(D_i f)P_t y_i(e)\right| < \infty$. Hence we deduce that $\sup_{t>0}|\mathcal{A}_t f| < \infty$ for all $f \in C_0^{2,R}$. But then

$$\sup_{t>0}||\mathcal{A}_t|| < \infty,$$

by the uniform boundedness theorem.

2. It follows by Proposition 5.3.2 (3) that $C_0^{2,R}(G) \cap \mathcal{D}_{\mathcal{L}}$ is dense in $\mathcal{D}_{\mathcal{L}}$ (in the $||\cdot||_\infty$ topology), and that $C_0^{2,R}(G)\cap\mathcal{D}_{\mathcal{L}} \subseteq \mathcal{D}_{\mathcal{A}}$, and so $\lim_{t\to 0}\mathcal{A}_t f = \mathcal{A}f$ for all $f \in C_0^{2,R}(G)\cap\mathcal{D}_{\mathcal{L}}$. But we have just seen that $(\mathcal{A}_t, t > 0)$ is uniformly bounded. So we deduce by a straightforward density argument that $\lim_{t\to 0}\mathcal{A}_t f = \mathcal{A}f$ for all $f \in C_0^{2,R}(G)$ and that the limit is finite.
3. Let $(\widetilde{P}_t, t \geq 0)$ be the Hunt semigroup and $\widetilde{\mathcal{A}}$ be the generating functional of $(\widetilde{\mu}_t, t \geq 0)$. Let $f \in C_0^{2,L}(G)$. Then $\widetilde{f} \in C_0^{2,R}(G)$. It is straightforward algebra to verify that $\widetilde{P}_t\widetilde{f}(e) = P_t f(e)$ for all $t \geq 0$, and so

$$\mathcal{A}f = \lim_{t\to 0}\frac{P_t f(e) - f(e)}{t} = \lim_{t\to 0}\frac{\widetilde{P}_t\widetilde{f}(e) - \widetilde{f}(e)}{t} = \widetilde{\mathcal{A}}f.$$

Since $f \in \mathrm{Dom}(\widetilde{\mathcal{A}})$ by (2) above we deduce that $f \in \mathcal{D}_{\mathcal{A}}$ as was required. □

We will not require right invariant vector fields again in this section, and so henceforth we will write $X = X^L$ for $X \in \mathfrak{g}$ and $C_0^{(2)}(G) := C_0^{2,L}(G)$.[1]

5.3.5 Hunt's Theorem

We are now ready to state and prove Hunt's theorem [103] for the classification of $\mathcal{L}$.

Theorem 5.3.3 (Hunt's theorem) *If $(\mu_t, t \geq 0)$ is a convolution semigroup of measures in G with generator $\mathcal{L}$, then*

(a) $C_0^{(2)}(G) \subseteq \mathrm{Dom}(\mathcal{L})$.
(b) For each $\sigma \in G$, $f \in C_0^{(2)}(G)$,

[1] We use the superscript (2) to distinguish the space $C_0^{(2)}(G)$ from $C_0^2(G)$ which is the subspace of $C_0(G)$ comprising all twice (continuously) differentiable functions.

$$\mathcal{L}f(\sigma) = \sum_{i=1}^{d} b_i X_i f(\sigma) + \frac{1}{2} \sum_{i,j=1}^{d} a_{ij} X_i X_j f(\sigma)$$

$$+ \int_G (f(\sigma\tau) - f(\sigma) - \sum_{i=1}^{d} x_i(\tau) X_i f(\sigma)) \nu(d\tau), \qquad (5.3.8)$$

where $b = (b_1, \ldots, b_d) \in \mathbb{R}^d$, $a = (a_{ij})$ is a non-negative definite, symmetric $d \times d$ real-valued matrix and ν is a Lévy measure on G.

Conversely, any linear operator with a representation as in (5.3.8) is the restriction to $C_0^{(2)}(G)$ of the Hunt generator corresponding to a unique convolution semigroup of probability measures.

Proof We only give the main points in the proof of the first part of the theorem here and we direct the reader to the literature (Hunt [103], Heyer [95] and Liao [132]) for the second part. As this argument is quite involved we split it into sections.

1. *Construction of the Lévy Measure ν.*

   Let W be an open subset of G such that $e \notin \overline{W}$ and choose $f \in \mathcal{D}_{\mathcal{A}} \cap C_c(W)$. Then $\mathcal{A}f = \lim_{t\to\infty} \frac{1}{t} \int_W f(\tau)\mu_t(d\tau)$. By Lemma 5.3.4,

$$|\mathcal{A}f| \leq ||f||_\infty \sup_{t>0} \frac{1}{t} \mu_t(W) < \infty.$$

   Hence $f \to \mathcal{A}f$ defines a positive bounded linear functional on $\mathcal{D}_{\mathcal{A}} \cap C_c(W)$, which extends by density (in the $|| \cdot ||_\infty$ norm) to a positive bounded linear functional $\mathcal{A}_W$ on $C_c(W)$. Thus by the Riesz representation theorem, there exists a finite Borel measure ν_W on W so that

$$\mathcal{A}_W f = \int_W f(\tau)\nu_W(d\tau)$$

   for all $f \in C_c(W)$. Using the second countability of G, we can find a sequence of open sets $(U_n, n \in \mathbb{N})$, with $\overline{U_{n+1}} \subseteq U_n$ for all $n \in \mathbb{N}$, that form a basis for the topology at e, in that if V is any neighbourhood of e then $V \subseteq U_n$ for some $n \in \mathbb{N}$. Define $W_n = \overline{U_n}^c$. Then $(W_n, n \in \mathbb{N})$ is a sequence of open sets that increases to $G \setminus \{e\}$. By the construction given above, we obtain an associated sequence of bounded linear functionals $(\mathcal{A}_n, n \in \mathbb{N})$ which are compatible in the sense that if $f \in C_c(W_n)$ with $\text{supp}(f) \subseteq W_m$ for some $m < n$, then $\mathcal{A}_n f = \mathcal{A}_m f_m$, where f_m is the restriction of f to W_m. Given any $f \in C_c(G \setminus \{e\})$ we must have $\text{supp}(f) \subseteq W_n$ for some $n \in \mathbb{N}$. We obtain a positive linear functional $\mathcal{I}$ on $C_c(G \setminus \{e\})$ by defining

$$\mathcal{I}(f) = \mathcal{A}_n(f_n)$$

where n is any integer for which $\text{supp}(f) \subseteq W_n$. Hence by the Riesz representation theorem again, there exists a regular Borel measure ν on $G \setminus \{e\}$ so that

$$\mathcal{I}(f) = \int_{G\setminus\{e\}} f(\tau)\nu(d\tau).$$

We define $\nu(\{e\}) = 0$ and show that ν is a Lévy measure. For each $n \in \mathbb{N}$, we denote by ν_n, the restriction of the measure ν to W_n.

It is clear that $\nu(U^c) < \infty$ for any neighbourhood U of e since, by construction, $U^c \subseteq W_n$ for some $n \in \mathbb{N}$. We choose a sequence $(\phi_n, n \in \mathbb{N})$ where for each $n \in \mathbb{N}$, $\phi_n \in C_c(W_n)$ with $0 \leq \phi_n \leq 1$ and $\{\tau \in G, \phi_n(\tau) = 1\} \uparrow G - \{e\}$ as $n \to \infty$. Then

$$\begin{aligned}
\int_G \left(\sum_{i=1}^d x_i(\tau)^2\right)\nu(d\tau) &= \lim_{n\to\infty} \int_{W_n} \left(\sum_{i=1}^d x_i(\tau)^2\right)\phi_n(\tau)\nu_n(d\tau) \\
&= \lim_{n\to\infty}\lim_{t\to 0}\frac{1}{t}\int_{W_n}\left(\sum_{i=1}^d x_i(\tau)^2\right)\phi_n(\tau)\mu_t(d\tau) \\
&\leq \sup_{t>0}\frac{1}{t}\int_G\left(\sum_{i=1}^d x_i(\tau)^2\right)\mu_t(d\tau) < \infty,
\end{aligned}$$

by Corollary 5.3.1.

2. *Construction of the matrix a.*

For each $1 \leq i, j \leq d$ we define

$$a_{ij} = \mathcal{A}(x_i x_j) - \int_G x_i(\tau)x_j(\tau)\nu(d\tau).$$

Since ν is a Lévy measure it is clear that a_{ij} is finite, and it is immediate that the $d \times d$ matrix $a = (a_{ij})$ is symmetric. To show that a is non-negative definite, we follow Liao [132] pp. 57–58 and define $h \in C_c(G)$ by

$$h := \left|\sum_{i=1}^d \xi_i x_i\right|^2,$$

where $\xi_i \in \mathbb{R}$, $1 \leq i \leq d$. Then $h \in \mathcal{D}_{\mathcal{A}}$ and

$$\mathcal{A}h = \lim_{t\to 0}\frac{1}{t}\int_G h(\tau)\mu_t(d\tau)$$

$$\begin{aligned}&\geq \lim_{t\to 0}\frac{1}{t}\int_G h(\tau)\phi_n(\tau)\mu_t(d\tau)\\&= \int_G h(\tau)\phi_n(\tau)\nu(d\tau) \to \int_G h(\tau)\nu(d\tau),\end{aligned}$$

as $n \to \infty$ and so,

$$\begin{aligned}\sum_{i,j=1}^{d} a_{ij}\xi_i\overline{\xi_j} &= \sum_{i,j=1}^{d} \xi_i\overline{\xi_j}\left(\mathcal{A}(x_ix_j) - \int_G x_i(\tau)x_j(\tau)\nu(d\tau)\right)\\&= \mathcal{A}h - \int_G h(\tau)\nu(d\tau) \geq 0,\end{aligned}$$

as was required.

3. *Computation of the Action of $\mathcal{A}$ on $C_0^{(2)}(G)$.*

Let $f \in C_0^{(2)}(G)$. By Taylor's theorem, for all $\tau \in G$

$$f(\tau) = f(e) + \sum_{i=1}^{d}(D_i f)x_i(\tau) + \frac{1}{2}\sum_{j,k=1}^{d}(D_{jk}f)x_j(\tau)x_k(\tau) + R_f(\tau),$$

where $R_f(\tau) := \frac{1}{2}\sum_{j,k=1}^{d} x_j(\tau)x_k(\tau)(X_jX_kf(\tau') - D_{jk}f)$. By Theorem 5.3.1 (3) we may compute $\mathcal{A}f$ and so obtain

$$\mathcal{A}f = \sum_{i=1}^{d}(D_i f)\mathcal{A}x_i + \frac{1}{2}\sum_{j,k=1}^{d}(D_{jk}f)\mathcal{A}x_jx_k + \mathcal{A}R_f.$$

We next show that

$$\mathcal{A}R_f = \int_G R_f(\tau)\nu(d\tau). \tag{5.3.9}$$

First we observe that there exists $c_f \in C(G)$ with $c_f \geq 0$ and $c_f(\tau) \to 0$ as $\tau \to e$ so that $|R_f(\tau)| \leq c_f(\tau)\sum_{i=1}^{d} x_i(\tau)^2$ for $\tau \in G$. It follows that R_f is ν-integrable. To establish (5.3.9), we see that

$$\begin{aligned}\left|\mathcal{A}R_f - \int_{W_n} R_f(\tau)\nu(d\tau)\right| &\leq \sup_{t>0}\frac{1}{t}\int_{W_n^c}|R_f(\tau)|\mu_t(d\tau)\\&\leq \sup_{t>0}\frac{1}{t}\int_G\sum_{i=1}^{d} x_i(\tau)^2\mu_t(d\tau)\ \sup_{\tau\in W_n^c} c_f(\tau)\\&\to 0 \text{ as } n \to \infty.\end{aligned}$$

We thus obtain

$$\mathcal{A}R_f = \frac{1}{2}\sum_{j,k=1}^{d}\int_G x_j(\tau)x_k(\tau)\nu(d\tau)(X_jX_kf(\tau') - D_{jk}f).$$

Writing $b_i := \mathcal{A}x_i$ for each $1 \leq i \leq d$, we then obtain

$$\mathcal{A}f = \sum_{i=1}^{d} b_i(D_if) + \frac{1}{2}\sum_{j,k=1}^{d} a_{ij}(D_{jk}f) + \frac{1}{2}\sum_{j,k=1}^{d}\int_G x_j(\tau)x_k(\tau)\nu(d\tau)X_jX_kf(\tau'),$$

and by Taylor's theorem again we conclude that

$$\mathcal{A}f = \sum_{i=1}^{d} b_i(D_if) + \frac{1}{2}\sum_{j,k=1}^{d} a_{ij}(D_{jk}f) + \int_G\left[f(\tau) - f(e) - \sum_{i=1}^{d} x_i(\tau)D_if\right]\nu(d\tau). \quad (5.3.10)$$

4. *Computation of the action of $\mathcal{L}$ on $C_0^{(2)}(G)$.*

Once we have shown that $C_0^{(2)}(G) \subseteq \mathcal{D}_{\mathcal{L}}$, we can obtain (5.3.8) directly from (5.3.10) by using (5.3.6). To establish the former, we introduce the resolvent $R_\lambda := (\lambda I - \mathcal{L})^{-1}$ and recall that for $\lambda > 0$ it is a bounded linear operator on $C_0(G)$ whose range lies in $\mathcal{D}_{\mathcal{L}}$. The required result follows by showing that for each $f \in C_0^{(2)}(G)$, $\lambda > 0$,

$$R_\lambda(\lambda - \mathcal{B}f) = f$$

where

$$\mathcal{B}f := \mathcal{A}(f \circ L_{\sigma^{-1}}).$$

This follows by using the fact that

$$R_\lambda f = \int_0^\infty e^{-\lambda t}P_tf dt.$$

Details may be found in Liao [132], p. 60. □

Note that the representation of the generator $\mathcal{L}$ appearing in (5.3.8) is uniquely determined by the triple (b, a, ν) which we call the *characteristics* of the convolution semigroup $(\mu_t, t \geq 0)$. Indeed, as was shown in the proof of Theorem 5.3.3 these are given by the following expressions. For $1 \leq i, j, k \leq d$,

$$b_i = \mathcal{A}(x_i),\ a_{jk} = \mathcal{A}(x_j x_k) - \int_G x_j(\tau)x_k(\tau)\nu(d\tau), \tag{5.3.11}$$

$$\int_G f(\tau)\nu(d\tau) = \lim_{t\to 0}\frac{1}{t}\int_G f(\tau)\mu_t(d\tau) \tag{5.3.12}$$

for all $f \in C_c(G)$ with f having support outside a neighbourhood of e.

The next result of this section will be to show that $C_c^\infty(G)$ is a core for $\mathcal{L}$. Here we follow the proof given by Breuillard [35] which is based on earlier work of Hirsch and Roth [102] (Théorèm 13) and Hazod [84] (Satz 4.1) (see also Sato [177] pp. 208–211 for the case where $G = \mathbb{R}^d$ and Theorem 20.3 (c) of Hazod and Siebert [85] for general locally compact groups). We use the fact that the core property is equivalent to requiring that $\lambda I - \mathcal{L}$ maps $C_c^\infty(G)$ to a dense linear manifold in $C_0(G)$ for some $\lambda \in \rho_{\mathcal{L}}$, the resolvent set of $\mathcal{L}$. This is proved in Appendix A.8. In our case, since $\mathcal{L}$ generates a C_0-semigroup, we have $(0,\infty) \subseteq \rho_{\mathcal{L}}$ and we will choose $\lambda = 1$ for convenience.

Theorem 5.3.4 *$C_c^\infty(G)$ is a core for $\mathcal{L}$.*

Proof Let ρ be a positive continuous linear functional on $C_c^\infty(G)$. Then it is a consequence of the Riesz representation theorem that ρ may be identified with a finite Borel measure on G. Using the result of Appendix A.8, we see that the required result is established if we can show that $\rho((I - \mathcal{L})f) = 0$ for all $f \in C_c^\infty(G) \Rightarrow \rho = 0$. Choose an arbitrary $f \in C_c^\infty(G)$ with $f \geq 0$ and consider the convolution operator P_ρ. By continuity we can find $\sigma_0 \in G$ so that $(P_\rho f)(\sigma_0) = ||P_\rho f||_\infty$. Since $L_{\sigma_0^{-1}} f \in C_c^\infty(G)$ we have $\rho((I - \mathcal{L})L_{\sigma_0^{-1}} f) = 0$. From this we see that $h_f(e) = \mathcal{L}h_f(e) = ||h_f||_\infty$ where $h_f := P_\rho L_{\sigma_0^{-1}} f$. But

$$\mathcal{L}h_f(e) = \lim_{t\to 0}\frac{1}{t}\int_G (h_f(\tau) - h_f(e))\mu_t(d\tau) \leq 0,$$

and so $||h_f||_\infty \leq 0 \Rightarrow h_f = 0$. So in particular $h_f(\sigma_0^{-1}) = \rho(f) = 0$ and hence $\rho = 0$ as was required. □

We say that the Lévy measure ν has a *finite first moment* if

$$\int_G |x_i(\sigma)|\nu(d\sigma) < \infty$$

for all $1 \leq i \leq d$. This condition is always satisfied when the measure ν is finite.

In this case we find that the Hunt generator has a particularly simple form.

Corollary 5.3.2 *If $(\mu_t, t \geq 0)$ is a convolution semigroup of measures on G whose Lévy measure ν has a finite first moment, then for all $f \in C_0^{(2)}(G)$, $\sigma \in G$*

$$\mathcal{L}f(\sigma) = \sum_{i=1}^{d} b_i^{(0)} X_i f(\sigma) + \frac{1}{2} \sum_{i,j=1}^{d} a_{ij} X_i X_j f(\sigma) + \int_G (f(\sigma\tau) - f(\sigma))\nu(d\tau), \tag{5.3.13}$$

where $b_i^{(0)} := b_i - \int_G x_i(\tau)\nu(d\tau)$ $(1 \leq i \leq d)$.

The proof is immediate.

Returning to the general case, the Hunt semigroup associated to the convolution semigroup $(\widetilde{\mu_t}, t \geq 0)$ is denoted by $(\widetilde{P}_t, t \geq 0)$ and its generator is $\widetilde{\mathcal{L}}$, so for all $t \geq 0, f \in C_0(G), \sigma \in G$

$$(\widetilde{P}_t f)(\sigma) = \int_G f(\sigma\tau^{-1})\mu_t(d\tau). \tag{5.3.14}$$

Note that we can always choose the co-ordinate functions x_i, $1 \leq i \leq d$ to satisfy $x_i(\sigma) = -x_i(\sigma^{-1})$ for all $\sigma \in G$ and from now on we will make this assumption.

Proposition 5.3.3 *$\mathcal{L}$ has characteristics (b, a, ν) if and only if $\widetilde{\mathcal{L}}$ has characteristics $(-b, a, \widetilde{\nu})$.*

Proof The fact that the Lévy measure corresponding to $(\widetilde{\mu_t}, t \geq 0)$ is $\widetilde{\nu}$ follows from (5.3.12). Let $\widetilde{\mathcal{A}}$ be the generating function of $(\widetilde{\mu_t}, t \geq 0)$. Then using (5.3.11) we have for all $1 \leq i \leq d$,

$$\widetilde{\mathcal{A}}x_i = \lim_{t \to 0} \frac{1}{t} \int_G x_i(\sigma)\widetilde{\mu_t}(d\sigma) = \lim_{t \to 0} \frac{1}{t} \int_G x_i(\sigma^{-1})\mu_t(d\sigma) = -\mathcal{A}(x_i) = -b_i.$$

Finally we can similarly check that for all $1 \leq i, j \leq d$,

$$a_{ij} = \widetilde{\mathcal{A}}(x_i x_j) - \int_G x_i(\tau)x_j(\tau)\widetilde{\nu}(d\tau). \qquad \square$$

For all $t \geq 0$, let $u(t) = P_t f$ where the initial condition $f \in C_0(G)$ is arbitrary. Since P_t maps $C_0(G)$ into $\text{Dom}(\mathcal{L})$ for all $t \geq 0$ we have the partial differential equation:

$$\frac{\partial u_t}{\partial t} = \mathcal{L}u(t).$$

We can dualise this equation to obtain the distribution-sense equation

$$\frac{\partial \mu_t}{\partial t} = \mathcal{L}^\dagger \mu_t,$$

where $\mathcal{L}^\dagger$ is the formal adjoint to $\mathcal{L}$ (see e.g. Applebaum [7], Sect. 3.5.3) and if μ_t has a density f_t for all $t > 0$ which is differentiable with respect to t and twice differentiable with respect to the group variable, then we have the *Kolmogorov forward* or *Fokker-Planck equation*:

$$\frac{\partial f_t}{\partial t} = \mathcal{L}^\dagger f_t,$$

(see also Corollary 2 in Lo and Ng [136]). In the case where $\nu = 0$ and $\mathcal{L}$ is a diffusion operator, engineering applications of this equation within the Lie group framework are developed in Chap. 16 of Chirikjian and Kyatko [48].

Remark More generally, Kolmogorov (or Fokker-Planck) equations apply to the transition probabilities $P(Z(t) \in A|Z(s) = \rho)$ (where $0 \leq s \leq t < \infty, \rho \in G, A \in \mathcal{B}(G)$) of a Markov process $Z = (Z(t), t \geq 0)$ taking values in G. In the case under discussion, Z is a G-valued (left)-Lévy process wherein $Z(t)$ has law μ_t for all $t \geq 0$ and $P(Z(t) \in A|Z(s) = \rho) = \mu_{t-s}(\rho^{-1}A)$ (see Proposition 5.9.1 in Sect. 5.9).

5.4 L^2-Properties of Hunt Semigroups

So far we have considered $(P_t, t \geq 0)$ as a contraction semigroup acting in $C_0(G)$, but by using similar arguments to those in e.g. the proof of Theorem 3.4.2 in Applebaum [7], pp. 173–174, we may also realise it as a contraction semigroup acting in $L^2(G)$ (equipped with a right invariant Haar measure). Note that the semigroup is *Markovian* in that if $f \in L^2(G)$ with $0 \leq f \leq 1$ (a.s.), then $0 \leq P_t f \leq 1$ (a.s.) for all $t \geq 0$. Define $C_c^{(2)}(G) := C_0^{(2)}(G) \cap C_c(G)$. Then since $C_c^\infty(G) \subseteq C_c^{(2)}(G)$ it follows that $C_c^2(G)$ is dense in $L^2(G)$. Let $\mathcal{L}'$ denote the infinitesimal generator of $(P_t, t \geq 0)$ in $L^2(G)$.

Proposition 5.4.1 *$C_c^{(2)}(G) \subseteq Dom(\mathcal{L}')$ and $\mathcal{L}' f = \mathcal{L} f$ for all $f \in C_c^{(2)}(G)$.*

Proof Let $f \in C_c^{(2)}(G)$. By a direct computation using (5.3.8) and Taylor's theorem on Lie groups, we have $\mathcal{L} f \in L^2(G)$. It is sufficient to show that

$$\lim_{t \to 0} \int_G \left(\frac{1}{t}(P_t f(\sigma) - f(\sigma)) - \mathcal{L} f(\sigma) \right) h(\sigma) d\sigma = 0,$$

for all $h \in L^1(G) \cap L^2(G)$, as weak and strong generators coincide (see e.g. Pazy [158] Theorem 1.4, pp. 43–44). This follows by dominated convergence. To see that such an argument is valid, first note that for all $t > 0$

$$P_t f - f = \int_0^t P_s \mathcal{L} f ds,$$

and so, using the contraction property of the semigroup, we have

$$\left|\left| \frac{1}{t}(P_t f - f) \right|\right|_\infty \leq ||\mathcal{L} f||_\infty.$$

We then find that for all $\sigma \in G$,

$$\left| \frac{1}{t}(P_t f(\sigma) - f(\sigma)) - \mathcal{L} f(\sigma)) h(\sigma) \right| \leq 2||\mathcal{L} f||_\infty |h(\sigma)|,$$

and the result follows. □

We remark that the proof of Proposition 5.4.1 is much simpler when G is compact, as readers can quickly verify. To ease notation we will henceforth write $\mathcal{L}$ for $\mathcal{L}'$.

Proposition 5.4.2 *For all $t \geq 0$, $\widetilde{P}_t = P_t^*$.*

Proof Using Fubini's theorem and the right invariance of m we deduce that for all $t \geq 0$, $f, g \in L^2(G)$,

$$\begin{aligned} \langle P_t f, g \rangle &= \int_G \int_G f(\sigma\tau) g(\sigma) \mu_t(d\tau) d\sigma \\ &= \int_G \int_G f(\sigma) g(\sigma\tau^{-1}) \mu_t(d\tau) d\sigma \\ &= \langle f, \widetilde{P}_t g \rangle, \end{aligned}$$

by (5.3.14). □

It follows immediately from Proposition 5.4.2 that $\widetilde{\mathcal{L}} = \mathcal{L}^*$, and so in particular we have that $C_c^{(2)}(G) \subseteq \text{Dom}(\mathcal{L}^*)$. Moreover, by Proposition 5.3.3 for all $f \in C_c^{(2)}(G), \sigma \in G$,

$$\begin{aligned} \mathcal{L}^* f(\sigma) = &- \sum_{i=1}^d b_i X_i f(\sigma) + \frac{1}{2} \sum_{i,j=1}^d a_{ij} X_i X_j f(\sigma) \\ &+ \int_G \left(f(\sigma\tau^{-1}) - f(\sigma) + \sum_{i=1}^d x_i(\tau) X_i f(\sigma) \right) \nu(d\tau). \end{aligned} \quad (5.4.15)$$

We now investigate some properties of the semigroup and its generator that are specific to the L^2 action.

5.4.1 Self-adjointness

The following is the main result of this section (see Kunita [128], Applebaum [9]):

Theorem 5.4.1 *The following are equivalent.*

(i) The convolution semigroup $(\mu_t, t \geq 0)$ is symmetric.
(ii) $P_t = P_t^*$ for each $t \geq 0$.
(iii) $\mathcal{L} = \mathcal{L}^*$.
(iv) $b = 0, \nu = \widetilde{\nu}$.

(v) For all $f \in C_c^{(2)}(G)$,

$$\mathcal{L}f(\sigma) = \frac{1}{2}\sum_{i,j=1}^{d} a_{ij}X_iX_jf(\sigma) + \frac{1}{2}\int_G (f(\sigma\tau) - 2f(\sigma) + f(\sigma\tau^{-1}))\nu(d\tau). \tag{5.4.16}$$

Proof (i) $\Rightarrow$ (ii) follows from standard properties of convolution operators. To see that (ii) $\Rightarrow$(i) observe that for all $A \in \mathcal{B}(G)$, by Proposition 5.4.2

$$\mu_t(A) = P_t 1_A(e) = P_t^* 1_A(e) = \widetilde{P}_t 1_A(e) = \widetilde{\mu_t}(A).$$

(ii) $\Leftrightarrow$ (iii) is standard semigroup theory (see e.g. Davies [53, 54]). To verify that (i) $\Rightarrow$ (iv), choose $f \in \mathrm{Dom}(\mathcal{A})$ which vanishes in a neighbourhood of e. Then by (5.3.12), we have

$$\begin{aligned}\int_G f(\sigma)\nu(d\sigma) &= \lim_{t\to 0}\frac{1}{t}\int_G f(\sigma)\mu_t(d\sigma)\\ &= \lim_{t\to 0}\frac{1}{t}\int_G f(\sigma)\widetilde{\mu_t}(d\sigma)\\ &= \int_G f(\sigma)\widetilde{\nu}(d\sigma).\end{aligned}$$

Hence $\nu = \widetilde{\nu}$. The fact that $b = 0$ then follows from (iii) when we compare expressions for $\mathcal{L}$ and $\mathcal{L}^*$ as given by (5.3.8) and (5.4.15) respectively.

(iv) $\Rightarrow$ (v) follows from the fact that if $\nu = \widetilde{\nu}$, then for each $f \in C_c^{(2)}(G)$,

$$\begin{aligned}&\int_G \left(f(\sigma\tau) - f(\sigma) - \sum_{i=1}^{d} x_i(\tau)X_if(\sigma) \right)\nu(d\tau)\\ &= \int_G \left(f(\sigma\tau^{-1}) - f(\sigma) + \sum_{i=1}^{d} x_i(\tau)X_if(\sigma) \right)\nu(d\tau),\end{aligned}$$

where we recall that we have chosen $x_1, \ldots, x_d$ to be such that $x_i(\sigma) = -x_i(\sigma^{-1})$ for each $\sigma \in G, 1 \le i \le d$.

Finally for (v) $\Rightarrow$ (i), if (v) holds then $\mathcal{L}f = \mathcal{L}^*f$ for each $f \in C_c^{(2)}(G)$. To see this, it suffices to observe that for all $\sigma \in G$,

$$\begin{aligned}&\int_G (f(\sigma\tau) + f(\sigma\tau^{-1}) - 2f(\sigma))\nu(d\tau)\\ &= \int_G (R_\tau - 2I + R_{\tau^{-1}})f(\sigma)\nu(d\tau)\\ &= \int_G (R_\tau - 2I + R_\tau^*)f(\sigma)\nu(d\tau),\end{aligned}$$

where $R_\tau f(\sigma) = f(\sigma\tau)$ for each $\tau \in G$. By a density argument in $C_0(G)$ we then deduce that $\mathcal{L}f = \widetilde{\mathcal{L}}f$ for each $f \in C^{(2)}(G)$. Now since by Theorem 5.3.3, $\mathcal{L}|_{C^{(2)}(G)}$ uniquely determines $(\mu_t, t \geq 0)$ and $\widetilde{\mathcal{L}}|_{C^{(2)}(G)}$ uniquely determines $(\widetilde{\mu_t}, t \geq 0)$, we find that $\mu_t = \widetilde{\mu_t}$ for all $t \geq 0$, as was required. □

Now suppose that $\mathcal{L}$ is a self-adjoint Hunt generator in $L^2(G, m)$. For each $f \in C_c^{(2)}(G)$ we write

$$\mathcal{L}f = \mathcal{L}_l f + \mathcal{L}_n f,$$

where $\mathcal{L}_l f = \frac{1}{2}\sum_{i,j=1}^d a_{ij} X_i X_j f$. We call $\mathcal{L}_l$ the *local part* of the Hunt generator while $\mathcal{L}_n$ is called the *non-local part*. We now motivate this terminology by examining the Dirichlet form $\mathcal{E}$ associated to $\mathcal{L}$ by the prescription

$$\mathcal{E}(f, g) := -\langle f, \mathcal{L}g\rangle,$$

for each $f, g \in \mathrm{Dom}(\mathcal{L})$. The next result gives a classical Beurling-Deny representation for $\mathcal{E}$ (see e.g. Theorem 3.2.1 in Fukushima et. al. [69]).

Theorem 5.4.2 *For each $f, g \in C_c^{(2)}(G)$,*

$$\mathcal{E}(f, g) = \mathcal{E}_l(f, g) + \mathcal{E}_n(f, g),$$

where

$$\mathcal{E}_l(f, g) := -\langle f, \mathcal{L}_l g\rangle = \frac{1}{2}\sum_{i,j=1}^d a_{ij}\int_G (X_i f)(\sigma)(X_j g)(\sigma)d\sigma$$

and

$$\mathcal{E}_n(f, g) := -\langle f, \mathcal{L}_n g\rangle = \frac{1}{2}\int_{(G\times G)} (f(\rho) - f(\sigma))(g(\rho) - g(\sigma))\mu(d\rho, d\sigma),$$

where $\mu(d\rho, d\sigma) := \nu(\sigma^{-1}d\rho)d\sigma$.

Proof The result for $\mathcal{E}_l$ is an easy consequence of the fact that the X_i's are skew-adjoint linear operators in $L^2(G)$. Let $(U_n, n \in \mathbb{N})$ be a decreasing sequence of Borel neighbourhoods of e such that $U_n \downarrow \{e\}$ as $n \to \infty$. By (5.4.16), for each $f, g \in C_c^{(2)}(G)$,

$$\begin{aligned}\mathcal{E}_n(f, g) &= -\frac{1}{2}\int_G\int_G f(\sigma)[g(\sigma\tau) + g(\sigma\tau^{-1}) - 2g(\sigma)]\nu(d\tau)d\sigma \\ &= \lim_{m\to\infty}\frac{1}{2}\int_G\int_{U_m^c} f(\sigma)[g(\sigma) - g(\sigma\tau) + g(\sigma) - g(\sigma\tau^{-1})]\nu(d\tau)d\sigma.\end{aligned}$$

Now use right invariance of Haar measure and Fubini's theorem to see that for each $m \in \mathbb{N}$, $\int_G\int_{G\setminus U_m} f(\sigma)g(\sigma)\nu(d\tau)d\sigma = \int_G\int_{G\setminus U_m} f(\sigma\tau)g(\sigma\tau)\nu(d\tau)d\sigma$ and

$\int_G \int_{G\setminus U_m} f(\sigma)g(\sigma\tau^{-1})\nu(d\tau)d\sigma = \int_G \int_{G\setminus U_m} f(\sigma\tau)g(\sigma)\nu(d\tau)d\sigma$. On passage to the limit we thus obtain

$$\mathcal{E}_n(f, g) = \frac{1}{2}\int_G \int_G (f(\sigma\tau) - f(\sigma))(g(\sigma\tau) - g(\sigma))\nu(d\tau)d\sigma,$$

and the result follows. □

5.4.2 Injectivity

If P_t is self-adjoint for all $t \geq 0$, it is easy to see that it is injective in $L^2(G)$. To verify this, fix $t \geq 0$ and assume that $f \in L^2(G)$ is such that $P_t f = 0$. Then

$$0 = \langle P_t f, f\rangle = ||P_{\frac{t}{2}} f||^2,$$

and so $P_{\frac{t}{2}} f = 0$. Iterating this argument yields $P_{\frac{t}{2^n}} f = 0$ for all $n \in \mathbb{N}$, and so $f = \lim_{n\to\infty} P_{\frac{t}{2^n}} f = 0$ by strong continuity. Of course this argument is quite general and doesn't make any use of the probabilistic context of these operators. The fact that P_ts associated to arbitrary convolution semigroups are injective is proved by Kunita [128] using a much more sophisticated argument.

5.4.3 Analyticity

If P_t is self-adjoint for all $t \geq 0$, then it has a spectral decomposition

$$P_t = \int_0^\infty e^{-t\lambda} E(d\lambda),$$

where $E(\cdot)$ is the projection-valued measure associated to the self-adjoint generator $\mathcal{L}$. It is then easy to see that the semigroup has an analytic extension to the half-plane $\{z \in \mathbb{C}; z = t + i\alpha, t \geq 0\}$ given by

$$P_{t+i\alpha} = \int_0^\infty e^{-(t+i\alpha)\lambda} E(d\lambda),$$

for each $t \geq 0, \alpha \in \mathbb{R}$. Moreover, it is straightforward to verify that $P_{t+i\alpha}$ is a contraction (see Stein [192], pp. 67–68). A finer result is due to Kunita [128], which we state w ithout proof. First we need some terminology. We regard the matrix a as a linear map from $\mathbb{R}^d$ to $\mathbb{R}^d$ with range $\text{Im}(a)$. Let

$$\Sigma_a := \left\{ \sigma \in G; \sigma = \exp\left(\sum_{i=1}^{d} y_i(\sigma) X_i \right), y_i(\sigma) \in \text{Im}(a), 1 \leq i \leq d \right\}.$$

Theorem 5.4.3 (Kunita) *If $b \in Im(a)$ and $supp(\nu - \widetilde{\nu}) \subseteq \Sigma_a$, then $(P_t, t \geq 0)$ is analytic.*

5.4.4 Trace Class

Finally we turn our attention to tracial properties of semigroups. We summarise our main result in the next theorem which was obtained in Applebaum [9]. We use without comment basic facts about trace class and Hilbert-Schmidt operators that can be found in Appendix A.6.

Theorem 5.4.4 *Let $(\mu_t, t \geq 0)$ be a convolution semigroup acting on a compact Lie group G and $(P_t, t \geq 0)$ be the associated Hunt semigroup acting in $L^2(G)$. The following are equivalent:*

1. *μ_t has a square integrable density for all $t > 0$.*
2. *P_t is Hilbert-Schmidt for all $t > 0$.*
3. *P_t is trace class for all $t > 0$.*

Proof The equivalence of (1) and (2) is an immediate consequence of Theorem 4.7.1. The equivalence of (2) and (3) is a consequence of general semigroup theory, indeed (3) $\Rightarrow$ (2) as the linear space of all trace class operators in $L^2(G)$ is a subspace of the linear space of all Hilbert-Schmidt operators. To see that (2) $\Rightarrow$ (3), choose an arbitrary $t > 0$, then write $P_t = P_{\frac{t}{2}} P_{\frac{t}{2}}$ and use the fact that the product of two Hilbert-Schmidt operators is trace-class. □

In the sequel, we will use Tr to denote the trace of a trace class operator acting in $L^2(G)$.

5.5 The Lévy-Khintchine Formula on Compact Lie Groups

The central result in the theory of infinitely divisible measures on $\mathbb{R}^d$ (and more generally real Banach spaces—see e.g. Linde [135]) is the celebrated Lévy-Khintchine formula. To some degree, the appropriate generalisation of this to Lie groups is Hunt's theorem (Theorem 5.3.3), but (as a corollary to this) we can also obtain a direct analogue of the Lévy-Khintchine formula. We present this result, for compact Lie groups only, again at the level of convolution semigroups. Observe that we hereby characterise the family of compatible matrices that appeared in (5.1.1). Readers may wish to recall basic facts about derived representations from Sect. 2.5.1.

Theorem 5.5.1 *Let $(\mu_t, t \geq 0)$ be a family of probability measures on a compact connected Lie group G with $\mu_0 = \delta_e$. Then $(\mu_t, t \geq 0)$ is a convolution semigroup if and only if for all $t \geq 0, \pi \in \widehat{G}$,*

$$\widehat{\mu_t}(\pi) = e^{t\mathcal{A}_\pi},$$

where

$$\mathcal{A}_\pi = \sum_{i=1}^{d} b_i d\pi(X_i) + \frac{1}{2}\sum_{j,k=1}^{d} a_{jk} d\pi(X_j) d\pi(X_k) + \int_G \left(\pi(\tau) - I_\pi - \sum_{i=1}^{d} x_i(\tau) d\pi(X_i)\right)\nu(d\tau), \tag{5.5.17}$$

(where b, a, ν and $x_i (1 \leq i \leq d)$ are as in Theorem 5.3.3).

Proof It follows from Theorem 2.5.1 that for all $1 \leq i, j \leq d_\pi$, $\pi_{ij} \in C^\infty(G) \subseteq C^{(2)}(G)$. In the following we will write $P_t\pi(\sigma)$ and $\mathcal{L}\pi(\sigma)$ for the matrices $(P_t\pi_{ij}(\sigma))$ and $(\mathcal{L}\pi_{ij}(\sigma))$ (respectively), where $\sigma \in G$.

First suppose that $(\mu_t, t \geq 0)$ is a convolution semigroup. Then we have

$$\begin{aligned}\frac{d}{dt}\widehat{\mu_t}(\pi) &= \frac{d}{dt}P_t\pi(e)\\ &= P_t\mathcal{L}\pi(e)\\ &= \int_G \mathcal{L}(\pi)(\sigma)\mu_t(d\sigma).\end{aligned}$$

We compute $\mathcal{L}(\pi)(\sigma)$ using (5.3.8). First observe that for each $X, Y \in \mathfrak{g}$:

$$\begin{aligned}X\pi(\sigma) &= \left.\frac{d}{du}\pi(\sigma \exp(uX))\right|_{u=0}\\ &= \pi(\sigma)\left.\frac{d}{du}\pi(\exp(uX))\right|_{u=0}\\ &= \pi(\sigma)d\pi(X),\end{aligned}$$

and by a similar argument,

$$\begin{aligned}XY\pi(\sigma) &= \left.\frac{\partial^2}{\partial u \partial v}\pi(\sigma \exp(uX)\exp(vY))\right|_{u=v=0}\\ &= \pi(\sigma)d\pi(X)d\pi(Y).\end{aligned}$$

We also note that for all $\tau \in G$,

$$\pi(\sigma\tau) - \pi(\tau) = \pi(\sigma)(\pi(\tau) - I_\pi).$$

Thus we conclude that

$$\mathcal{L}(\pi)(\sigma) = \pi(\sigma)\mathcal{A}_\pi,$$

and hence

$$\frac{d}{dt}\widehat{\widetilde{\mu_t}}(\pi) = \widehat{\widetilde{\mu_t}}(\pi)\mathcal{A}_\pi.$$

Solving the matrix-valued differential equation with the initial condition $\widehat{\widetilde{\mu_0}}(\pi) = I_\pi$ yields the required result.

The converse follows from Proposition 5.1.2. □

Note The use of $\widetilde{\mu}$ rather than μ is a consequence of our definition of the Fourier transform. If we had used μ, then the characteristics (b, a, ν) in (5.5.17) would have to be replaced by $(-b, a, \widetilde{\nu})$.

Corollary 5.5.1 *If G is a compact, connected Lie group then a probability measure μ on G is infinitely divisible if and only*

$$\widehat{\mu}(\pi) = e^{\mathcal{A}_\pi},$$

for all $\pi \in \widehat{G}$.

Proof This is a direct consequence of Theorems 5.5.1 and 5.2.1 using the fact that μ is infinitely divisible if and only if $\widetilde{\mu}$ is. □

We say that the convolution semigroup $(\mu_t, t \geq 0)$ has a *standard Gaussian component* if its characteristics are of the form (b, cI, ν) where $c > 0$. Let G be compact. In this case it follows easily from (5.5.17) that for all $\pi \in \widehat{G}, t \geq 0$

$$\widehat{\mu_t}(\pi) = e^{-ct\kappa_\pi}\widehat{\mu_t^{(0)}}(\pi),$$

where $(\mu_t^{(0)}, t \geq 0)$ is a convolution semigroup having characteristics $(b, 0, \nu)$. It then follows directly from Corollary 4.5.1 that μ_t has a smooth density for all $t > 0$.

We now turn our attention to the central case where we can derive a simple and elegant form of the Lévy-Khintchine formula for a rich class of Lie groups. We recall that $(\mu_t, t \geq 0)$ is a *central convolution semigroup* if μ_t is central for all $t \geq 0$. It follows from (5.3.12) that the Lévy measure of a central convolution semigroup is itself central.

We recall the group character $\chi_\pi(\cdot) := \text{tr}(\pi(\cdot))$, for each $\pi \in \widehat{G}$ and we define the normalised character $\varrho_\pi(\cdot) := \dfrac{1}{d_\pi}\chi_\pi(\cdot)$. The next result is due to Applebaum and Bañuelos [13]. The complex numbers $\{\alpha_\pi, \pi \in \widehat{G}\}$ which appear here are those defined in Theorem 5.1.2.

Theorem 5.5.2 *Let $(\mu_t, t \geq 0)$ be a central convolution semigroup on a compact, connected semi-simple Lie group.*

1. *If the Lévy measure ν has a finite first moment, i.e. $\int_G |x_i(\sigma)|\nu(d\sigma) < \infty$ for all $1 \leq i \leq n$, then for all $\pi \in \widehat{G}$,*

$$\alpha_\pi = -c\kappa_\pi + \int_G (\overline{\varrho_\pi(\tau)} - 1)\nu(d\tau), \tag{5.5.18}$$

where $c \geq 0$.

2. *If the central convolution semigroup is also symmetric, then for all $\pi \in \widehat{G}$,*

$$\alpha_\pi = -c\kappa_\pi + \int_G (\Re(\varrho_\pi(\tau)) - 1)\nu(d\tau),$$

where $c \geq 0$.

Proof

1. It is shown in Proposition 4.4 of Liao [132] p. 99 that under the given conditions we must have $a = cI$ for some $c \geq 0$. The argument of Proposition 4.5 in the same source shows that we must also have $b = 0$. If we assume that the Lévy measure ν has a finite first moment, then by Corollary 5.3.2 we have for all $f \in C_0^{(2)}(G), \sigma \in G$,

$$\mathcal{L}f(\sigma) = c\Delta f(\sigma) + \int_G (f(\sigma\tau) - f(\sigma))\nu(d\tau).$$

By Theorem 5.5.1, $\widehat{\mu_t}(\pi) = e^{t\mathcal{A}_\pi}$ for all $t \geq 0$, where

$$\mathcal{A}_\pi = -c\kappa_\pi I_\pi + \int_G (\pi(\tau) - I_\pi)\nu(d\tau).$$

Comparing this identity with (5.1.2) and using Theorem 4.2.1(4), we deduce that

$$\alpha_\pi I_\pi = -c\kappa_\pi I_\pi + \int_G (\pi(\tau^{-1}) - I_\pi)\nu(d\tau),$$

and we obtain the required result on taking the trace in V_π of both sides of this last equation, using the fact that $\chi_\pi(\tau^{-1}) = \overline{\chi_\pi(\tau)}$ for all $\tau \in G$.

2. In this case for all $\pi \in \widehat{G}$, $\alpha_\pi \in \mathbb{R}$. By (5.4.16) and the representation of $\mathcal{L}$ given in (1), we find that for all $\sigma \in G$, $f \in C_0^{(2)}(G)$ we have

$$\mathcal{L}f(\sigma) = c\Delta f(\sigma) + \frac{1}{2}\int_G (f(\sigma\tau) - 2f(\sigma) + f(\sigma\tau^{-1}))\nu(d\tau),$$

and so for all $\pi \in \widehat{G}$

$$\mathcal{L}(\pi) = -c\kappa_\pi I_\pi + \frac{1}{2}\int_G (\pi(\tau) - 2I_\pi + \pi(\tau^{-1}))\nu(d\tau).$$

The result then follows by the argument of (1). □

Remark It has been pointed out to the author by Ming Liao that the representation (5.5.18) holds without the constraint on the first moment if the integral is interpreted as a principal value.

5.6 Gaussian Measures and the Central Limit Theorem

A convolution semigroup $(\mu_t^G, t \geq 0)$ is said to be *Gaussian* if

$$\lim_{t \to 0} \frac{1}{t} \mu_t(V^c) = 0$$

for every neighbourhood V of e. It then follows from (5.3.12) that the Lévy measure $\nu \equiv 0$, and so the associated Hunt semigroup $(P_t^G, t \geq 0)$ has generator $\mathcal{L}_G$ whose action on $f \in C_0^{(2)}(G)$ is given by

$$\mathcal{L}_G f(\sigma) = \sum_{i=1}^{d} b_i X_i f(\sigma) + \frac{1}{2} a_{ij} X_i X_j f(\sigma)$$

for each $\sigma \in G$, so $\mathcal{L}_G$ is a *diffusion operator* with constant coefficients on G.

An important special case is obtained by taking $b_i = 0$ $(1 \leq i \leq d)$ and $a_{ij} = 2\delta_{ij}$ $(1 \leq i, j \leq d)$. If we also choose the basis $\{X_1, \ldots, X_d\}$ to be orthonormal with respect to some given left invariant Riemannian metric on G, then we have

$$\mathcal{L}_G = \sum_{i=1}^{d} X_i^2 = \Delta,$$

and in this case $(\mu_t^G, t \geq 0)$ is the standard Gaussian semigroup that we have already met several times in this book.

A measure $\rho \in \mathcal{P}(G)$ is said to be *Gaussian* if it can be embedded into some Gaussian convolution semigroup, i.e. there exists a Gaussian $(\mu_t^G, t \geq 0)$ for which $\rho = \mu_1$. Note that this precisely captures the usual notion of Gaussianity when $G = \mathbb{R}^d$.

Arguably the most important result concerning Gaussian measures on $\mathbb{R}^d$ is the central limit theorem; indeed this underpins a great deal of modelling in statistical mechanics and is the central plank of statistical inference. We give a statement and proof of this result which mainly relies on functional analytic techniques and which follows closely the presentation of Breuillard [35]. We begin with two lemmas giving some general results about bounded approximations to Gaussian semigroups and their generators and resolvents. The central limit theorem is a simple consequence of these. Our set up is as follows. Let $(\rho_n, n \in \mathbb{N})$ be a sequence of measures in $\mathcal{P}(G)$

for which there exists $b \in \mathbb{R}^d$ and a non-negative definite symmetric $d \times d$ matrix $a = (a_{ij})$ so that for all $1 \leq i, j \leq d$, the following hold as $n \to \infty$:

A(i) $\int_G x_i(\sigma)\rho_n(d\sigma) = \frac{b_i}{n} + o\left(\frac{1}{n}\right)$,

A(ii) $\int_G x_i(\sigma)x_j(\sigma)\rho_n(d\sigma) = \frac{a_{ij}}{n} + o\left(\frac{1}{n}\right)$,

A(iii) $\rho_n(V^c) = o\left(\frac{1}{n}\right)$, for every neighbourhood V of the identity.

For all $n \in \mathbb{N}$ we introduce the linear operator on $C_0(G)$ defined by

$$\mathcal{L}_n = n(P_{\rho_n} - I).$$

It is clear that $||\mathcal{L}_n||_\infty \leq 2n$. In fact for all $f \in C_0(G), \sigma \in G$ we have

$$(\mathcal{L}_n f)(\sigma) = \int_G (f(\sigma\tau) - f(\sigma))n\rho_n(d\tau),$$

and so $\mathcal{L}_n$ is itself the infinitesimal generator of a Hunt semigroup associated to a compound Poisson convolution semigroup. So for each $t \in \mathbb{R}$ we may define the operators

$$e^{t\mathcal{L}_n} := \sum_{m=0}^{\infty} \frac{t^m}{m!}\mathcal{L}_n^m.$$

Then $(e^{t\mathcal{L}_n}, t \in \mathbb{R})$ is a strongly continuous one-parameter group of bounded linear operators on $C_0(G)$ and $e^{t\mathcal{L}_n}$ is a contraction for each $t \geq 0$. It follows that 1 is in the resolvent set and we write the corresponding resolvent as $R_n = (I - \mathcal{L}_n)^{-1}$. We also write $R_G = (I - \mathcal{L}_G)^{-1}$. Note that R_G and R_n are contractions for all $n \in \mathbb{N}$.[2]

Lemma 5.6.1 *Under the assumptions A(i), A(ii) and A(iii), we have*

$$\lim_{n\to\infty} \mathcal{L}_n f = \mathcal{L}_G f,$$

for all $f \in C_c^\infty(G)$.

Proof By Taylor's theorem (Theorem 5.3.1), we can find a neighbourhood U of e so that for $\sigma \in G, \tau \in U$,

$$f(\sigma\tau) - f(\sigma) = \sum_{i=1}^{d} x_i(\tau)X_i f(\sigma) + \frac{1}{2}\sum_{i,j=1}^{d} x_i(\tau)x_j(\tau)X_iX_j f(\sigma\tau'),$$

[2] For basic facts about resolvents of semigroups, see e.g. Chap. 8 of Davies [54] pp. 210–227.

where $\tau' \in U$. Now

$$(\mathcal{L}_n f)(\sigma) = n \int_U (f(\sigma\tau) - f(\sigma))\rho_n(d\tau) + \Psi_n(\sigma),$$

where $(\Psi_n f)(\sigma) := n \int_{U^c}(f(\sigma\tau) - f(\sigma))\rho_n(d\tau)$. By assumption A(iii) we see that for all $\sigma \in G$,

$$||\Psi_n f||_\infty \leq 2n||f||_\infty \rho_n(U^c) \to 0 \text{ as } n \to \infty.$$

We also have (uniformly in $\sigma \in G$) by assumption A(i),

$$n \sum_{i=1}^d (X_i f)(\sigma) \int_U x_i(\tau)\rho_n(d\tau) \to \sum_{i=1}^d b_i X_i f(\sigma) \text{ as } n \to \infty,$$

and by assumption A(ii)

$$\frac{1}{2} n \sum_{i,j=1}^d (X_i X_j f)(\sigma) \int_U x_i(\tau) x_j(\tau)\rho_n(d\tau) \to \frac{1}{2} \sum_{i,j=1}^d a_{ij} X_i X_j f(\sigma) \text{ as } n \to \infty.$$

For each $\tau \in G$, let $X(\tau)f(\sigma) := \sum_{i=1}^d x_i(\tau)X_i f(\sigma)$. We are done if we can show that

$$n \int_U (X(\tau)^2 f(\sigma\tau') - X(\tau)^2 f(\sigma))\rho_n(d\tau) \to 0 \text{ as } n \to \infty,$$

uniformly in $\sigma \in G$. By Taylor's theorem again, there exists $\tau'' \in U$ so that

$$X(\tau)^2 f(\sigma\tau') - X(\tau)^2 f(\sigma) = X(\tau)^3 f(\sigma\tau''),$$

and so

$$\begin{aligned} |X(\tau)^2 f(\sigma\tau') - X(\tau)^2 f(\sigma)| &\leq ||X(\tau)^3 f||_\infty \\ &\leq \sum_{i,j,k=1}^d x_i(\tau)x_j(\tau)x_k(\tau)||X_i X_j X_k f||_\infty. \end{aligned}$$

Hence we see that we need only show that each of the terms $n \int_U x_i(\tau)x_j(\tau)x_k(\tau)\rho_n(d\tau)$ can be made sufficiently small as n increases. Let $\epsilon > 0$. By continuity, we can and will replace U with a neighbourhood of the identity V such that $V \subseteq U$ and

$$\max_{1 \leq i \leq d} \sup_{\tau \in V} |x_i(\tau)| < \frac{\epsilon}{\alpha},$$

where $\alpha := \max_{1\leq i\leq d}(1+a_{ii}) > 0$. Then by assumption A(ii), for all $1 \leq i, j, k \leq d$, as $n \to \infty$,

$$\begin{aligned} n\left|\int_V x_i(\tau)x_j(\tau)x_k(\tau)\rho_n(d\tau)\right| &< \frac{\epsilon}{\alpha}n\int_V |x_j(\tau)x_k(\tau)|\rho_n(d\tau) \\ &\leq \frac{\epsilon}{\alpha}\left(n\int_V x_j(\tau)^2\rho_n(d\tau)\right)^{\frac{1}{2}}\left(n\int_V x_k(\tau)^2\rho_n(d\tau)\right)^{\frac{1}{2}} \\ &\leq \frac{\epsilon}{\alpha}a_{jj}^{\frac{1}{2}}a_{kk}^{\frac{1}{2}} \\ &< \epsilon, \end{aligned}$$

and the result follows. □

Lemma 5.6.2 *Under the assumptions A(i), A(ii) and A(iii), we have for all $f \in C_0(G)$,*

1. $\lim_{n\to\infty} R_n f = R_G f$,
2. $\lim_{n\to\infty} e^{t\mathcal{L}_n} f = P_t^G f$ *for all* $t \geq 0$,
3. $\lim_{n\to\infty} ||P_{\rho_n^{*n}} f - e^{\mathcal{L}_n} f||_\infty = 0$.

Proof

1. As R_n (for all $n \in \mathbb{N}$) and R_G are bounded operators, it suffices to prove the result on a dense subspace of $C_0(G)$. By Theorem 5.3.4 and Theorem A.8.1 (in Appendix A.8), we see that $(I - \mathcal{L}_G)C_c^\infty(G)$ is dense in $C_0(G)$. By Lemma 5.6.1, for all $f \in C_c^\infty(G)$ we have

$$(I - \mathcal{L}_n)f \to (I - \mathcal{L}_G)f$$

as $n \to \infty$. Applying R_n and taking limits we deduce that

$$\lim_{n\to\infty} R_n(I - \mathcal{L}_G)f = f.$$

Now put $g = (I - \mathcal{L}_G)f$ to conclude that $\lim_{n\to\infty} R_n g = R_G g$, as required.

2. For all $t \geq 0$, $n \in \mathbb{N}$, $f \in \text{Dom}(\mathcal{L}_G)$, we have

$$\begin{aligned} P_t^G f - e^{t\mathcal{L}_n} f &= \int_0^t \frac{d}{ds}(e^{(t-s)\mathcal{L}_n} P_s^G) f ds \\ &= e^{t\mathcal{L}_n}\int_0^t e^{-s\mathcal{L}_n}(\mathcal{L}_G - \mathcal{L}_n)P_s^G f ds. \end{aligned}$$

Since $R_G : C_0(G) \to \text{Dom}(\mathcal{L}_G)$ is surjective, we can write $f = R_G g$ for some $g \in C_0(G)$. Now using the fact that R_n is bounded, that $R_n e^{t\mathcal{L}_n} = e^{t\mathcal{L}_n} R_n$ for all $n \in \mathbb{N}$ and that $P_t^G R_G = R_G P_t^G$ for all $t \geq 0$, we find that

$$\begin{aligned} R_n(P_t^G - e^{t\mathcal{L}_n})R_G g &= e^{t\mathcal{L}_n}\int_0^t e^{-s\mathcal{L}_n}R_n(\mathcal{L}_G - \mathcal{L}_n)R_G P_s^G g ds \\ &= e^{t\mathcal{L}_n}\int_0^t e^{-s\mathcal{L}_n}(R_G - R_n)P_s^G g ds, \end{aligned}$$

since $R_n(\mathcal{L}_G - \mathcal{L}_n)R_G = -R_n((I - \mathcal{L}_G) - (I - \mathcal{L}_n))R_G = R_G - R_n$. Using the result of (1) and dominated convergence, we deduce that

$$\lim_{n\to\infty} R_n(P_t^G - e^{t\mathcal{L}_n})R_G g = 0.$$

Then by a density argument we see that $\lim_{n\to\infty} R_n(P_t^G - e^{t\mathcal{L}_n})f = 0$ for all $f \in C_0(G)$. But

$$R_n(P_t^G - e^{t\mathcal{L}_n}) = e^{t\mathcal{L}_n}(R_G - R_n) - (R_G - R_n)P_t^G + (P_t^G - e^{t\mathcal{L}_n})R_G,$$

and so by another use of (1), we deduce that $\lim_{n\to\infty}(e^{t\mathcal{L}_n} - P_t^G)R_G f = 0$, and the result follows by another density argument.

3. For all $n \in \mathbb{N}$ we have the following telescopic sum of bounded linear operators:

$$P_{\rho_n^{*n}} - e^{\mathcal{L}_n} = (P_{\rho_n} - e^{\frac{\mathcal{L}_n}{n}})\sum_{k=0}^{n-1} P_{\rho_n^{*(n-1-k)}} e^{\frac{k\mathcal{L}_n}{n}}.$$

Then for each $f \in C_0(G)$ we have

$$||(P_{\rho_n^{*n}} - e^{\mathcal{L}_n})f||_\infty \leq n||(P_{\rho_n} - e^{\frac{\mathcal{L}_n}{n}})f||_\infty.$$

However (by definition of $\mathcal{L}_n$), $P_{\rho_n}f = f + \frac{1}{n}\mathcal{L}_n f$ and so

$$(P_{\rho_n} - e^{\frac{\mathcal{L}_n}{n}})f = \sum_{k=0}^{\infty}\frac{1}{(k+2)!}\left(\frac{\mathcal{L}_n}{n}\right)^k\left(\frac{\mathcal{L}_n}{n}\right)^2 f.$$

Using the fact that $||\mathcal{L}_n|| \leq 2n$, we then find that

$$||(P_{\rho_n} - e^{\frac{\mathcal{L}_n}{n}})f||_\infty \leq e^2\frac{||\mathcal{L}_n^2 f||_\infty}{n^2}.$$

Now replace f by $R_n^2 f$. Then

$$||\mathcal{L}_n^2 R_n^2||_\infty \leq ||\mathcal{L}_n R_n||_\infty^2 = ||I - R_n||^2 \leq 4$$

since R_n is a contraction. Hence

$$\lim_{n\to\infty} n||(P_{\rho_n} - e^{\frac{\mathcal{L}_n}{n}})R_n^2 f||_\infty = 0,$$

and so

$$\lim_{n\to\infty} ||(P_{\rho_n^{*n}} - e^{\mathcal{L}_n})R_n^2 f||_\infty = 0.$$

But by (1) we have $R_n^2 f \to R_G^2 f$ as $n \to \infty$, and the range of R_G^2 is dense in $C_0(G)$. The required result follows. □

Theorem 5.6.1 (Central Limit Theorem) *If* $(\rho_n, n \in \mathbb{N})$ *is a sequence of Borel probability measures on G for which assumptions A(i), A(ii) and A(iii) are satisfied, then the sequence* $(\rho_n^{*n}, n \in \mathbb{N})$ *converges vaguely as* $n \to \infty$ *to a Gaussian measure on G.*

Proof This follows directly from Lemma 5.6.2 (ii) and (iii), since for all $f \in C_0(G)$ we have

$$\begin{aligned}\left|\int_G f(\sigma)\rho_n^{*n}(d\sigma) - \int_G f(\sigma)\mu_1(d\sigma)\right| &= |P_{\rho_n^{*n}} f(e) - P_1^G f(e)| \\ &\leq ||P_{\rho_n^{*n}} f - P_1^G f||_\infty \to 0 \text{ as } n \to \infty.\end{aligned}$$ □

We say that a Gaussian convolution semigroup $(\mu_t^G, t \geq 0)$ is *non-degenerate* if the matrix a is positive definite (so that $\det(a) > 0$). In this case we can always find a $d \times d$ matrix c so that $c^T c = a$. Then the operator

$$\mathcal{L}_G := \sum_{i,j=1}^d a_{ij} X_i X_j = \sum_{i=1}^d Y_i^2,$$

where $Y_i = \sum_{j=1}^d c_{ij} X_j$ for $1 \leq i \leq d$. This is the well-known *elliptic* case, where standard pde methods can be applied to show that μ_t^G has a smooth density for all $t > 0$ (see e.g. Treves [205], Chap. 36, pp. 347–353).

In the degenerate case where a is non-negative definite with $\det(a) = 0$, we can still find an $m \times d$ matrix f so that $f^T f = a$ and then $\mathcal{L}_G = \sum_{i=1}^m Z_i^2$ where $Z_i = \sum_{j=1}^d f_{ij} X_j$ for $1 \leq i \leq d$. The operator $\mathcal{L}_G$ is said to be *hypo-elliptic* if the Lie algebra generated by $\{Z_1, \ldots, Z_m\}$ is the whole of $\mathfrak{g}$. If hypo-ellipticity holds, then it is a consequence of Hörmander's theorem (Hörmander [101]) that μ_t^G has a smooth density for all $t > 0$.

5.7 Subordination

Subordination is a key technique that enables us to construct new and interesting examples of convolution semigroups. First we introduce a *subordinator*. This is precisely a convolution semigroup $(\rho_t, t \geq 0)$ of probability measures on $\mathbb{R}$ wherein each $\text{supp}(\rho_t) \subseteq [0, \infty)$. Recall that $h \in C^\infty((0, \infty))$ is called a *Bernstein function* if $h \geq 0$ and $(-1)^n h^{(n)} \leq 0$ for all $n \in \mathbb{N}$. Every Bernstein function has the representation

$$h(x) = a + bx + \int_{(0,\infty)} (1 - e^{-ux})\lambda(du), \tag{5.7.19}$$

for all $x \in (0, \infty)$ where $a, b \geq 0$ and λ is a Borel measure on $(0, \infty)$ for which $\int_{(0,\infty)} (1 \wedge x)\lambda(dx) < \infty$. To every subordinator $(\rho_t, t \geq 0)$ there corresponds exactly one Bernstein function h with $\lim_{t\to\infty} h(t) = 0$ (and so $a = 0$ in (5.7.19)) given by

$$\int_{(0,\infty)} e^{-xy} \rho_t(dy) = e^{-th(x)} \tag{5.7.20}$$

for all $x > 0, t \geq 0$ (see e.g. Schilling et al. [178] for details). From now on we will find it convenient to write $(\rho_t^h, t \geq 0)$ for the subordinator whose Bernstein function (as given in (5.7.20)) is h.

Now let $(\mu_t, t \geq 0)$ be a convolution semigroup on a Lie group G and $(\rho_t^h, t \geq 0)$ be a subordinator with Bernstein function h. By the Riesz representation theorem, for each $t \geq 0$ there exists a Borel probability measure μ_t^h on G so that for all $f \in C_c(G)$,

$$\int_G f(\sigma)\mu_t^h(d\sigma) = \int_{(0,\infty)} \int_G f(\sigma)\mu_s(d\sigma)\rho_t^h(ds). \tag{5.7.21}$$

The relationship between the three families of measures is frequently expressed using the *vague integral*

$$\mu_t^h(A) = \int_{(0,\infty)} \mu_s(A)\rho_t^h(ds) \tag{5.7.22}$$

for all $t \geq 0$, $A \in \mathcal{B}(G)$, where the meaning of (5.7.22) is as described in (5.7.21).

Theorem 5.7.1 *$(\mu_t^h, t \geq 0)$ is a convolution semigroup of measures on G.*

Proof We verify the conditions (i), (ii)* and (iii) from the definition of convolution semigroups given in Sect. 5.1.

(i) For all $s, t \geq 0$, $f \in C_c(G)$, using Fubini's theorem we have

$$\begin{aligned}\int_G f(\sigma)\mu_{s+t}^h(d\sigma) &= \int_{(0,\infty)} \int_G f(\sigma)\mu_r(d\sigma)\rho_{s+t}^h(dr) \\ &= \int_{(0,\infty)} \int_{(0,\infty)} \int_G f(\sigma)\mu_{r+u}(d\sigma)\rho_s^h(dr)\rho_t^h(du)\end{aligned}$$

$$= \int_{(0,\infty)} \int_{(0,\infty)} \int_G \int_G f(\sigma\tau)\mu_r(d\sigma)\mu_u(d\tau)\rho_s^h(dr)\rho_t^h(du)$$
$$= \int_{(0,\infty)} \int_G \left(\int_{(0,\infty)} \int_G (R_\tau f)(\sigma)\mu_r(d\sigma)\rho_s^h(dr) \right) \mu_u(d\tau)\rho_t^h(du)$$
$$= \int_G \int_G f(\sigma\tau)\mu_s^h(d\sigma)\mu_t^h(d\tau)$$
$$= \int_G f(\sigma)(\mu_s^h * \mu_t^h)(d\sigma),$$

and so we see that $\mu_{s+t}^h = \mu_s^h * \mu_t^h$.
(ii) For all $f \in C_b(G)$,

$$\lim_{t\to 0} \int_G f(\sigma)\mu_t^h(d\sigma) = \lim_{t\to 0} \int_{(0,\infty)} \left(\int_G f(\sigma)\mu_s(d\sigma) \right) \rho_t^h(ds)$$
$$= \int_G f(\sigma)\mu_0(d\sigma) = f(e).$$

(iii) is obvious. □

Corollary 5.7.1 *1. If $(\mu_t, t \geq 0)$ is a symmetric convolution semigroup of measures on G, then so is $(\mu_t^h, t \geq 0)$.*
2. If $(\mu_t, t \geq 0)$ is a central convolution semigroup of measures on G, then so is $(\mu_t^h, t \geq 0)$.

Proof Immediate from (5.7.22) (or use (5.7.21)). □

Let $(P_t, t \geq 0)$ be the usual Hunt semigroup corresponding to $(\mu_t, t \geq 0)$ and let $(P_t^h, t \geq 0)$ be the semigroup corresponding to $(\mu_t^h, t \geq 0)$. Then we have

Proposition 5.7.1 *For all $t \geq 0$, $f \in C_0(G)$, $\sigma \in G$,*

$$P_t^h f(\sigma) = \int_{(0,\infty)} P_s f(\sigma)\rho_t^h(ds).$$

Proof This essentially follows from (5.7.21), since

$$P_t^h f(\sigma) = \int_G f(\sigma\tau)\mu_t^h(d\tau)$$
$$= \int_{(0,\infty)} \int_G f(\sigma\tau)\mu_s(d\tau)\rho_t^h(ds)$$
$$= \int_{(0,\infty)} P_s f(\sigma)\rho_t^h(ds),$$

as required. □

If G is compact, then for all $t \geq 0$, $\pi \in \widehat{G}$, $1 \leq i, j \leq d_\pi$ we have

$$\begin{aligned}\widehat{\mu_t^h}(\pi)_{ij} &= \int_G \pi_{ij}(\tau^{-1})\mu_t^h(d\tau) \\ &= \int_{(0,\infty)}\int_G \pi_{ij}(\tau^{-1})\mu_s(d\tau)\rho_t^h(ds) \\ &= \int_{(0,\infty)} \widehat{\mu_s}(\pi)_{ij}\rho_t^h(ds). \end{aligned} \tag{5.7.23}$$

The equation is particularly valuable when $(\mu_t, t \geq 0)$ is central. In that case, for all $\pi \in G$ there exist continuous functions $c_\pi : \mathbb{R}^+ \to \mathbb{C}$ and $c_\pi^h : \mathbb{R}^+ \to \mathbb{C}$ so that for all $t \geq 0$, $\widehat{\mu_t}(\pi) = c_\pi(t)I_\pi$ and $\widehat{\mu_t^h}(\pi) = c_\pi^h(t)I_\pi$. Then (5.7.23) tells us that

$$c_\pi^h(t) = \int_{(0,\infty)} c_\pi(s)\rho_t^h(ds). \tag{5.7.24}$$

For many important applications, we will take $(\mu_t, t \geq 0)$ to be a standard Gaussian convolution semigroup, so that for all $\pi \in \widehat{G}, t \geq 0, c_\pi(t) = e^{-t\kappa_\pi}$. Substituting into (5.7.24) and using (5.7.20), we then find that

$$c_\pi^h(t) = e^{-th(\kappa_\pi)}. \tag{5.7.25}$$

We now explore these results at the semigroup level. By standard functional calculus, $h(-\Delta)$ is a positive self adjoint operator in $L^2(G)$ with dense domain $\{f \in L^2(G), ||h(-\Delta)f|| < \infty\}$ (c.f. Sect. 3.1.1). By functional calculus again the operators $Q_t^h := e^{-th(-\Delta)}$, where $t \geq 0$, form a self-adjoint C_0-semigroup of contractions in $L^2(G)$.

Theorem 5.7.2 *If G is a compact Lie group, then for all $t \geq 0$, $P_t^h = Q_t^h$.*

Proof It follows from standard spectral theory that for all $t \geq 0, \pi \in \widehat{G}, 1 \leq i, j \leq d_\pi$,

$$Q_t^h \pi_{ij} = e^{-th(\kappa_\pi)}\pi_{ij}.$$

But by Theorem 4.7.2 and (5.7.24)

$$P_t^h \pi_{ij} = e^{-th(\kappa_\pi)}\pi_{ij}.$$

The result now follows by a straightforward density argument, using the form of the Peter-Weyl theorem given in Corollary 2.2.4. □

It follows that the generator $\mathcal{L}^h$ of the Hunt semigroup associated to the subordinated convolution semigroup $(\mu_t^h, t \geq 0)$ is $-h(-\Delta)$. The result of Theorem 5.7.2 also holds on general Lie groups, where it is a consequence of Phillip's theorem (see e.g. Sato [177], Theorem 32.1 pp. 212–217).

Now suppose that $(\phi(t), t \geq 0)$ is a Lévy process on the group G so that $\phi(t)$ has law μ_t for all $t \geq 0$. A subordinating process $(T(t), t \geq 0)$ is a real-valued Lévy process that is almost surely non-decreasing. Then for each $t \geq 0$, $T(t)$ takes values (almost surely) in $[0, \infty)$ and its law at time t is precisely $\rho^h(t)$, where the Bernstein function h is such that $\mathbb{E}(e^{uT(t)}) = e^{-th(u)}$ for all $t \geq 0, u > 0$. If the processes $(\phi(t), t \geq 0)$ and $(T(t), t \geq 0)$ are defined on the same probability space and are independent, then the subordinated process $(\phi_T(t), t \geq 0)$ is a Lévy process on G where $\phi_T(t) := \phi(T(t))$ for all $t \geq 0$. Furthermore the law of $\phi_T(t)$ is $\mu^h(t)$ for each $t \geq 0$ (see e.g. Applebaum [6]).

5.8 Examples of Subordination and Regularity of Densities

Let $(\mu_t, t \geq 0)$ be a convolution semigroup of measures on G and $(\rho_t^h, t \geq 0)$ be a subordinator with associated Bernstein function h. Let $(\mu_t^h, t \geq 0)$ be the associated subordinated convolution semigroup. It follows easily from (5.7.21) and Fubini's theorem that if μ_t has a density g_t for all $t > 0$, then μ_t^h has a density g_t^h for all $t > 0$ and for all $\sigma \in G$,

$$g_t^h(\sigma) = \int_{(0,\infty)} g_s(\sigma)\rho_t^h(ds). \tag{5.8.26}$$

In particular this always holds when $(\mu_t, t \geq 0)$ is a standard Gaussian semigroup. Indeed in that case we can easily deduce that the mapping g_t^h is central. From now on in this section we will always make this Gaussian assumption. We will also assume that G is compact and connected so that we can apply the machinery developed in earlier chapters to investigate the regularity of g_t^h (see also Applebaum [10]).

Example 1 Stable-Type Convolution Semigroups

In this case we take $h(s) = s^\alpha$ for $s > 0$ where $0 < \alpha < 1$. Thus the Hunt semigroup $(P_t^h, t \geq 0)$ has infinitesimal generator $-(-\Delta)^\alpha$. This case is of particular interest for applications when $G = \mathbb{R}^d$. In that case the corresponding subordinated Lévy process enjoys a self-similarity property that is useful for mathematical modelling. On Lie groups it is known that such a self-similarity property can hold if and only if the group G is nilpotent and simply connected. This implies that as a manifold it is diffeomorphic to Euclidean space. For details and developments in this case see Theorem 2.2 in Applebaum [6], Cohen [49] and Kunita [125–127].

The compact case cannot provide the requisite structure for self-similarity, which is why we use the (perhaps overworked) terminology "stable-type" rather than "stable" in this case. Note that by (5.7.25), for all $\pi \in \widehat{G}$,

$$c_\pi^h(t) = e^{-t(\kappa_\pi)^\alpha}. \tag{5.8.27}$$

Proposition 5.8.1 *If $0 < \alpha < 1$, then the α-stable type subordinated measure μ_t^h has a smooth density for all $t > 0$.*

Proof By Theorem 4.5.3 it is sufficient to check that the Fourier transform (as given by (5.8.27)) lies in Sugiura space. Using Corollaries 2.5.1 and 2.5.2 we find that for all $k \in \mathbb{N}$

$$\begin{aligned}\limsup_{|\lambda|\to\infty} |\lambda|^k |||\widehat{\mu_t^h}(\lambda)|||_{HS} &= \limsup_{|\lambda|\to\infty} |\lambda|^k d_\lambda^{\frac{1}{2}} e^{-t(\kappa_\pi)^\alpha} \\ &\leq C \limsup_{|\lambda|\to\infty} |\lambda|^{k+\frac{m}{2}} e^{-t|\lambda|^{2\alpha}} \\ &= 0,\end{aligned}$$

and the result follows. □

Example 2 The Relativistic Schrödinger Convolution Semigroup

In this case our Bernstein function is $h(s) = \sqrt{s+m^2} - m$ for $s > 0$, where the parameter $m > 0$ (c.f. Applebaum [7] pp. 40–42). The interest in this case derives from the fact that when $G = \mathbb{R}^d$, the generator $\mathcal{L}^h = -\sqrt{-\Delta + m^2 I} - mI$ is the relativistic Schrödinger operator which has important applications to the stability of matter (see e.g. Lieb and Yau [134]). In order to study the density on a compact connected Lie group G, we easily compute

$$c_\pi^h(t) = e^{t(m - \sqrt{\kappa_\pi + m^2})},$$

and we may then use similar calculations to those in the proof of Proposition 5.8.1 to see that the measures have a smooth density for all $t > 0$.

Example 3 The Student t-Distribution

Here we follow the research report by Hurst [104] (but see also Seneta [182]). The Student t-distribution is very well known on the real line, where it plays an important role in statistical inference. In that context it has density

$$g_\nu(x) = \frac{\Gamma(\frac{1}{2}(1+\nu))}{\sqrt{\nu\pi}\,\Gamma(\frac{\nu}{2})}\left(1 + \frac{x^2}{\nu}\right)^{-\frac{1}{2}(1+\nu)},$$

for $x \in \mathbb{R}$, where $\nu \in \mathbb{N}$ indicates the degrees of freedom. The corresponding convolution semigroup of measures can be obtained by subordinating a standard Gaussian with a generalised inverse Gaussian subordinator. We omit the details and simply write down the characteristic function for the distribution on the real line as this is the main object of interest. Indeed (see Hurst [104] and also p. 186 of Seneta [182] and references therein), we have

$$\int_{\mathbb{R}} e^{-iux} g_\nu(x)dx = \frac{K_{\frac{\nu}{2}}(\sqrt{\nu}|x|)(\sqrt{\nu}|x|)^{\frac{\nu}{2}}}{\Gamma(\frac{\nu}{2})2^{\frac{\nu}{2}-1}},$$

where for $\lambda \in \mathbb{R}$, K_λ denote a Bessel function of the third kind. It then follows that on a compact, connected Lie group G, the distribution of interest μ^ν is that for which for all $\pi \in \widehat{G}$, $\widehat{\mu^\nu}(\pi) = \alpha_\pi^\nu I_\pi$, where

$$\alpha_\pi^\nu = \frac{K_{\frac{\nu}{2}}(\sqrt{\nu\kappa_\pi})(\sqrt{\nu\kappa_\pi})^{\frac{\nu}{2}}}{\Gamma(\frac{\nu}{2})2^{\frac{\nu}{2}-1}}. \tag{5.8.28}$$

Proposition 5.8.2 *The Student t-distribution on a compact Lie group has a C^∞ density.*

Proof We use the fact that as $|x| \to \infty$, $K_\lambda(x) \sim \sqrt{\frac{\pi}{2x}}e^{-x}$ (see Watson [215], Sect. A.3), and so as $\kappa_\pi \to \infty$,

$$\alpha_\pi^\nu \sim \frac{\sqrt{\pi}}{\Gamma(\frac{\nu}{2})2^{\frac{1}{2}(\nu-1)}} \sqrt{\nu\kappa_\pi}^{\frac{1}{2}(\nu-1)} e^{-\sqrt{\nu\kappa_\pi}}.$$

Then following the argument of Proposition 5.8.1 we can assert that there exists $L > 0$ so that for all $k \in \mathbb{N}$:

$$\begin{aligned}\limsup_{|\lambda|\to\infty} |\lambda|^k |||\widehat{\mu^\nu}(\lambda)|||_{HS} &\leq L\frac{\sqrt{\pi}}{\Gamma(\frac{\nu}{2})2^{\frac{1}{2}(\nu-1)}} |\lambda|^{k+\frac{m}{2}}\nu^{\frac{1}{4}(\nu-1)}(1+|\lambda|)^{\frac{1}{4}(\nu-1)}e^{-\sqrt{\nu}|\lambda|} \\ &\to 0.\end{aligned}$$

□

Example 4 The Laplace Distribution

In this case (see [12]) we take $h(s) = \log(1+\beta^2 s)$, where $s > 0$ and the parameter $\beta > 0$. We then obtain the Laplace (or double-exponential distribution) μ_β at time 1 and check that in this case for all $\pi \in \widehat{G}$,

$$\widehat{\mu_\beta}(\pi) = (1+\beta^2\kappa_\pi)^{-1} I_\pi$$

This distribution is important because of applications to statistical inference on compact Lie groups (see Chap. 6). However, we can see quite easily that $\widehat{\mu_\beta}$ is not an element of Sugiura space, and so the distribution fails to have a C^∞ density. We can check that it has an L^2-density for some low values of d. To see this, we apply Theorem 4.5.1 and compute

$$\sum_{\pi\in\widehat{G}-\{\pi_0\}} \frac{d_\pi^2}{(1+\beta^2\kappa_\pi)^2} \leq \frac{N}{\beta^2} \sum_{\lambda\in D-\{0\}} \frac{|\lambda|^{2m}}{|\lambda|^4} = \frac{N}{\beta^2}\zeta(2-m),$$

where $N > 0$.

By Sugiura's convergence result (Theorem 3.2.1) for ζ, we see that a sufficient condition for convergence is $m < 2 - \frac{r}{2}$. Hence $d \in \{1, 2, 3\}$. So for example, the Laplace distribution has a square-integrable density on the groups SO(3), SU(2) and Sp(1), each of which has dimension 3 and rank 1.

5.9 Small Time Asymptotics for Subordinated Gaussian Semigroups

We begin with some motivation. A Lévy process on a Lie group $(\phi(t), t \geq 0)$ is in particular a (homogeneous) Markov process and so we may consider its *transition probability* $P(\phi(t) \in A|\phi(0) = \sigma)$ for $A \in \mathcal{B}(G), \sigma \in G, t \geq 0$. In general, a time homogeneous G-valued Markov process $(Y(t), t \geq 0)$ has a *transition density* if there exists $\rho : G \times G \times (0, \infty) \to \mathbb{R}^+$ so that

$$P(Y(t) \in A|Y(0) = \sigma) = \int_A \rho(\sigma, \tau, t) d\tau,$$

for all $A \in \mathcal{B}(G), \sigma \in G, t > 0$.

Proposition 5.9.1 *If $(\phi(t), t \geq 0)$ is a Lévy process for which $\phi(t)$ has a density g_t for all $t > 0$, then it has a transition density ρ and*

$$\rho(\sigma, \tau, t) = g_t(\sigma^{-1}\tau),$$

for all $t > 0, \sigma, \tau \in G$.

Proof For each $\sigma \in G$ we have

$$\begin{aligned} P(\phi(t) \in A|\phi(0) = \sigma) &= P(\sigma\phi(t) \in A) \\ &= P(\phi(t) \in \sigma^{-1}A) \\ &= \int_{\sigma^{-1}A} g_t(\tau) d\tau \\ &= \int_A g_t(\sigma^{-1}\tau) d\tau, \end{aligned}$$

and the result follows. □

Note that in particular, for all $\sigma \in G$, $\rho(\sigma, \sigma, t) = g_t(e)$ is the density of transitions from a point on the group back to the same point in time t. The next result show that for at least one interesting class of processes, this quantity is closely related to the trace of the associated semigroup of operators. It was established in Applebaum [10] using the proof given below, but it can also be obtained as a consequence of Mercer's theorem (see Proposition 5.6.9 in Davies [54], pp. 156–157). In fact Mercer's theorem is more general, as it doesn't require the uniform convergence assumption that we will impose in order to prove the result using Fourier series.

Theorem 5.9.1 *Let $(\mu_t, t \geq 0)$ be a central symmetric convolution semigroup on a compact Lie group G for which μ_t has a continuous density g_t for all $t > 0$. Then for all $t > 0$,*

$$Tr(P_t) = g_t(e).$$

Proof To provide a swift Fourier-analytic proof, we make the assumption that g_t has a uniformly convergent Fourier series expansion as in Proposition 4.5.1. If g_t is continuous it is square-integrable, and hence P_t is trace class by Theorem 5.4.4. By Theorem 5.1.2, for each $\pi \in \widehat{G}$ there exists $\alpha_\pi \in \mathbb{R}$ so that $\widehat{\mu_t}(\pi) = e^{t\alpha_\pi} I_\pi$. Then by Theorem 4.7.3 we have $P_t \pi_{ij} = e^{t\alpha_\pi} \pi_{ij}$ for all $1 \leq i, j \leq d_\pi$. Since the multiplicity of the eigenvalue $e^{t\alpha_\pi}$ is d_π^2 in the eigenspace V_π, we deduce that

$$\mathrm{Tr}(P_t) = \sum_{\pi \in \widehat{G}} d_\pi^2 e^{t\alpha_\pi}.$$

On the other hand, by the Fourier expansion (2.3.7) and (4.5.11), for all $\sigma \in G$

$$\begin{aligned} g_t(\sigma) &= \sum_{\pi \in \widehat{G}} d_\pi \mathrm{tr}(\widehat{\mu_t}(\pi)\pi(\sigma)) \\ &= \sum_{\pi \in \widehat{G}} d_\pi e^{t\alpha_\pi} \chi_\pi(\sigma), \end{aligned}$$

and so, since $\chi_\pi(e) = d_\pi$, $g_t(e) = \sum_{\pi \in \widehat{G}} d_\pi^2 e^{t\alpha_\pi}$ and the result follows. □

The material that follows is based on work by Bañuelos and Baudoin [18] (but note that they work on manifolds and use a more general class of Bernstein functions). It gives a positive solution to a conjecture posed by the author in Applebaum [10], where it is verified in some examples. Let G be a compact connected Lie group that is equipped with some Riemannian metric, and h be the Bernstein function for which $h(s) = Cs^\alpha$ for all $s > 0$, where $C > 0$ and $0 < \alpha < 1$. Note that for all $t > 0$, the measure μ_t^h has a C^∞ density k_t^h by the argument of the proof of Proposition 5.8.1. Consider the subordinated semigroup $(P_t^h, t \geq 0)$ whose generator is $-h(-\Delta)$. Recall that $-\Delta$ has eigenvalues $\{\kappa_\pi, \pi \in \widehat{G}\}$ and rewrite this spectrum as $\{\kappa_n, n \in \mathbb{N}\}$, where $\kappa_{n+1} \geq \kappa_n$ for all $n \in \mathbb{N}$. Then the spectrum of $h(-\Delta)$ is $\{h(\kappa_n), n \in \mathbb{N}\}$.

For $\kappa > 0$, let $N(\kappa) := \#\{i \in \mathbb{N}; \kappa_i \leq \kappa\}$. Then a classic result due to Hermann Weyl (see e.g. Chavel [44], pp. 31–32) tells us that as $\kappa \to \infty$

$$N(\kappa) \sim \frac{\mathrm{Vol}(G)}{\Gamma(\frac{d}{2}+1)(4\pi)^{\frac{d}{2}}} \kappa^{\frac{d}{2}}, \tag{5.9.29}$$

where Vol(G) is the Riemannian volume of G (an explicit formula for this quantity is given in Theorem 12.7 on page 109 of Fegan [66]).

For $\kappa > 0$ define $N_h(\kappa) := \#\{i \in \mathbb{N}; h(\kappa_i) \leq \kappa\}$. Then using (5.9.29) we see that as $\kappa \to \infty$,

$$\begin{aligned} N_h(\kappa) &\sim \# \left\{ i \in \mathbb{N}; \kappa_i \leq \left(\frac{\kappa}{C}\right)^{\frac{1}{\alpha}} \right\} \\ &\sim \frac{\text{Vol}(G)}{\Gamma(\frac{d}{2}+1)(4\pi)^{\frac{d}{2}}} \left(\frac{\kappa}{C}\right)^{\frac{d}{2\alpha}}. \end{aligned} \tag{5.9.30}$$

By Theorem 5.9.1 (or Mercer's theorem) for all $t > 0$,

$$k_t^h(e) = \text{Tr}(P_t^h) = \int_0^\infty e^{-th(\kappa)} dN_h(\kappa), \tag{5.9.31}$$

(see Kac [111] and Blumenthal and Getoor [32] for pioneering use of this technique).

Now we recall Karamata's Tauberian theorem (see e.g. Bingham et al. [30], pp. 37–38), which tells us that given a non-negative measurable function F defined on $\mathbb{R}^+$, we have

$$F(x) \sim \frac{M}{\Gamma(1+\rho)} x^\rho$$

as $x \to \infty$ (where $M, \rho > 0$) if and only if

$$\widehat{F}(s) \sim M s^{-\rho}$$

as $s \to \infty$, where

$$\widehat{F}(s) := \int_0^\infty e^{-sx} dF(x)$$

is the Laplace transform. From (5.9.31) and the asymptotics of the spectrum (5.9.30) we then deduce the short time asymptotics of the density at the neutral element as $t \downarrow 0$:

$$k_t^h(e) \sim \frac{\text{Vol}(G)}{(4\pi)^{\frac{d}{2}} C^{\frac{d}{2\alpha}}} \cdot \frac{\Gamma(1+\frac{d}{2\alpha})}{\Gamma(1+\frac{d}{2})} t^{-\frac{d}{2\alpha}}.$$

Note that for the heat kernel the fact that there exists $K > 0$ so that as $t \downarrow 0$

$$k_t(e) \sim K t^{-\frac{d}{2}} e^{t|\rho|^2},$$

can be obtained by using properties of weights and Hermite polynomials (see Chap. 12 of Fegan [66] and also Fegan [64, 65]). Here ρ is the usual half-sum of positive roots. The author cannot resist mentioning the celebrated "strange formula" of Freudenthal and de Vries which states that, in the case where G is also semi-simple and the Riemannian metric is induced by the Killing form,

$$||\rho||^2 = \frac{d}{24}.$$

This is established using heat kernel asymptotics in Fegan [66] p. 169. For an alternative elementary proof, see Burns [39].

5.10 Martingale Representation of Lévy Processes

We briefly describe the stochastic integral representation of Lévy processes due to Kunita and the author [15] (see also Liao [132] Sect. 3.3, pp. 70–77).

Let $\phi = (\phi(t), t \geq 0)$ be a Lévy process on a Lie group G. We will assume that it is càdlàg, i.e. that $P(\Omega' = 1)$, where $\Omega' \in \mathcal{F}$ is the set of all $\omega \in \Omega$ for which the mapping $t \to \phi(t)(\omega)$ is right continuous with left limits at every point. The process ϕ is adapted to its natural filtration and is a Feller-Markov process. As pointed out in Sect. 5.2, the laws of ϕ form a convolution semigroup and Theorem 5.3.3 classifies the infinitesimal generator $\mathcal{L}$ by means of its characteristics (b, a, ν). Then for all $f \in C_0^{(2)}(G), t \geq 0$ we have that $(M_f(t), t \geq 0)$ is a martingale (again, with respect to the natural filtration of the process), where for each $t \geq 0$,

$$M_f(t) := f(\phi(t)) - f(e) - \int_0^t \mathcal{L}f(\phi(s))ds. \tag{5.10.32}$$

Define the jump in the process at time t to be $\Delta\phi(t) := \phi(t-)^{-1}\phi(t)$, where $\phi(t-) := \lim_{s \uparrow t} \phi(s)$ is the left limit. Then the prescription

$$N(t, E) = \#\{0 \leq s \leq t; \Delta\phi(s) \in E\}$$

gives rise to a Poisson random measure N on $\mathbb{R}^+ \times G$, and it can be shown that

$$\mathbb{E}(N(t, E)) = t\nu(E),$$

where ν is the Lévy measure associated to the process through the representation (5.3.8). Hence, at least for those $E \in \mathcal{B}(G)$ which are complements of neighbourhoods of the neutral element, we may define the compensated Poisson random measure $\widetilde{N}(t, E) = N(t, E) - t\nu(E)$. In Applebaum and Kunita [15] it is shown that there exists a Brownian motion $B = (B_1, \ldots, B_d)$ on $\mathbb{R}^d$, which is independent of N, having autocovariance $\text{Cov}(B_i(s), B_j(t)) = \mathbb{E}(B_i(s)B_j(t)) = a_{ij}s \wedge t$ for all $s, t \geq 0, 1 \leq i, j \leq d$ such that (in the sense of Itô integrals)

$$M_f(t) = \int_0^t (X_i f)(\phi(s-))dB_i(s) + \int_0^t \int_G (f(\phi(s-)\tau) - f(\phi(s-)))\widetilde{N}(ds, d\tau). \tag{5.10.33}$$

The Eqs. (5.10.32) and (5.10.33) together yield a stochastic integral representation for the Lévy process ϕ, and conversely, any solution to the corresponding stochastic integral equation is unique and is a Lévy process on G. A special case of considerable interest is *Brownian motion on G*. This is obtained by equipping G with a Riemannian metric, choosing $\{X_1, \ldots, X_d\}$ to be an orthonormal basis of $\mathfrak{g}$ and taking $b = \nu = 0$ and $a = I_d$. Then $\mathcal{L} = \frac{1}{2}\Delta$ and the convolution semigroup formed by the laws of ϕ is a standard Gaussian semigroup (with $\sigma^2 = 1/2$). In this case the stochastic integral equation for ϕ is equivalent to the Stratonovitch stochastic differential equation:

$$d\phi(t) = X_i(\phi(t)) \circ dB^i(t). \tag{5.10.34}$$

5.11 Notes and Further Reading

The study of convolution semigroups on general Lie groups began with the pioneering work of Hunt [103] in 1956. Ramaswami's paper [165] helped to make Hunt's proof more accessible and his results were incorporated into the monograph treatment in Heyer [95] (Chap. 4). The most recent version, and the one on which the account here is based, is due to Liao [132]. But it should be emphasised that all of these proofs are variations on Hunt's. In Euclidean space, Hunt's theorem is essentially equivalent to the classical Lévy-Khintchine formula (see e.g. Chapter 3 of Applebaum [7]), and a direct generalisation of that formula to locally compact abelian groups can be found in Chap. 7 of Parthasarathy [155]. Generalisations of Hunt's theorem to locally compact groups have been obtained where the generator is obtained as the sum of three terms corresponding to "drift", "diffusion" and "jumps" (see Heyer [95], Sects. 4.4 and 4.5). An expression for the generator that is closer to the spirit of (5.3.8) may be found in Born [34]. There are also interesting generalisations of these ideas to hypergroups (see Chap. 4 of Bloom and Heyer [31]) and quantum groups due to Schürmann [181]. The representation theoretic version of the Lévy-Khintchine formula first appeared in Heyer [94] and then was rediscovered by Lo and Ng [136]. It is generalised to general Lie groups in Siebert [185] p. 123 (see also Applebaum [5] for a different perspective).

The central limit theorem (CLT) was originally due to Donald Wehn and appears in his PhD thesis. It was never published and the version of it here is presented by Breuillard [35] who writes "(Wehn)'s Yale dissertation, made under the supervision of S.Kakutani in 1959, is only accessible through the archives department of the Sterling Memorial Library at Yale university, and it bears a special notice prohibiting its copying." An alternative approach to the CLT based on solving a martingale problem can be found in Stroock and Varadhan [196]. Chapter 6 of Heyer's [95] monograph is devoted to investigating the CLT on a general locally compact group. Mello's paper

[148] gives a physicist's perspective on CLT's on groups. For an account of CLT's in locally compact abelian groups, see Bingham [28]. A very nice survey of results about Gaussian semigroups on compact groups, with many open problems, can be found in Saloff-Coste [175].

Stochastic analysis on Lie groups is an enormous area which merges quickly with that on general Riemannian manifolds. Brownian motion on a Lie group was first investigated by the great Kiyoso Itô [105] in 1950. There has been a great amount of research on this topic and associated stochastic differential equations. For orientation see Elworthy's [62] St Flour lecture notes or Chap. XI of Malliavan's monograph [145]. An interesting paper by Voit [212] uses representation theory in the spirit of the current volume to generalise Paul Lévy's classic martingale characterisation of Brownian motion to a Lie group. We also mention extensions of Brownian motion to loop groups by Driver [59] and also Driver and Lohrenz [60]. Finally there has been important work by Bendikov and Saloffe-Coste [23–25] on studying Gaussian processes on general locally compact groups, including absolute continuity and sample path properties. One of the most important Gaussian processes in Euclidean space that can be built from Brownian motion is the Ornstein-Uhlenbeck process, and a generalisation of this to Lie groups may be found in Baudoin et al. [20]. Stochastic partial differential equations is, at the time of writing, an area of considerable current activity. There have so far been one or two works that have investigated such equations on Lie groups using Gaussian space-time white noise. The paper by Tindel and Viens [204] is particularly interesting as the group is compact and many of the techniques used there will be familiar to readers of this book. Peszat and Tindel [160] go beyond the compact case. For an introduction to some ideas of Malliavin calculus on Lie groups, see Pontier and Üstünel [162]. Smoothness of densities for solutions of stochastic differential equations on manifolds and Lie groups driven by Lévy processes have been recently studied by Picard and Savona [161] in the pure jump case, and by Kunita [129] in the more general case with a Hörmander condition imposed on both the diffusion and jump parts.

The work of Applebaum and Kunita [15] on Lévy processes on Lie groups was influenced by the ideas of Holevo [100]. A great deal about this topic can be learned from the monograph of Liao [132]. For recent work and references, see e.g. Applebaum [10]. Applications of Lie group valued Lévy processes to develop operators of interest to harmonic analysts can be found in Applebaum and Bañuelos [13]. As mentioned at the end of the previous chapter, the use of pseudo-differential operator techniques to study Feller-Markov processes has become important in recent years and on compact groups a suitable technology based on representation theoretic ideas has been developed by Ruzhansky and Turunen [172]. Applications to Lévy processes on Lie groups and more general group-valued Markov processes can be found in Applebaum [10, 11].

Since any symmetric space M is a quotient of a semisimple Lie group G by a compact subgroup H, Lie group techniques play a key role in studying probability in symmetric spaces e.g. Brownian motions or Lévy processes on M are the images of those in G under the natural surjection. A Lévy-Khintchine type formula in symmetric spaces was first studied by Gangolli [70] in 1965. This was restricted to spherically symmetric distributions, i.e. the probability laws are bi-invariant under the action of K. Further work was carried out by the author [4], Liao and Wang [133] and using subordination, Albeverio and Gordina [1]. Recent work by Applebaum and Dooley [14] has enabled the spherical symmetry assumption to be dropped. Brownian motion on symmetric spaces has also been intensely studied, see e.g. Malliavin and Malliavin [143] for a study of its asymptotic behaviour.

Random fields on groups (and manifolds carrying a group action) is a subject that is currently attracting considerable attention, not least because of applications to cosmology and to earthquake modelling (see Malyenko [146] and Marinucci and Peccati [147]).

Chapter 6
Deconvolution Density Estimation

Abstract We review the scheme introduced by Kim and Richards for *deconvolution density estimation* on compact Lie groups. The Fourier transform is used to express the problem in terms of products of matrices of irreducible representations. We introduce a sequence of consistent estimators, by taking cut-offs in a certain Fourier expansion. Some smoothness classes of noise are discussed and used to find optimal rates of convergence of the estimators.

In this chapter we will describe an application of probability theory on Lie groups to nonparametric statistics within the specific context of deconvolution density estimation. Most of the material is based on the paper by Kim and Richards [118].

6.1 The Statistical Model

We fix a probability space $(\Omega, \mathcal{F}, P)$ and a compact connected Lie group G. We will assume that G is equipped with some Riemannian metric. Let $X : \Omega \to G$ be a random variable, which we call the *signal*. We assume that X has a density $f_X \in L^2(G)$, which we regard as an unknown. Let $\epsilon : \Omega \to G$ be another random variable that we treat as *noise* which distorts the signal. It is also assumed to have a density f_ϵ. We assume that X and ϵ are independent. Define $Y : \Omega \to G$ by $Y = \epsilon X$, so that $Y(\omega) = \epsilon(\omega)X(\omega)$ for all $\omega \in \Omega$. We call Y the *observed random variable*. We do not have direct knowledge of Y but we make observations $Y_1, \ldots, Y_n$ which are i.i.d. copies of Y. Note that Y has a density f_Y given by

$$f_Y = f_\epsilon * f_X \tag{6.1.1}$$

The problem of *deconvolution* is to invert this equation to find f_X. Note that for all $\pi \in \widehat{G}$ we have (by Theorem 4.2.1 (2))

$$\widehat{f_Y}(\pi) = \widehat{f_X}(\pi)\widehat{f_\epsilon}(\pi). \tag{6.1.2}$$

D. Applebaum, *Probability on Compact Lie Groups*, Probability Theory and Stochastic Modelling 70, DOI: 10.1007/978-3-319-07841-0_6,

We make the following:

Assumption 1 For all $\pi \in \widehat{G}$, the matrix $\widehat{f_\epsilon}(\pi)$ is invertible.

Then for all $\pi \in \widehat{G}$, we have from (6.1.2) that

$$\widehat{f_X}(\pi) = \widehat{f_Y}(\pi)\widehat{f_\epsilon}(\pi)^{-1}. \tag{6.1.3}$$

As an estimator of $\widehat{f_Y}(\pi)$ based on our observations, we define the *empirical characteristic function*

$$\widehat{f_Y^{(n)}}(\pi) := \frac{1}{n}\sum_{i=1}^{n} \pi(Y_i^{-1}), \tag{6.1.4}$$

for all $\pi \in \widehat{G}, n \in \mathbb{N}$. Observe that $\widehat{f_Y^{(n)}}(\pi)$ is itself a matrix-valued random variable and we have, as usual, suppressed the ω dependence.

The sequence $(\widehat{f_Y^{(n)}}(\pi), n \in \mathbb{N})$ is easily seen to be an unbiased estimator in that $\mathbb{E}(\widehat{f_Y^{(n)}}(\pi)) = \widehat{f_Y}(\pi)$ for all $\pi \in \widehat{G}, n \in \mathbb{N}$. This follows from the fact that for all $1 \leq i \leq n$

$$\mathbb{E}(\pi(Y_i^{-1})) = \int_G \pi(\sigma^{-1}) f_Y(\sigma) d\sigma = \widehat{f_Y}(\pi). \tag{6.1.5}$$

We can also show that it is a consistent estimator. Indeed this follows from the following.

Proposition 6.1.1 *For all* $\pi \in \widehat{G}, 1 \leq j, k \leq d_\pi$, $\widehat{f_Y^{(n)}}(\pi)_{jk} \to \widehat{f_Y}(\pi)_{jk}$ *a.s. as* $n \to \infty$.

Proof The random variables $(\pi_{jk}(Y_n), n \in \mathbb{N})$ are i.i.d. and so by the strong law of large numbers, almost surely,

$$\begin{aligned} \widehat{f_Y^{(n)}}(\pi)_{jk} &= \frac{1}{n}\sum_{i=1}^{n} \pi(Y_i^{-1})_{jk} \\ &\to \mathbb{E}(\pi_{jk}(Y^{-1})) = \widehat{f_Y}(\pi)_{jk} \text{ as } n \to \infty, \end{aligned}$$

where we have used (6.1.5). □

Using Fourier inversion (Theorem 2.3.1 (1)) and (6.1.3), we have (in the L^2-sense)

$$f_X = \sum_{\pi \in \widehat{G}} d_\pi \text{tr}(\widehat{f_Y}(\pi)\widehat{f_\epsilon}(\pi)^{-1}\pi).$$

Our first step in obtaining an estimator of f_X is to replace $\widehat{f_Y}(\pi)$ by $\widehat{f_Y^{(n)}}(\pi)$ in this formula. Our second step is to introduce the sequence of *smoothing parameters* $(T_n, n \in \mathbb{N})$, where $T_n > 0$ for all $n \in \mathbb{N}$ and $T_n \to \infty$ as $n \to \infty$. We define a sequence $(f_X^{(n)}, n \in \mathbb{N})$ of estimators of f_X, each of which is a finite sum, as follows:

$$f_X^{(n)}(\sigma) = \sum_{\pi \in \widehat{G}; \kappa_\pi \leq T_n} d_\pi \mathrm{tr}(\widehat{f_Y^{(n)}(\pi) \widehat{f_\epsilon}(\pi)^{-1} \pi(\sigma)}), \tag{6.1.6}$$

for all $n \in \mathbb{N}, \sigma \in G$. Note that $f_X^{(n)}(\cdot, \omega) \in C^\infty(G)$ for all $n \in \mathbb{N}, \omega \in \Omega$. Define the *kernel*

$$K_n^\epsilon(\sigma) = \sum_{\pi \in \widehat{G}; \kappa_\pi \leq T_n} d_\pi \mathrm{tr}(\widehat{f_\epsilon}(\pi)^{-1} \pi(\sigma)), \tag{6.1.7}$$

for $n \in \mathbb{N}, \sigma \in G$. Then $f_X^{(n)}$ is an example of a *kernel density estimator*, in that

$$f_X^{(n)}(\sigma) = \frac{1}{n} \sum_{i=1}^{n} K_n^\epsilon(\sigma Y_i^{-1}), \tag{6.1.8}$$

for each $n \in \mathbb{N}, \sigma \in G$.[1]

6.2 Asymptotic Behaviour of the Estimator

Our arguments will be much simpler if we can use pointwise convergent Fourier series. In fact we'll assume the following:

Assumption 2 f_X has a uniformly convergent Fourier series expansion.

Our first result show that $f_X^{(n)}$ is a consistent estimator.

Proposition 6.2.1 *For all* $\sigma \in G$,

$$\lim_{n \to \infty} \mathbb{E}(f_X^{(n)}(\sigma)) = f_X(\sigma).$$

Proof Using (6.1.8), (6.1.7) and (6.1.3), we obtain

$$\begin{aligned} \mathbb{E}(f_X^{(n)}(\sigma)) &= \mathbb{E}(K_n^\epsilon(\sigma Y^{-1})) \\ &= \int_G \sum_{\pi \in \widehat{G}; \kappa_\pi \leq T_n} d_\pi \mathrm{tr}(\widehat{f_\epsilon}(\pi)^{-1} \pi(\sigma \tau^{-1})) f_Y(\tau) d\tau \\ &= \sum_{\pi \in \widehat{G}; \kappa_\pi \leq T_n} d_\pi \mathrm{tr}(\widehat{f_Y}(\pi) \widehat{f_\epsilon}(\pi)^{-1} \pi(\sigma)) \\ &= \sum_{\pi \in \widehat{G}; \kappa_\pi \leq T_n} d_\pi \mathrm{tr}(\widehat{f_X}(\pi) \pi(\sigma)) \end{aligned}$$

[1] For background on kernel density estimation, see e.g. Silverman [186].

$$\to \sum_{\pi \in \widehat{G}} d_\pi \mathrm{tr}(\widehat{f_X}(\pi)\pi(\sigma)) \text{ as } n \to \infty$$
$$= f_X(\sigma).$$
□

In order to make further progress, we need to make a smoothness assumption on the density f_X that we seek to estimate. We recall the Sobolev space $H_s(G)$ of order $s > 0$ which we introduced in Chap. 3. From now on we will make the following:

Assumption 3 $f_X \in H_s(G)$ for some $s > \frac{d}{2}$.

If Assumption 3 is satisfied, then $f_X \in L^\infty(G)$ by Theorem 3.2.2. It follows that $f_Y \in L^\infty(G)$. Indeed, we easily deduce from (6.1.1) that $||f_Y||_\infty \leq ||f_X||_\infty$. Assumption 3 also entails that $f_X \in \mathrm{Dom}(\Delta^{\frac{s}{2}})$, where Δ is the usual Laplacian on G. If we take s in Assumption 3 to be sufficiently large, then Assumption 2 is automatically satisfied. This follows by using the Sobolev embedding theorem (Theorem 3.1.3) and Theorem 3.3.1(ii).

In the next theorem, $||\cdot||_2$ denotes the norm in $L^2(G)$ and the estimate should be understood as pointwise in $\omega \in \Omega$:

Proposition 6.2.2 *For all $n \in \mathbb{N}$,*

$$||f_X - \mathbb{E}(f_X^{(n)})||_2 \leq T_n^{-\frac{s}{2}} ||\Delta^{\frac{s}{2}} f_X||_2. \tag{6.2.9}$$

Proof From the proof of Proposition 6.2.1, we see that for all $\sigma \in G$,

$$f_X(\sigma) - \mathbb{E}(f_X^{(n)}(\sigma)) = \sum_{\pi \in \widehat{G}; \kappa_\pi \geq T_n} d_\pi \mathrm{tr}(\widehat{f_X}(\pi)\pi(\sigma)),$$

and so by the Parseval-Plancherel identity (2.3.8),

$$\begin{aligned} ||f_X - \mathbb{E}(f_X^{(n)})||_2^2 &= \sum_{\pi \in \widehat{G}; \kappa_\pi \geq T_n} d_\pi \mathrm{tr}(\widehat{f_X}(\pi)\widehat{f_X}(\pi)^*) \\ &\leq \sum_{\pi \in \widehat{G}; \kappa_\pi \geq T_n} d_\pi \frac{\kappa_\pi^s}{T_n^s} \mathrm{tr}(\widehat{f_X}(\pi)\widehat{f_X}(\pi)^*) \\ &\leq T_n^{-s} ||\Delta^{\frac{s}{2}} f_X||_2. \end{aligned}$$
□

The *integrated variance bias* at $n \in \mathbb{N}$ of the estimator is the quantity $\mathbb{E}(||f_X - f_X^{(n)}||_2^2)$, which we now estimate.

Corollary 6.2.1 *For all $n \in \mathbb{N}$,*

$$\mathbb{E}(||f_X - f_X^{(n)}||_2^2) \leq \int_G \mathrm{Var}(f_X^{(n)}(\sigma))d\sigma + T_n^{-s}||\Delta^{\frac{s}{2}} f_X||_2^2. \tag{6.2.10}$$

Proof For each $\omega \in \Omega$, it is clear that the $f_X^{(n)} - \mathbb{E}(f_X^{(n)})$ and $f_X - \mathbb{E}(f_X^{(n)})$ are orthogonal in $L^2(G)$, and so by Proposition 6.2.2,

$$\begin{aligned} ||f_X - f_X^{(n)}||_2^2 &= ||f_X^{(n)} - \mathbb{E}(f_X^{(n)})||_2^2 + ||f_X - \mathbb{E}(f_X^{(n)})||_2^2 \\ &\leq ||f_X^{(n)} - \mathbb{E}(f_X^{(n)})||_2^2 + T_n^{-s}||\Delta^{\frac{s}{2}} f_X||_2^2. \end{aligned}$$

The result follows upon taking expectations of both sides of the last inequality. □

The last result of this section is a useful bound on the integral of the variance of our estimator.[2]

Theorem 6.2.1 *For all $n \in \mathbb{N}$*

$$\int_G \mathrm{Var}(f_X^{(n)}(\sigma))d\sigma \leq \frac{||f_X||_\infty}{n} \int_G |K_n^\epsilon(\sigma)|^2 d\sigma + O\left(\frac{1}{n}\right). \tag{6.2.11}$$

Proof Using (6.1.8),

$$\begin{aligned} \int_G \mathrm{Var}(f_X^{(n)}(\sigma))d\sigma &= \int_G \mathrm{Var}\left(\frac{1}{n}\sum_{i=1}^n K_n^\epsilon(\sigma Y_i^{-1})\right)d\sigma \\ &= \frac{1}{n}\int_G \mathrm{Var}(K_n^\epsilon(\sigma Y^{-1}))d\sigma \\ &= \frac{1}{n}\int_G \mathbb{E}(|K_n^\epsilon(\sigma Y^{-1})|^2)d\sigma - \frac{1}{n}\int_G |\mathbb{E}(K_n^\epsilon(\sigma Y^{-1}))|^2 d\sigma. \end{aligned}$$

By the proof of Proposition 6.2.1, for all $\sigma \in G, n \in \mathbb{N}$,

$$\mathbb{E}(K_n^\epsilon(\sigma Y^{-1})) = \sum_{\pi \in \widehat{G}; \kappa_\pi \leq T_n} d_\pi \mathrm{tr}(\widehat{f_X}(\pi)\pi(\sigma)).$$

[2] Recall that if $Z : \Omega \to \mathbb{C}$ is a complex-valued random variable, then its variance, $\mathrm{Var}(Z) := \mathbb{E}(|Z|^2) - |\mathbb{E}(Z)|^2$.

Hence by the Parseval-Plancherel identity (2.3.8),

$$\begin{aligned}\sup_{n\in\mathbb{N}}\int_G |\mathbb{E}(K_n^{\epsilon}(\sigma Y^{-1}))|^2 d\sigma &= \sup_{n\in\mathbb{N}} \sum_{\pi\in\widehat{G};\kappa_\pi\le T_n} d_\pi \mathrm{tr}(\widehat{f_X}(\pi)\widehat{f_X}(\pi)^*)\\ &\le \sum_{\pi\in\widehat{G}} d_\pi \mathrm{tr}(\widehat{f_X}(\pi)\widehat{f_X}(\pi)^*)\\ &= ||f_X||_2^2.\end{aligned}$$

By Fubini's theorem, for all $n \in \mathbb{N}$

$$\begin{aligned}\int_G \mathbb{E}(|K_n^{\epsilon}(\sigma Y^{-1})|^2)d\sigma &= \int_G\int_G |K_n^{\epsilon}(\sigma\tau^{-1})|^2 f_Y(\tau)d\tau d\sigma\\ &\le ||f_Y||_\infty \int_G\int_G K_n^{\epsilon}(\sigma\tau^{-1})|^2 d\sigma d\tau\\ &\le ||f_X||_\infty \int_G |K_n^{\epsilon}(\sigma)|^2 d\sigma,\end{aligned}$$

and the result follows. □

6.3 Smoothness Classes of Noise

In this section we discuss some smoothness classes of noise that were introduced by Kim and Richards [118]. Let $(\kappa_\pi, \pi \in \widehat{G})$ be the usual Casimir spectrum of the group.

The noise ϵ is said to be *statistically super-smooth* of order $\beta > 0$ if there exists $\gamma > 0$ and $a_1, a_2 \ge 0$ such that

$$||\widehat{f_\epsilon}(\pi)^{-1}||_{op} = O(\kappa_\pi^{a_1}\exp(\gamma\kappa_\pi^{\beta})) \text{ and } ||\widehat{f_\epsilon}(\pi)||_{op} = O(\kappa_\pi^{a_2}\exp(-\gamma\kappa_\pi^{\beta}))$$

as $\kappa_\pi \to \infty$. For example the standard Gaussian and α-stable type densities $(0 < \alpha < 1)$ are statistically super-smooth. It is natural to ask how this concept relates to our usual notion of smoothness.

Proposition 6.3.1 *If f is statistically super-smooth, then it is smooth.*

Proof For sufficiently large κ_π and using (2.5.23) and Corollary 2.5.2 we find that there exist $N, C, K > 0$ such that

$$\begin{aligned}
||\widehat{f}(\pi)||_{HS} &\leq ||\widehat{f}(\pi)||_{\text{op}}|||I_\pi||| \\
&= d_\pi^{\frac{1}{2}}||\widehat{f}(\pi)||_{\text{op}} \\
&\leq N^{\frac{1}{2}}|\lambda_\pi|^{\frac{m}{2}} \cdot C\kappa_\pi^{a_2}\exp(-\gamma\kappa_\pi^{\beta}) \\
&\leq K|\lambda_\pi|^{\frac{m}{2}}(1+|\lambda_\pi|^2)^{a_2}\exp(-\gamma|\lambda_\pi|^{2\beta}),
\end{aligned}$$

from which it follows that $\widehat{f} \in \mathcal{S}(D)$ and the result follows by Theorem 3.4.2. □

The noise ϵ is said to be *statistically smooth* of order $\delta > 0$ if

$$||\widehat{f}_\epsilon(\pi)^{-1}||_{op} = O(\kappa_\pi^{\delta}) \text{ and } ||\widehat{f}_\epsilon(\pi)||_{op} = O(\kappa_\pi^{-\delta})$$

as $\kappa_\pi \to \infty$. There is no analogue of Proposition 6.3.1 in this context. Indeed the Laplace distribution is an example of a density which is statistically smooth but not smooth in the usual sense (c.f. Example 4 in Sect. 5.8).

A third example of a smoothness class is the *statistically log-super-smooth class.* We will not write about this class here but refer the interested reader to Kim and Richards [118]. At the time of writing the only known example in this class is the *von Mises-Fisher distribution* on $SO(n)$. This distribution is described by two parameters—the concentration parameter $\kappa > 0$ and the mean direction $\mu \in SO(n)$. The density has the form

$$f_{\kappa,\mu}(\sigma) = c(\kappa)e^{\kappa \text{tr}(\mu^{-1}\sigma)},$$

for $\sigma \in G$, where $c(\kappa)$ is a normalisation constant. Infinite divisibility of this distribution has been established for all κ by Kent [117].

6.4 Optimal Rates of Convergence

The following definitions are based on work of Stone [195] and are adapted from Kim and Richards [118]. Consider a sequence $(f_n, n \in \mathbb{N})$ where $f_n \in L^2(\Omega \times G)$. We say that $(f_n, n \in \mathbb{N})$ is an *L^2-estimator* of f_X if $||f_n - f_X||_2$ converges to zero in probability. Such an estimator is said to be a *L^2-estimator based on* $Y_1, \ldots, Y_n$ if f_n is $\sigma\{Y_1, \ldots, Y_n\}$ measurable for all $n \in \mathbb{N}$. Let $(b_n, n \in \mathbb{N})$ be a sequence of positive numbers. We say that it is an *lower bound on convergence of estimators* if for every sequence $(f_n, n \in \mathbb{N})$ of L^2-estimators of f where f_n is based on $Y_1, \ldots, Y_n$ we have

$$\lim_{c\to 0}\liminf_{n\to\infty} P(||f_n - f_X||_2 > cb_n) = 1 \qquad (6.4.12)$$

and an *upper bound on convergence of estimators* if there exists a sequence $(f_n^{(0)}, n \in \mathbb{N})$ of L^2-estimators of f where $f_n^{(0)}$ is based on $Y_1, \ldots, Y_n$ so that

$$\lim_{c \to \infty} \limsup_{n \to \infty} P(||f_n^{(0)} - f_X||_2 > cb_n) = 0. \tag{6.4.13}$$

We say that $(b_n, n \in \mathbb{N})$ achieves the *optimal rate of convergence* if it is both a lower and an upper bound. The key theorem is due to Kim and Richards [118].

Theorem 6.4.1 (Kim, Richards) *Suppose that $f_X \in H_s(G)$ where $s > \frac{d}{2}$.*

1. *If ϵ is statistically super-smooth of order $\beta > 0$, then the optimal rate of convergence of $(f_X^{(n)}, n \in \mathbb{N})$ to f_X is $b_n = (\log(n))^{-\frac{s}{2\beta}}$ and $f_n^{(0)} = f_X^{(n)}$ for all $n \in \mathbb{N}$.*
2. *If ϵ is statistically smooth of order $\delta > 0$, then the optimal rate of convergence of $(f_X^{(n)}, n \in \mathbb{N})$ to f_X is $b_n = n^{-(2s+4\delta+d)}$ and $f_n^{(0)} = f_X^{(n)}$ for all $n \in \mathbb{N}$.*

We will only give a taste of the proof of the theorem by showing how to obtain the upper bound in case (1). Readers who want the whole story are directed to the paper by Kim and Richards [118]. Before we present the analysis we mention a key asymptotic result that plays a key role there:

For any κ_0 in the Casimir spectrum of G we have (see e.g. Minakshisundaram and Pleijel [150]) that for all $\sigma \in G$

$$\lim_{n \to \infty} T_n^{-\frac{d}{2}} \sum_{\pi \in \widehat{G}; \kappa_0 \le \kappa_\pi < T_n} \sum_{1 \le i,j \le d_\pi} d_\pi \pi_{ij}(\sigma)^2 = \frac{\mathrm{Vol}(G)}{(2\sqrt{\pi})^d \Gamma(1 + d/2)}.$$

When we evaluate this at $\sigma = e$ we find that

$$\lim_{n \to \infty} T_n^{-\frac{d}{2}} \sum_{\pi \in \widehat{G}; \kappa_0 \le \kappa_\pi < T_n} d_\pi^2 = \frac{\mathrm{Vol}(G)}{(2\sqrt{\pi})^d \Gamma(1 + d/2)}. \tag{6.4.14}$$

Proof To find the upper bound in (1).

By the Parseval-Plancherel identity (2.3.8), for each $n \in \mathbb{N}$

$$\begin{aligned} \int_G |K_n^\epsilon(\sigma)|^2 d\sigma &= \sum_{\pi \in \widehat{G}: \kappa_\pi < T_n} d_\pi ||\widehat{f_\epsilon}(\pi)^{-1}||_{HS}^2 \\ &\le \sum_{\pi \in \widehat{G}: \kappa_\pi < T_n} d_\pi^2 ||\widehat{f_\epsilon}(\pi)^{-1}||_{op}^2. \end{aligned}$$

Now since ϵ is statistically super-smooth, we may find κ_0 so that for $\kappa_\pi \ge \kappa_0$,

$$||\widehat{f_\epsilon}(\pi)^{-1}||_{op} \le D_1 \kappa_\pi^{a_1} \exp(\gamma \kappa_\pi^\beta),$$

for some $D_1 > 0$. Then by (6.4.14) there exist $D_2(\epsilon), D_3 \ge 0$ so that for sufficiently large n,

$$\int_G |K_n^\epsilon(\sigma)|^2 d\sigma \le D_2(\epsilon) + D_1 \sum_{\pi\in\widehat{G};\kappa_0\le\kappa_\pi<T_n} d_\pi^2 \kappa_\pi^{2a_1}\exp(2\gamma\kappa_\pi^\beta)$$

$$= D_2(\epsilon) + D_1 T_n^{-2a_1} \sum_{\pi\in\widehat{G};\kappa_0\le\kappa_\pi<T_n} d_\pi^2 \left(\frac{\kappa_\pi}{T_n}\right)^{2a_1}\exp(2\gamma\kappa_\pi^\beta)$$

$$\le D_2(\epsilon) + D_1 T_n^{\frac{d}{2}-2a_1}\exp(2\gamma T_n^\beta) T_n^{-\frac{d}{2}} \sum_{\pi\in\widehat{G};\kappa_0\le\kappa_\pi<T_n} d_\pi^2$$

$$\le D_2(\epsilon) + D_3 T_n^{\frac{d}{2}-2a_1}\exp(2\gamma T_n^\beta).$$

Now by (6.2.10) and (6.2.11), we have

$$\mathbb{E}(||f_X - f_X^{(n)}||_2^2) \le T_n^{-s}||\Delta^{\frac{s}{2}} f_X||_2^2 + \frac{||f_X||_\infty}{n}\int_G |K_n^\epsilon(\sigma)|^2 d\sigma + \mathrm{O}\left(\frac{1}{n}\right)$$

$$\le T_n^{-s}||\Delta^{\frac{s}{2}} f_X||_2^2 + \frac{||f_X||_\infty}{n}\left(D_2(\epsilon) + D_3 T_n^{\frac{d}{2}-2a_1}\exp(2\gamma T_n^\beta)\right) + \mathrm{O}\left(\frac{1}{n}\right).$$

The bound is optimised when the exponential growth term is linearised, and this is achieved by choosing $T_n = K(\log n)^{\frac{1}{\beta}}$ where $K > 0$. We then find that there are $C_1(f_X), C_2(f_X) > 0$ so that

$$\mathbb{E}(||f_X - f_X^{(n)}||_2^2) \le C_1(f_X)(\log n)^{-\frac{s}{\beta}} + C_2(f_X)(\log n)^{\frac{\frac{d}{2}-2a_1}{\beta}} + \mathrm{O}\left(\frac{1}{n}\right).$$

By Chebychev's inequality, for any $c > 0$

$$P(||f_X - f_X^{(n)}||_2 \ge c(\log(n))^{-\frac{s}{2\beta}}) \le \frac{C_1(f_X)}{c^2} + \frac{C_2(f_X)}{c^2}(\log n)^{\frac{\frac{d}{2}-s-2a_1}{\beta}} + \frac{1}{c^2}\mathrm{O}\left(\frac{1}{n}\right),$$

and the result follows. □

6.5 Decompounding

We briefly mention an interesting extension of the deconvolution concept. Suppose that $X = (X(t), t \ge 0)$ is a compound Poisson process taking values in the compact Lie group G that is built from a sequence $(Z_n, n \in \mathbb{N})$ of i.i.d. G-valued random variables and an independent Poisson process, as described in Chap. 5. We assume that the Z_ns have a common density f_Z. Let $\epsilon = (\epsilon(t), t \ge 0)$ be a noise process on G that is independent of X, for example ϵ might be a group valued Brownian motion. Then suppose that we make observations on the process $\Phi = (\Phi(t), t \ge 0)$ where

$\Phi(t) = \epsilon(t)X(t)$ for each $t \geq 0$. The problem of *decompounding* is to estimate f_X from these observations. This problem is studied in Said et al. [173] where applications (in the case $G = SO(3)$) are developed to the multiple scattering of waves under a statistical isotropy assumption on the spherical symmetry of scattering events about the direction of propagation.

6.6 Statistics and Differential Geometry

The applications of differential geometry to statistical inference first began receiving detailed attention in the 1980s; see in particular the survey by Barndorff-Nielsen et al. [19] and the more recent monograph by Amari [2]. We emphasise two distinct aspects:

1. The development of statistical methodology to investigate data that already has an inherently geometric character. The work described above fits into this scheme.
2. The use of differential geometric ideas within statistical inference in the usual sense, without any geometric assumptions on the data.

One of the most interesting examples of (2) is the circle of ideas that has now become known as *information geometry*. Let $\mathcal{M}$ be a parametric statistical model described by a family of probability densities $f(\cdot, \boldsymbol{\theta})$ on $\mathbb{R}^n$ where $\boldsymbol{\theta} = (\theta_1, \ldots, \theta_r)$ are the parameters. We consider the usual log-likelihood $L_{\cdot} := \log(f(\cdot, \boldsymbol{\theta}))$. We assume that f varies smoothly with $\theta_1, \ldots, \theta_r$ and define the *Fisher information matrix*

$$I(\boldsymbol{\theta})_{ij} := \mathbb{E}((\partial_i L_{\boldsymbol{\theta}})(\partial_j L_{\boldsymbol{\theta}})),$$

for $1 \leq i, j \leq r$. Then $I(\boldsymbol{\theta})$ is a Riemannian metric on $\mathcal{M}$. From this metric, a family of connections may be constructed which are called *α-connections*. These may be applied to e.g. the higher-order statistical inference of certain exponential models (see e.g. Barndorff-Nielsen et al. [19]).

For another example of aspect (1), we consider a general scheme for density estimation developed by Hendriks [90]. Let M be a d-dimensional Riemannian manifold having volume measure μ. We seek information on an unknown quantity represented by the M-valued random variable Y, and we assume that Y has a density $f_Y \in L^2(M, \mu)$. The data are given by an i.i.d. sample $Y_1, \ldots, Y_n$. Recall that the Laplace-Beltrami operator Δ is essentially self-adjoint on $C_c(M)$ and that $-\Delta$ has a discrete spectrum $\lambda_1 < \cdots < \lambda_n \to \infty$ as $n \to \infty$. Let ϕ_j be the eigenvector associated to λ_j where $j \in \mathbb{N}$. Then for fixed $T > 0$ we define the spectral function:

$$k(x, y; T) := \sum_{\lambda_k < T} \phi_k(x)\overline{\phi_k(y)},$$

for each $x, y \in M$. It is easily verified that P_T is the orthogonal projection in $L^2(M, \mu)$ onto the direct sum of the eigenspaces corresponding to eigenvalues less than T, where for $g \in L^2(M, \mu)$,

$$(P_T g)(x) = \int_M k(x, y; T) g(y) \mu(dy),$$

and $\lim_{T \to \infty} ||P_T g - g||_2 = 0$. Note that $(P_T f_Y)(x) = \mathbb{E}(k(x, Y; T))$ for all $x \in M$. As an estimate for the density f_Y we define

$$f_{Y,T}^{*(N)}(x) := \frac{1}{N} \sum_{j=1}^{N} k(x, Y_j; T),$$

where $x \in M$, $N \in \mathbb{N}$. The following key theorem was established by Hendriks [90]. Here we require Sobolev spaces $H^s(M)$ on general Riemannian manifolds for $s > 0$ (see Hebay [87]).

Theorem 6.6.1 *If $f_Y \in H^s(M)$ and $T_0 > 0$, then there exist $A, B > 0$ so that for $T > T_0$,*

$$\mathbb{E}(||f_Y - f_{Y,T}^{*(N)}||_2^2) \leq A \cdot \frac{T^{\frac{d}{2}}}{N} + BT^{-s}.$$

Furthermore, for suitable choice of T,

$$\mathbb{E}(||f_Y - f_{Y,T}^{*(N)}||_2^2) = O(N^{\frac{2s}{2s+d}})$$

as $N \to \infty$.

A similar theorem is proven giving estimates and large N asymptotics for $\mathbb{E}(||f_Y - f_{Y,T}^{*(N)}||_\infty^2)$. It is interesting that the proofs require knowledge of the rate of convergence of the zeta function:

$$\zeta(x, s) = \sum_{n \in \mathbb{N}} |\phi_n(x)|^2 \lambda_n^{-s},$$

where $x \in M$ and $s \in \mathbb{C}$. This series converges absolutely if $\Re(s) > \frac{d}{2}$ and has a meromorphic extension to the complex plane.

6.7 Notes and Further Reading

This section will be shorter than in earlier chapters as so much of the key literature has already been mentioned in the text of this short chapter. Also the subject of statistics on Lie groups is still very much in its infancy. Deconvolution is a growing concern that has been intensively developed by P. Kim, J. Y. Koo and collaborators. We mention in particular the development of asymptotic minimax bounds in the compact case [123], work on the two-sphere [86] and the important case of the three-dimensional Euclidean group [138]. See Yazici [219] for an overview of engineering applications of deconvolution on groups with a particular emphasis on second-order stationary processes. For a different application of estimation on compact Lie groups within the context of pattern theory, see Bigot et al. [27]. Estimation of parameters for Gaussian distribution on compact groups and applications to signal processing are discussed in Said and Manton [174]. In Marinucci and Peccati [147] residual cosmic microwave background radiation from the Big Bang is described using random fields on the two sphere (see also Chap. 5 of Malyarenko [146]). Representations of $SO(3)$ play an important role in this theory and statistical tests are developed for non-Gaussianity.

Appendices

A.1 Appendix 1: Basic Concepts of Topology

This section is a quick reminder of basic ideas in topology. Readers can find more details and proofs in any standard text such as Kelley [116] or Simmons [187].

Let S be a set. A *topology* on S is a collection τ of subsets of S which has the following properties:

- $S, \emptyset \in \tau$,
- τ is closed under arbitrary unions and finite intersections.

The pair (S, τ) is then called a *topological space*, and the sets in τ are called the *open sets* for the topology. A *basis* for a topology τ is a subcollection $\sigma \subset \tau$ so that every set in τ is a union of sets from σ. The space (S, τ) is said to be *second countable* if it has a countable basis.

Let $A \subset S$. The *induced, relative* or *subspace* topology for A is the collection $\tau_A = \{A \cap U; U \in \tau\}$. Then (A, τ_A) is itself a topological space which is called a (topological) *subspace* of (S, τ).

Suppose that S is equipped with topologies τ_1 and τ_2 wherein $\tau_1 \subset \tau_2$. In this case we say that τ_1 is *weaker* (or *coarser*) that τ_2. Equivalently τ_2 is *stronger* (or *finer*) than τ_1. The *discrete topology* τ_D on an arbitrary set S is that topology for which every subset of S is open. It is the strongest topology on S in that if τ is any other topology, then $\tau \subset \tau_D$. If S is equipped with the discrete topology we call it a *discrete space*.

Let (S_1, τ_1) and (S_2, τ_2) be topological spaces. A mapping $f : S_1 \to S_2$ is said to be *continuous* if $f^{-1}(U) \in \tau_1$ whenever $U \in \tau_2$. A mapping f is said to be a *homeomorphism* between (S_1, τ_1) and (S_2, τ_2) if it is bijective and continuous and if f^{-1} is also continuous.

If $x \in S$, a *neighbourhood* of x is a set $N \subseteq S$ such that there exists some $U \in \tau$ for which $x \in U \subseteq N$. If N is a neighbourhood of x and $N \in \tau$, we call it an *open neighbourhood*. The space (S, τ) is said to be *Hausdorff* if for all distinct $x, y \in S$ there exist open neighbourhoods U_x of x and V_y of y so that $U_x \cap V_y = \emptyset$.

D. Applebaum, *Probability on Compact Lie Groups*, Probability Theory and Stochastic Modelling 70, DOI: 10.1007/978-3-319-07842-7,

A sequence $(x_n, n \in \mathbb{N})$ of points in S is said to *converge* to a *limit* $x \in S$ if given any neighbourhood U of x there exists $N \in \mathbb{N}$ so that $x_n \in U$ for all $n > N$. If (S, τ) is Hausdorff, then convergent sequences in S have unique limits. If (S_i, τ_i) for $i = 1, 2$ are topological spaces wherein (S_1, τ_1) is second countable, then $f : S_1 \to S_2$ is continuous if and only if given any sequence $(x_n, n \in \mathbb{N})$ in S_1 that converges to $x \in S_1$, then the sequence $(f(x_n), n \in \mathbb{N})$ converges to $f(x)$ in S_2.

A collection $\mathcal{U} := \{U_i, i \in I\}$ of open sets in τ is an *open cover* for S if $S = \bigcup_{i \in I} U_i$. A *subcover* of a cover $\mathcal{U}$ is a subcollection of $\mathcal{U}$ which is itself an open cover for S. The space (S, τ) is said to be *compact* if every open cover of S has a *finite* subcover. The continuous image of a compact set is compact. A discrete space is compact if and only if it has a finite number of elements. If (S, τ) is locally compact and Hausdorff, its *one-point compactification* is the compact Hausdorff space (S_∞, τ_∞) where $S_\infty = S \cup \{\infty\}$ (with $\infty \notin S$), and τ_∞ comprises the open sets of S (now regarded as subsets of S_∞), S_∞ itself, and the complements in S_∞ of compact subsets of S.

A subset of S is said to be *closed* if its complement is open. A closed subset of a compact space is compact. Every compact subset of a Hausdorff space is closed. In particular, if (S, τ) is Hausdorff, then $\{x\}$ is closed for all $x \in S$.

If A is a subset of S, then its *closure* $\overline{A}$ is the intersection of all closed sets containing A. We always have $A \subseteq \overline{A}$ but $A = \overline{A}$ if and only if A is closed. A subset A of S is said to be *relatively compact* if $\overline{A}$ is compact. The space (S, τ) is said to be *locally compact* if every point in S has a neighbourhood which has compact closure. The set A is said to be *dense* in (S, τ) if $\overline{A} = S$. The space (S, τ) is said to be *separable* if S contains a countable dense subset. If (S, τ) is second countable, then it is separable.

A space (S, τ) is *connected* if we cannot write $S = A \cup B$ for two non-empty, disjoint open sets A and B. The continuous image of a connected set is connected. If (S, τ) is an arbitrary topological space and $x \in S$, the *component* of x is the largest connected subset of S that contains x. Components are always closed sets. The set of all components $\{A_i, i \in I\}$ in S forms a *partition* of S in that $A_i \cap A_j = \emptyset$ whenever $i \neq j$ and $S = \bigcup_{i \in I} A_i$. The corresponding equivalence relation $\sim$ on S is given by $x \sim y$ if there exists a connected set containing both x and y. A space (S, τ) is said to be *locally connected* if every neighbourhood of every point in S contains a connected neighbourhood of that point.

A property of a space (S, τ) is said to be a *topological property* if it is preserved under homeomorphisms. The Hausdorff property, compactness, local compactness, connectedness and local connectedness are all topological properties.

If $\{(S_i, \tau_i), i = 1, 2, \ldots, n\}$ are a (finite) collection of topological spaces, the *product topology* τ on the Cartesian product $S := S_1 \times S_2 \times \cdots \times S_n$ is defined via the basis $U_1 \times U_2 \times \cdots \times U_n$ where $U_i \in \tau_i (1 \leq i \leq n)$. The space (S, τ) is compact (respectively, connected, Hausdorff, locally compact, locally connected) if (S_i, τ_i) is for all $i = 1, 2, \ldots, n$. The product topology can be extended to arbitrary collections $\{(S_\alpha, \tau_\alpha), \alpha \in I\}$, where I is some index set, but the construction is different from that given for finite index sets. We won't give the details here (they can be found in the standard texts) but we'll quote the famous *Tychonoff theorem* which states that the product of an arbitrary collection of compact spaces is itself compact. Kelley (p. 143) writes that this "is unquestionably the most useful theorem on compactness. It is probably the most important single theorem of general topology." In relation to the examples discussed in Sect. 1.1, we see that the infinite torus group Π^{∞} (which is the product of countably many tori) is compact.

Finally we state a corollary of the celebrated *Urysohn lemma* which says that if X is a locally compact Hausdorff space, K is a compact subset of X and U is an open set with $K \subset U$, then there exists a function $h \in C_c(X, \mathbb{R})$ for which $1_K \leq h \leq 1_U$. From this we can deduce the following:

Proposition A.0.1 *If X is a locally compact Hausdorff space then*

1. *$C_c(X, \mathbb{R})$ is dense in $C_0(X, \mathbb{R})$.*
2. *$C_c(X, \mathbb{C})$ is dense in $C_0(X, \mathbb{C})$.*

Proof

1. Let $f \in C_0(X, \mathbb{R})$. Then given $\epsilon > 0$ there exists a compact set K with $||f 1_{K^c}||_\infty < \frac{\epsilon}{2}$. Now apply the corollary to Urysohn's lemma, taking U to be the union of open neighbourhoods of each point in K. Thus we can assert the existence of $h \in C_c(X, \mathbb{R})$ for which $1_K \leq h \leq 1_U$. In particular, it follows that $hf \in C_c(X, \mathbb{R})$ and

$$\begin{aligned} ||f - hf||_\infty &\leq ||f - 1_K f||_\infty + ||1_K f - hf||_\infty \\ &\leq ||f - 1_K f||_\infty + ||1_K f - 1_U f||_\infty \\ &= ||f - 1_K f||_\infty + ||1_{U \setminus K} f||_\infty \\ &\leq 2||f 1_{K^c}||_\infty < \epsilon. \end{aligned}$$

2. Follows easily from (1) by approximating real and imaginary parts. □

A.2 Appendix 2: Concerning Left/Right Uniform Continuity

We begin by discussing some useful facts about topological groups G. First some notation. If $A, B \subset G$ we write $AB := \{gh; g \in A, h \in B\}$. In particular if $g \in G$, then $gA := \{g\}A$ and $Ag := A\{g\}$. We also write $A^{-1} := \{g \in G; g^{-1} \in A\}$.

Fact 1. It is pointed out in Sect. 1.1 that for each $g \in G$, left translation l_g and right translation r_g are homeomorphisms of G. It follows that if U is a neighbourhood of e, then both gU and Ug are neighbourhoods of g.

We say that $V \subset G$ is *symmetric* if $V^{-1} = V$.

Fact 2. Given any neighbourhood U of e there exists a symmetric neighbourhood v of e so that $VV \subseteq U$. To see this it's sufficient to use continuity of the group operation from $G \times G$ to G to deduce that there exist neighbourhoods W_1 and W_2 of e so that $W_1 W_2 \subseteq U$. Now take $V = W_1 \cap W_1^{-1} \cap W_2 \cap W_2^{-1}$.

Fact 3. Let $f : G \to \mathbb{R}$ be continuous and fix $g \in G$. Then given any $\epsilon > 0$ there exists a neighbourhood U_g of g so that so that $g' \in U_g \Rightarrow |f(g) - f(g')| < \epsilon$. By Fact 1 we can find a neighbourhood W_g of e so that $gW_g \subseteq U_g$ and for all $h \in W_g$, $|f(g) - f(gh)| < \epsilon$.

Now we are ready to prove the main result.

Theorem A.0.1 *Every continuous function of compact support on G is both left and right uniformly continuous.*

Proof We follow Folland [68], Proposition 2.6, p. 34 very closely (see also Higgins [98] Proposition 21, p. 67–68). We only deal with right uniform continuity here as the other case works in the same way. Let $f \in C_c(G)$ and $\epsilon > 0$. Write $K := \text{supp}(f)$ and let $g \in K$ be arbitrary. Then by Fact 3 we can find a neighbourhood W_g of e so that for all $h \in W_g$, $|f(g) - f(gh)| < \frac{\epsilon}{2}$. By Fact 2, there exists a symmetric neighbourhood V_g of e so that $V_g V_g \subseteq W_g$. By Fact 1, the sets $\{gV_g, g \in G\}$ cover K and so there exists $g_1, \ldots, g_N \in G$ so that $K \subseteq \bigcup_{j=1}^N g_j V_{g_j}$. Then there exists some j so that $g_j^{-1} g \in V_{g_j}$. Now for all $h \in V := \bigcap_{j=1}^N V_{g_j}$ we have $gh = g_j(g_j^{-1}g)h \in g_j W_{g_j}$ while $g = g_j(g_j^{-1}g) \in g_j W_{g_j}$. We then have

$$|f(gh) - f(g)| \leq |f(gh) - f(g_j)| + |f(g_j) - f(g)| < \frac{\epsilon}{2} + \frac{\epsilon}{2} = \epsilon.$$

A similar argument works when $gh \in K$ and the result follows. □

Corollary A.0.1 *Every continuous function on G that vanishes at infinity is both left and right uniformly continuous.*

Proof We only prove left uniform continuity as the same argument works in the other case. Let $f \in C_0(G)$. Since $C_c(G)$ is dense in $C_0(G)$, given any $\epsilon > 0$ there exists $h \in C_c(G)$ so that $||f - h||_\infty < \frac{\epsilon}{3}$. By Theorem A.2.1 there exists an open neighbourhood U of e so that $||L_g h - h||_\infty < \frac{\epsilon}{3}$ for all $g \in U$. It follows that for all $g \in U$,

$$\begin{aligned} ||L_g f - f||_\infty &\leq ||L_g f - L_g h||_\infty + ||L_g h - h||_\infty + ||h - f||_\infty \\ &= ||L_g h - h||_\infty + 2||h - f||_\infty \\ &< \frac{\epsilon}{3} + \frac{2\epsilon}{3} = \epsilon. \end{aligned}$$ □

A.3 Appendix 3: Manifolds

This appendix gathers together some useful facts about manifolds. For proofs and more detailed discussion we recommend a dedicated text on differential geometry such as Helgason [88] or Warner [214]. Spivak [189] gives an excellent quick introduction and contains a very nice account of integration. For Riemannian manifolds, we particularly recommend Chavel [43].

A.3.1 Manifolds, Tangent Spaces, Vector Fields

A Hausdorff, second countable topological space M is a *(real, topological) manifold* of dimension d if each point in M has an open neighbourhood that is homeomorphic to an open set in $\mathbb{R}^d$. It follows immediately that M is locally compact. If U is an open neighbourhood of $p \in M$ and ϕ is a homeomorphism from U to an open subset of $\mathbb{R}^d$, we call (U, ϕ) a *chart* or *co-ordinate system* at p. For each $q \in U$, we write $\phi(q) = (x_1(q), \ldots, x_d(q))$, and call $x_1, \ldots, x_d$ *local co-ordinates* at U. We will also use the notation $q_i = x_i(q)(1 \leq i \leq d)$.

An *atlas* is a collection of charts $\{(U_\alpha, \phi_\alpha); \alpha \in I\}$ so that $(U_\alpha, \alpha \in I)$ covers M. If M has an atlas for which the mappings $\phi_\alpha \circ \phi_\beta^{-1} : \mathbb{R}^d \to \mathbb{R}^d$ are C^∞ for all $\alpha, \beta \in I$, then M is said to be a C^∞-manifold. Henceforth whenever we use the word "manifold" we will always assume that it is C^∞.

A function $f : M \to \mathbb{R}$ is said to be *smooth* or C^∞ if $f \circ \phi_\alpha^{-1}$ is C^∞ from $\phi_\alpha(U_\alpha)$ to $\mathbb{R}$ for all $\alpha \in I$. Let $C^\infty(M, \mathbb{R})$ denote the set of all smooth real-valued functions on M. It is a (real) algebra under the usual pointwise operations.

A *tangent vector* $X(p)$ at $p \in M$ is a linear functional on $C^\infty(M, \mathbb{R})$ which satisfies the local derivation property

$$X(p)(fg) = f(p)X(p)(g) + (X(p)f)g(p), \tag{A.1}$$

for each $f, g \in C^\infty(M, \mathbb{R})$. In local co-ordinates we can write $X(p) = \sum_{i=1}^d X_i(p) \frac{\partial}{\partial x_j}$ where $X_1(p), \ldots, X_d(p) \in \mathbb{R}$. The set of all tangent vectors at p forms a d-dimensional real linear space called the *tangent space* to p at M. We denote it by $T_p(M)$. We can give the set $T(M) := \bigcup_{p \in M} T_p(M)$ the structure of a $2d$-dimensional manifold, and $T(M)$ is called the *tangent bundle* to M.

Let M_1 and M_2 be manifolds of dimension d_1 and d_2 (respectively) and let $\{(U_\alpha, \phi_\alpha); \alpha \in I\}$ be an atlas for M_1 and $\{(V_\beta, \psi_\beta); \beta \in J\}$ be an atlas for M_2. A mapping $f : M_1 \to M_2$ is said to be C^∞ if $\psi_\beta \circ f \circ \phi_\alpha^{-1}$ is C^∞ from $\mathbb{R}^{d_1}$ to $\mathbb{R}^{d_2}$ for all $\alpha \in I, \beta \in J$. A C^∞-mapping $f : M_1 \to M_2$ is said to be a *diffeomorphism* if it is a bijection and f^{-1} is also a C^∞-mapping.

Suppose that M_1 and M_2 are manifolds, and that $\Phi : M_1 \to M_2$ is a C^∞ mapping. Let $p \in M_1$ and $X_p \in T_p(M_1)$. Then if $q = \Phi(p)$, there exists $Y_q \in T_q(M_2)$ so that for all $f \in C^\infty(M_2, \mathbb{R})$,

$$Y_q f = X_p(f \circ \Phi).$$

We write $Y_q = d\Phi_p(X_p)$ and call the linear map $d\Phi_p : T_p(M_1) \to T_q(M_2)$ the *differential* of Φ (at p). It is common to write $d\Phi$ instead of $d\Phi_p$.

The following composition property is very important. If M_1, M_2 and M_3 are manifolds and $\Phi_1 : M_1 \to M_2$ and $\Phi_2 : M_2 \to M_3$ are C^∞ mappings, then for all $p \in M_1$:

$$d(\Phi_2 \circ \Phi_1)_p = (d\Phi_2)_q \circ (d\Phi_1)_p, \tag{A.2}$$

where $q = \Phi_1(p)$.

A C^∞ mapping X from M to $T(M)$ is said to be a *vector field* if $X(p) \in T_p(M)$ for all $p \in M$. In each local co-ordinate system (U, ϕ) we can write $X(p) = \sum_{i=1}^d X_i(p)\frac{\partial}{\partial x_j}$, as we did for individual tangent vectors but now X_i is a C^∞ function from $\phi(U)$ to $\mathbb{R}$. From an analytic point of view, a vector field X can be regarded as a first order differential operator mapping $C^\infty(M, \mathbb{R})$ to itself by the prescription $(Xf)(p) = X(p)f$ for $p \in M$. Indeed, the local derivation property (A.1) now extends to a global derivation property on the algebra $C^\infty(M, \mathbb{R})$:

$$X(fg) = X(f)g + fX(g).$$

The collection of all vector fields on M forms a real linear space $\mathcal{L}(M)$. It is also an (infinite-dimensional) real Lie algebra, where the Lie bracket is defined by

$$[X, Y]f = X(Yf) - Y(Xf),$$

for each $X, Y \in \mathcal{L}(M)$, $f \in C^\infty(M)$.

If M_1 and M_2 are manifolds, and $\Phi : M_1 \to M_2$ is a C^∞ mapping then the vector fields $X \in \mathcal{L}(M_1)$ and $Y \in \mathcal{L}(M_2)$ are said to be *Φ-related* if $d\Phi(X_p) = Y_{\Phi(p)}$ for all $p \in M$. If X_i is Φ-related to Y_i for $i = 1, 2$, then $[X_1, X_2]$ is Φ-related to $[Y_1, Y_2]$.

If M and N are manifolds and $\phi : M \to N$ is C^∞ we say that ϕ is an *immersion* if $d\phi_p$ is non-singular for each $p \in M$, and we say that (M, ϕ) is a *submanifold* of N if ϕ is an injective immersion.

If K and U are compact and open sets (respectively) with $K \subset U \subset M$, then there exists a mapping $\phi \in C_c^\infty(M)$ which is called a *cut-off function* for which $\text{supp}(\phi) \subset U, 0 \leq \phi \leq 1$ and $\phi(p) = 1$ for all $p \in K$. We can imitate the proof of Proposition A.1.1, using cut-off functions instead of Urysohn's lemma, to show that $C_c^\infty(M, \mathbb{R})$ is dense in $C_0(M, \mathbb{R})$, and hence that $C_c^\infty(M, \mathbb{C})$ is dense in $C_0(M, \mathbb{C})$.

A.3.2 Differential Forms

If $f \in C^\infty(M, \mathbb{R})$, then since $T_x(\mathbb{R}) \cong \mathbb{R}$ for all $x \in \mathbb{R}$, we have $df(X) = Xf$ for all $f \in C^\infty(M, \mathbb{R})$. The mapping $df : \mathcal{L}(M) \to C^\infty(M, \mathbb{R})$ is called a *one-form*.[1] If we localise at $p \in M$, then df_p is a linear mapping from $T_p(M)$ to $\mathbb{R}$, i.e. an element of the algebraic dual $T_p^*(M)$ to $T_p(M)$ which is called the *cotangent space* to M at p.

We can form tensor products of r tangent and s co-tangent spaces at a point $p \in M$ to obtain the linear space

[1] The most general one-forms are finite linear combinations of the df's.

$$\bigotimes_p^{(r,s)}(M) = \underbrace{T_p(M) \otimes \cdots \otimes T_p(M)}_{r \text{ copies}} \otimes \underbrace{T_p^*(M) \otimes \cdots \otimes T_p^*(M)}_{s \text{ copies}}.$$

The set $\bigotimes^{(r,s)}(M) := \bigcup_{p \in M} \bigotimes_p^{(r,s)}(M)$ can be given the structure of a real d^{r+s}-dimensional manifold, and is called the (r, s)-tensor bundle over M. An (r, s) *tensor field* is a C^∞ mapping F from M to $\bigotimes^{(r,s)}(M)$ such that $F(p) \in \bigotimes_p^{(r,s)}(M)$ for each $p \in M$. A tensor field is said to be *covariant* if $s = 0$ and *contravariant* if $r = 0$. So for example a vector field is covariant (with $r = 1$) and a one-form is contravariant (with $s = 1$).

Recall that if V is a real d-dimensional vector space and $V^{\otimes^r}$ is its r-fold tensor product, then the subspace $\Lambda^r(V) \subseteq V^{\otimes^r}$ of *alternating r-tensors* is spanned by

$$\mathrm{Alt}(v_1, \ldots, v_r) := \frac{1}{r!} \sum_{\sigma \in \Sigma(r)} \mathrm{sign}(\sigma) v_{\sigma(1)} \otimes \cdots \otimes v_{\sigma(r)},$$

where $v_1, \ldots, v_r \in V$ and $\Sigma(r)$ is the symmetric group on r letters.[2] We have $\Lambda^1(V) = V$ and $\Lambda^r(V) = \{0\}$ for $r > d$. Note also that $\dim(\Lambda^r(V)) = \binom{d}{r}$, and so in particular $\dim(\Lambda^d(V)) = 1$. It follows that $\Lambda^d(V) - \{0\}$ has two connected components, and an *orientation* on V is a specific choice of one of these.

If $v = \mathrm{Alt}(v_1, \ldots, v_r) \in \Lambda^r(V)$ and $w = \mathrm{Alt}(w_1, \ldots, w_s) \in \Lambda^s(V)$, their *wedge product* $v \wedge w \in \Lambda^{r+s}(V)$ is

$$v \wedge w := \frac{(r+s)!}{r!s!} \mathrm{Alt}(v_1, \ldots, v_r, w_1, \ldots w_s).$$

Furthermore,

$$v \wedge w = (-1)^{rs} w \wedge v.$$

Consider the linear space $\Lambda^r(T_p^*(M))$ of alternating r-tensors over the cotangent space at $p \in M$. We can again give a manifold structure to $\Lambda^r(M) := \bigcup_{p \in M} \Lambda^r(T_p^*(M))$ (indeed, $\Lambda^r(M)$ is a *sub-bundle* of $\bigotimes^{(0,r)}(M)$). A smooth *r-form* ω is a C^∞ mapping from M to $\Lambda^r(M)$ such that for each $p \in M$, $\omega_p \in \Lambda^r(T_p^*(M))$. In particular, if $X_1, \ldots, X_r \in \mathcal{L}(M)$,

$$\omega(X_1, \ldots, X_r)(p) = \omega_p(X_1(p), \ldots, X_r(p)).$$

In local co-ordinates (U, ϕ) at $p \in M$ an r-form ω can be written

$$\omega(p) = \sum_{\{i_1,\ldots,i_r\} \subseteq \{1,\ldots,d\}} \omega_{i_1,\ldots,i_r}(p) dx_{i_1} \wedge \cdots dx_{i_r},$$

[2] The *sign* $\mathrm{sign}(\sigma)$ of a permutation σ is defined to be 1 if it is even, and -1 if it is odd.

where $\omega_{i_1,\ldots,i_r}$ are smooth mappings from $\phi(U)$ to $\mathbb{R}$ and $dx_j\left(\frac{\partial}{\partial x_k}\right) = \delta_{jk}$.

We denote the linear space of all C^∞ r-forms on M by $\Omega^r(M)$. In fact $\Omega^r(M)$ is a left $C^\infty(M)$-module where for $f \in C^\infty(M), \omega \in \Omega^r(M)$, $f\omega$ is the r-form whose value at $p \in M$ is $f(p)\omega_p$.

If M_1 and M_2 are manifolds and $\Phi : M_1 \rightarrow M_2$ is a C^∞ mapping, then its *pullback* Φ^* is the linear mapping from $\Omega^r(M_2)$ to $\Omega^r(M_1)$ defined by

$$\Phi^*(\omega)(X_1, \ldots, X_r)(p) = \omega_{\Phi(p)}(d\Phi_p(X_1(p)), \ldots, d\Phi_p(X_r(p))),$$

for all $X_1, \ldots, X_r \in \mathcal{L}(M_1)$, $p \in M_1$.

A.3.3 Integration on Manifolds

Suppose that M is a connected d-dimensional manifold. Define $O \in \Lambda^d(M)$ by $O = \bigcup_{p \in M}\{0_p\}$ where 0_p is the zero vector in $\Lambda^d(T_p^*(M))$. Since $\Lambda^d(T_p^*(M)) - \{0_p\}$ has exactly two connected components for each $p \in M$, it follows that $\Lambda^d(M) - \{O\}$ has at most two of these. We say that M is *orientable* if $\Lambda^d(M) - \{O\}$ has exactly two connected components, and a choice of component is called an *orientation* on M. If M is not connected then it is orientable if each connected component is orientable in the above sense. Equivalently, the manifold M is orientable if and only if there is a nowhere-vanishing d-form on M. From now on we assume that M is *oriented*, i.e. it is equipped with an orientation. We say that a d-form ω is *positive* if (in each connected component of M) it belongs to the component corresponding to the chosen orientation.

Now let $f \in C_c^\infty(M, \mathbb{R})$ and assume that supp$(f) \subseteq U$ where (U, ϕ) is a chart. Let $\omega \in \Lambda^d(M)$ and suppose that $\omega(q) = h_\omega(x_1(q), \ldots, x_d(q))dx^1 \wedge \cdots \wedge dx^d$ for all $q \in U$. Then we may define the *integral* of the form $f\omega$ by

$$\int_M f\omega = \int_{\mathbb{R}^d} f \circ \phi^{-1}(x_1, \ldots, x_d)h_\omega(x_1, \ldots, x_d)dx_1 \ldots dx_d.$$

If ω is positive and (U, ϕ) is chosen (as it may be) so that $dx^1 \wedge \cdots \wedge dx^d$ is positive in $\phi(U)$, then $h_\omega > 0$ in $\phi(U)$. To define more general integrals we need a *partition of unity*, which is a collection of functions $(\psi_i, i \in I)$ in $C^\infty(M)$ for which

(i) At each $p \in M$ only finitely many of the ψ_is are non-zero,
(ii) For each $i \in I$ supp(ψ_i) is compact,
(iii) For each $p \in M, i \in I$, $\psi_i(p) \geq 0$ and $\sum_{i \in I} \psi_i(p) = 1$.

Now suppose that $(U_\alpha, \alpha \in J)$ is an open cover of M. We say that a partition of unity is *subordinate* to this cover if for each $i \in I$ there exists $\alpha \in J$ such that supp$(\psi_i) \subseteq U_\alpha$. The key fact we need is that given a manifold M with an atlas

$\{(U_\alpha, \phi_\alpha), \alpha \in J\}$ there exists a partition of unity that is subordinate to the cover $(U_\alpha, \alpha \in J)$.

For arbitrary $\omega \in \Lambda^d(M)$ we may now define

$$\int_M \omega = \sum_{i \in I} \int_M \phi_i \omega,$$

whenever the right hand side converges absolutely, where the integral $\int_M \phi_i \omega$ is defined by the prescription we gave earlier. In particular, if M is compact, then we can cover M with finitely many of the U_αs, and so the series is a finite one.

In the discussion so far we have insisted that ω be a smooth form, but this is not essential. In fact it is sufficient for it to be continuous. Now let $\omega \in \Gamma^d(M)$ be positive and let $f \in C_c(M, \mathbb{R})$. Then we may define $\int_M f\omega$ as above. But $f \to \int_M f\omega$ is then a linear functional defined on $C_c(M, \mathbb{R})$, which is positive in that $f \geq 0 \Rightarrow \int_M f\omega \geq 0$. Hence by the Riesz representation theorem (see Appendix A.5), there is a unique regular Borel measure μ_ω on M so that

$$\int_M f(x)\mu_\omega(dx) = \int_M f\omega,$$

for all $f \in C_c(M, \mathbb{R})$, where the integral on the left hand side is in the Lebesgue sense. We call μ_ω the measure induced by the form ω. We use this process to construct Haar measure on a Lie group at the end of Sect. 1.3.2.

Let $g \in \bigotimes^{0,2}(M)$. Then for each $p \in M$, g_p is a real bilinear form on $T_p(M)$ which varies smoothly with p. We say that g is a *Riemannian metric* on M if g_p is an inner product on $T_p(M)$ for all $p \in M$. The pair (M, g) is then called a *Riemannian manifold*. The metric g induces a metric (in the usual sense of "metric spaces") on M which we denote as ρ_g by the prescription:

$$\rho_g(p, q) = \inf_\gamma \int_0^1 g_{\gamma(t)}(d\gamma(t), d\gamma(t))^{\frac{1}{2}} dt,$$

where the infimum is taken over all paths γ from $[0, 1]$ to M such that $\gamma(0) = p$ and $\gamma(1) = q$. Note that paths are required to be continuous and at least piecewise differentiable. Every Riemannian manifold is orientable, and there exists a d-form called the *volume element* whose value at p (in local co-ordinates) is $\sqrt{g_p} dx_1 \wedge \cdots \wedge dx_d$, where g_p is the determinant of the matrix whose (i, j)th component is $g_p\left(\frac{\partial}{\partial i}, \frac{\partial}{\partial j}\right)$.

A.4 Appendix 4: The Symplectic and Spin Groups

The material in this appendix can be found in standard monographs on Lie groups such as Chevellay [45] or Bröcker and tom Dieck [36].

A.4.1 Quaternions and the Symplectic Group

The *quaternions* are the elements of the four dimensional real division algebra Q generated by $\{1, i, j, k\}$ satisfying the relations

$$i^2 = j^2 = k^2 = -1,$$

and

$$ij = -ji = k, jk = -kj = i, ki = -ik = j.$$

Any $q \in Q$ may be written $q = \alpha_0 + \alpha_1 i + \alpha_2 j + \alpha_3 k$, and its *conjugate* is $\tilde{q} = \alpha_0 - \alpha_1 i - \alpha_2 j - \alpha_3 k$. We obtain a norm $|\cdot|$ on Q by the prescription

$$|q| = (\tilde{q}q)^{\frac{1}{2}} = (\alpha_0^2 + \alpha_1^2 + \alpha_2^2 + \alpha_3^2)^{\frac{1}{2}}.$$

We may also regard Q as a two dimensional complex vector space with basis $\{1, j\}$ by writing

$$q = \alpha_0 + \alpha_1\sqrt{-1} + j(\alpha_2 - \alpha_3\sqrt{-1}). \tag{A.3}$$

We can then regard the Cartesian product Q^n as a $4n$-dimensional real vector space or as a $2n$-dimensional complex vector space. If $x = (x_1, \ldots, x_n)$ and $y = (y_1, \ldots, y_n) \in Q^n$ we define their *symplectic product* to be

$$x \cdot y = \sum_{i=1}^{n} x_i \tilde{y}_i.$$

The symplectic product is a (real) inner product on Q^n and we define the *sympletic group* $Sp(n)$ to be the set of all matrices A in $M_n(Q)$ for which

$$Ax \cdot Ay = x \cdot y,$$

for all $x, y \in Q^n$.

By using (A.3) we can equivalently characterise $Sp(n)$ as the group of all $A \in U(2n)$ for which $A^T J A = J$, where $J := \begin{pmatrix} 0 & I_n \\ -I_n & 0 \end{pmatrix}$, and this latter approach leads naturally to the description of the Lie algebra $\mathbf{sp(n)}$ given in Sect. 1.3.1.

Quaternions were invented by William Rowan Hamilton (1805–1865) to generalise complex numbers to higher dimensions. Hamilton was also responsible for the reformulation of classical mechanics that is now known as *Hamiltonian mechanics*. The symplectic group arises as the natural symmetry group in this context (see e.g. Dragt [58]).

A.4.2 Clifford Algebras and the Spin Group

Let V be a real vector space equipped with an inner product $\langle \cdot, \cdot \rangle$. The *Clifford algebra* $\mathrm{Cl}(V)$ over V is the real algebra generated by $\{q(v), v \in V\}$ and the identity I such that $q(\cdot)$ is linear and for each $u, v \in V$:

$$q(u)q(v) + q(v)q(u) = -2\langle u, v\rangle I. \tag{A.4}$$

Note that in particular $q(v)^2 = -||v||^2 I$ for each $v \in V$. If $\dim(V) = n$, then $\dim(\mathrm{Cl}(V)) = 2^n$.

From now on we take V to be $\mathbb{R}^n$ with its usual inner product and write $\mathrm{Cl}(n) := \mathrm{Cl}(\mathbb{R}^n)$. Then $\mathrm{Cl}(n)$ is generated by I and $\{q_1, \ldots, q_n\}$, where $q_i := q(e_i)$ and $\{e_1, \ldots, e_n\}$ is the natural basis in $\mathbb{R}^n$, i.e. for $1 \leq i \leq n$, $e_i = (0, \ldots, 0, \overset{(i)}{1}, 0, \ldots, 0)$. Indeed, if $v = \sum_{i=1}^n v_i e_i \in \mathbb{R}^n$, then $q(v) = \sum_{i=1}^n v_i q_i$. The generating relations (A.4) are then equivalent to

$$q_i^2 = -I \text{ and } q_j q_k + q_k q_j = 0 \tag{A.5}$$

for each $1 \leq i \leq n$, $1 \leq j < k \leq n$.

There is a natural vector space $\mathbb{Z}_2$ grading on $\mathrm{Cl}(n)$, so that we can write

$$\mathrm{Cl}(n) = \mathrm{Cl}(n)_+ \oplus \mathrm{Cl}(n)_-,$$

where $\mathrm{Cl}(n)_+$ is the vector space spanned by I and $\{q_{i_1} \cdots q_{i_{2m}}; 1 \leq i_1 < \cdots < i_{2m} \leq n, m \in \mathbb{N}\}$ and $\mathrm{Cl}(n)_-$ is the vector space spanned by $\{q_{j_1} \cdots q_{j_{2p-1}}; 1 \leq j_1 < \cdots < j_{2p-1} \leq n, p \in \mathbb{N}\}$. Note that $\mathrm{Cl}(n)_+$ is a subalgebra of $\mathrm{Cl}(n)$, but $\mathrm{Cl}(n)_-$ is not closed under products, indeed

$$\mathrm{Cl}(n)_+\mathrm{Cl}(n)_+ \subseteq \mathrm{Cl}(n)_+,\ \mathrm{Cl}(n)_+\mathrm{Cl}(n)_- \subseteq \mathrm{Cl}(n)_-,\ \mathrm{Cl}(n)_-\mathrm{Cl}(n)_+ \subseteq \mathrm{Cl}(n)_- \\ \text{and } \mathrm{Cl}(n)_-\mathrm{Cl}(n)_- \subseteq \mathrm{Cl}(n)_+.$$

We now define three groups, the third of which is the main object of interest:

$\mathrm{Cl}^\times(n)$ is the group of all invertible elements in $\mathrm{Cl}(n)$.

$$\text{Pin}(n) \text{ is the group generated by } \{q(v) \in \text{Cl}_n^{\times}; ||v|| = 1\}.$$

$$\text{Spin}(n) = \text{Pin}(n) \cap \text{Cl}(n)_+ \text{ is the } \textit{spin group}.$$

We will not give a full proof here that Spin(n) is the covering group of $SO(n)$ for $n \geq 3$, but we will demonstrate that there is indeed a natural relationship between these two groups. Let W be the real vector space spanned by $\{q_1, \ldots, q_n\}$ and consider the automorphism $T(g)(v) = gvg^{-1}$ for $g \in \text{Spin}(n)$, $v \in \text{Cl}(n)$. It can be shown that $T(g) : W \subseteq W$, and so there is an $n \times n$ matrix $(T_{ij}(g))$ such that $T(g)q_i = \sum_{j=1}^n T_{ij}(g)q_j$ for each $1 \leq i \leq n$. Hence for arbitrary $w \in W$ with $w = \sum_{i=1}^n w_i q_i$ we have

$$T(g)w = \sum_{j=1}^{n}\left(\sum_{i=1}^{n} w_i T_{ij}(g)\right).$$

Now since $T(g)$ is an automorphism we have $T(g)v^2 = T(g)v \cdot T(g)v$. Computing both sides of this identity and using (A.5), we obtain

$$\sum_{j=1}^{n}\left(\sum_{i=1}^{n} w_i T_{ij}(g)\right)^2 = \sum_{i=1}^{n} \alpha_i^2,$$

i.e. the matrix $T(g)$ is orthogonal. Now for each $g, h \in \text{Spin}(n)$, $T(gh) = T(g)T(h)$ hence $\det(T(gh)) = \det(T(g))\det(T(h))$, and so we must have $\det(T(g)) = 1$ for all $g \in \text{Spin}(n)$. Hence we deduce that $T : \text{Spin}(n) \to SO(n)$ is a homomorphism.

Spin groups and associated spin structures play a key role in differential geometry and its applications to quantum physics, particularly through the study of Dirac operators, and applications to index theory. For a nice introduction to this rich area, see the article by Boi [33] (it has a very large bibliography for those who want to explore further), or the book by Roe [168].

A.5 Appendix 5: Measures on Locally Compact Spaces

In this section we gather together some useful results about measures on locally compact spaces, most of which can be found in either Cohn [50] or Bauer [21].

Throughout this section X is always assumed to be locally compact topological space and $\mathcal{B}(X)$ is its Borel σ-algebra, i.e. the smallest σ-algebra of subsets of X containing all open sets of X. Let μ be a Borel measure on X, i.e. a measure defined on $(X, \mathcal{B}(X))$. We say that μ is *outer regular* if for all $A \in \mathcal{B}(X)$,

$$\mu(A) = \inf\{\mu(O); A \subset O, O \text{ is open in X}\}.$$

The measure μ is said to be *inner regular* if

$$\mu(A) = \sup\{\mu(C); C \subset A, C \text{ is compact in X}\}.$$

We say that μ is *regular* if it is both inner and outer regular, and also satisfies $\mu(C) < \infty$ for all compact sets C in X.

Here are some useful facts about regular measures:

- If X is second countable and Hausdorff, then every Borel measure defined on X that is finite on compact sets is regular.
- If X is second countable and Hausdorff, then every regular measure defined on X is σ-finite.

Proposition A.0.2 *If μ is a non-trivial regular Borel measure on a locally compact Hausdorf space X then*

1. *$C_c(X, \mathbb{R})$ is dense in $L^p(X, \mathcal{B}(X), \mu; \mathbb{R})$ for $1 \leq p < \infty$,*
2. *$C_c(X, \mathbb{C})$ is dense in $L^p(X, \mathcal{B}(X), \mu; \mathbb{C})$ for $1 \leq p < \infty$.*

Proof

1. If $E \in \mathcal{B}(X)$ with $\mu(E) < \infty$ then by regularity of μ, given any $\epsilon > 0$, we can find a compact K and an open U with $K \subset E \subset U$ and $\mu(U) < \mu(K) + \epsilon$. By a corollary to Urysohn's lemma (see Appendix A.1) there exists $h \in C_c(X, \mathbb{R})$ so that $1_K \leq h \leq 1_U$. Moreover
$$||1_E - h||_p^p \leq ||1_U - 1_K||_p^p = \mu(U \setminus K) < \epsilon,$$
and the result follows.
2. follows from (1) by approximating real and imaginary parts.

□

Note If M is a manifold, then the result of Proposition A.5.1 may be refined to prove that $C_c^\infty(M, \mathbb{R})$ is dense in $L^p(M, \mathcal{B}(M), \mu; \mathbb{R})$ (with an analogous result for the complex case) by using cut-off functions instead of Urysohn's lemma (see e.g. the argument in Grigor'yan [74] pp. 20–21).

A Borel measure μ on X is said to be *locally finite* if every point of X has an open neighbourhood U such that $\mu(U) < \infty$, and μ is said to be a *Radon measure* if it is both inner regular and locally finite.

Proposition A.0.3 *1. A Borel measure μ on a locally compact topological space X is locally finite if and only if it is finite on compacta.*

2. Every regular Borel measure on X is a Radon measure.

Proof

1. Suppose that μ is locally finite and let K be a compact subset of X. Then every point in K has an open neighbourhood that has finite measure, and the collection of all such neighbourhoods is an open cover for K. It follows that there are a finite number of these, say $V_1, \ldots, V_N$, which also cover K. But then

$$\mu(K) \leq \mu\left(\bigcup_{i=1}^{N} V_i\right) \leq \sum_{i=1}^{N} \mu(V_i) < \infty.$$

The converse is a consequence of the definition of local compactness.
2. This follows immediately from (1), and the definitions.

□

It can be shown that if X is second countable and Hausdorff (and not necessarily locally compact), then every inner regular Borel measure on X which is finite on compact sets is a Radon measure and so if X is also locally compact, then every Borel measure that is finite on compact sets is regular and Radon. In particular:

If M is a smooth manifold , then every Borel measure on M that is finite on compact sets is regular and Radon. In particular, every finite Borel measure on M is regular and Radon.

We now describe the Riesz representation theorem, which is used extensively in this book. A (real) linear functional I on $C_c(X, \mathbb{R})$ is said to be *positive* if $f \geq 0 \Rightarrow I(f) \geq 0$.

Theorem A.0.2 (Riesz representation theorem) *Let X be locally compact and Hausdorff. If I is a positive linear functional on $C_c(X, \mathbb{R})$, then there is a unique regular Borel measure μ on X such that for all $f \in C_c(X, \mathbb{R})$,*

$$I(f) = \int_X f(x)\mu(dx).$$

A *signed Borel measure* on X is a σ-additive set function ν defined on $\mathcal{B}(X)$ and taking values in $[-\infty, \infty]$ for which $\nu(\emptyset) = 0$. It is said to be *finite* if $|\nu(X)| \leq \infty$. The *Jordan decomposition* of a signed measure ν states that we have a unique decomposition $\nu = \nu_+ - \nu_-$, where ν_+ and ν_- are positive measures at least one of which is finite. The *total variation* of ν is defined to be $||\nu|| := \nu_+(X) + \nu_-(X)$. The concepts of (inner, outer) regularity as defined above extend to signed measures in an obvious way. Let $\mathcal{M}(X)$ denote the set of all regular signed measures on X which have finite total variation. It becomes a real vector space with respect to the operations

$$(\nu_1 + \nu_2)(A) = \nu_1(A) + \nu_2(A),$$

$$\text{and} \ \ (\lambda\nu)(A) = \lambda\nu(A),$$

for $\nu_1, \nu_2, \nu \in \mathcal{M}(X)$ and $\lambda \in \mathbb{R}$, and is a real Banach space with respect to the total variation norm.

If f is a measurable function defined on X and ν is a signed measure on X, we can define

$$\int_X f(x)\nu(dx) := \int_X f(x)\nu_+(dx) - \int_X f(x)\nu_-(dx),$$

provided at least one of the integrals appearing on the right hand side is finite.

The next result is a variant of the Riesz representation theorem. Let $C_0(X, \mathbb{R})$ be the real Banach space (under the supremum norm $||\cdot||_\infty$) of all real-valued continuous functions on X which vanish at infinity. Note that if ν is a finite signed measure on X, then I_ν is a bounded linear functional on $C_0(X, \mathbb{R})$, where for each $f \in C_0(X, \mathbb{R})$

$$I_\nu(f) := \int_X f(x)\nu(dx).$$

Indeed we easily see that $|I_\nu(f)| \leq ||f||_\infty ||\nu||$ and so $||I_\nu|| \leq ||\nu||$.

Theorem A.0.3 *If X is locally compact and Hausdorff, then the mapping $\nu \to I_\nu$ is an isometric isomorphism of $\mathcal{M}(X)$ onto the continuous dual of $C_0(X, \mathbb{R})$.*

An immediate consequence of Theorem A.5.2 is the useful identity:

$$||\nu|| = \sup\left\{\left|\int_X f(x)\nu(dx)\right| ; ||f||_\infty = 1\right\}.$$

A.6 Appendix 6: Compact, Hilbert-Schmidt and Trace-Class Operators

The material is this appendix is often founds in texts on functional analysis such as Sects. 4.2, 5.5 and 5.6 of Davies [54] and Sects. VI.5 and 6 of Reed and Simon [166]. We also found the on-line notes of Walter [213] to be particularly helpful.

A.6.1 Compact Operators

Let E and F be real or complex Banach spaces. A bounded linear operator $T : E \to F$ is said to be *compact* if it maps the unit ball in E to a compact set in F. Equivalently, whenever $(x_n, n \in \mathbb{N})$ is a bounded sequence in E, then $(Tx_n, n \in \mathbb{N})$

has a convergent subsequence in F. From now on we will take $E = F = H$ to be a complex separable Hilbert space.

Theorem A.0.4 *If $T : H \to H$ is compact, then there exist complete orthonormal bases $(e_n, n \in \mathbb{N})$ and $(f_n, n \in \mathbb{N})$ and a sequence of complex numbers $(c_n, n \in \mathbb{N})$ with $\lim_{n\to\infty} c_n = 0$ such that*

$$T = \sum_{n=1}^{\infty} c_n \langle \cdot, e_n \rangle f_n. \tag{A.6}$$

A bounded linear operator on H is of *finite rank* if it has finite range. Let $\mathcal{F}(H)$ be the linear space of all finite rank operators on H. Then Theorem A.6.1 expresses the fact that every compact operator in H is a uniform limit of finite rank operators.

Theorem A.0.5 (Riesz-Schauder) *If $T : H \to H$ is compact, then its spectrum $\sigma(T)$ is discrete and has no limit points except possibly $\{0\}$. If $\lambda \in \sigma(T)$, then λ is an eigenvalue of T having at most finite multiplicity.*

Theorem A.0.6 (Hilbert-Schmidt) *If $T : H \to H$ is compact and self-adjoint, then it has a complete orthonormal basis of eigenvectors, and the corresponding eigenvalues $\{\lambda_n, n \in \mathbb{N}\}$ satisfy $\lim_{n\to\infty} \lambda_n = 0$.*

In fact if T is both compact and self adjoint, then we can write the representation (A.6) as

$$T = \sum_{n=1}^{\infty} \lambda_n \langle \cdot, e_n \rangle e_n, \tag{A.7}$$

where e_n is the normalised eigenvector corresponding to $\lambda_n (n \in \mathbb{N})$.

Let $\mathcal{C}(H)$ be the linear space of all compact operators on H.

1. $\mathcal{C}(H)$ is a closed subspace of $\mathcal{L}(H)$.
2. $\mathcal{C}(H)$ is a (two-sided) *-ideal of $\mathcal{L}(H)$, i.e. $T \in \mathcal{C}(H)$ if and only if $T^* \in \mathcal{C}(H)$ and if $T \in \mathcal{C}(H)$, $X, Y \in \mathcal{L}(H)$, then $XTY \in \mathcal{C}(H)$.

A.6.2 Hilbert-Schmidt Operators

We say that a bounded linear operator T acting on H is Hilbert-Schmidt if $\sum_{n=1}^{\infty} ||Te_n||^2 < \infty$ for some (and hence all) complete orthonormal basis $(e_n, n \in \mathbb{N})$ in H. Let $\mathcal{I}_2(H)$ denote the linear space of all Hilbert-Schmidt operators on H. It is itself a complex Hilbert space with respect to the inner product

$$\langle S, T \rangle_2 = \sum_{n=1}^{\infty} \langle Se_n, Te_n \rangle,$$

where $S, T \in \mathcal{I}_2(H)$. The associated norm will be denoted $|| \cdot ||_2$.

1. $\mathcal{I}_2(H)$ is a (two-sided) *-ideal of $\mathcal{L}(H)$.
2. $\mathcal{I}_2(H) \subseteq \mathcal{C}(H)$ and if $T \in \mathcal{C}(H)$, then $T \in \mathcal{I}_2(H)$ if and only if $\sum_{n=1}^{\infty} |c_n|^2 < \infty$ in (A.6) and we then have

$$||T||_2 = \left(\sum_{n=1}^{\infty} |c_n|^2 \right)^{\frac{1}{2}}. \tag{A.8}$$

3. $\mathcal{F}(H)$ is $|| \cdot ||_2$-dense in $\mathcal{I}_2(H)$.

We have a very precise description of Hilbert-Schmidt operators acting in L^2 spaces.

Theorem A.0.7 *Let (S, Σ, μ) be a measure space wherein the σ-algebra Σ is countably generated. Let $H = L^2(S, \Sigma, \mu)$. Then $T \in \mathcal{I}_2(H)$ if and only if for all $f \in H, x \in S$*

$$(Tf)(x) = \int_S f(y)k(x, y)\mu(dy),$$

where the kernel $k \in L^2(S \times S, \Sigma \otimes \Sigma, \mu \times \mu)$. Moreover

$$||T||_2^2 = \int_S \int_S |k(x, y)|^2 \mu(dx)\mu(dy).$$

Proof (Sketch). First we consider sufficiency. Since Σ is countably generated, H is separable. Let $(e_n, n \in \mathbb{N})$ be a complete orthonormal basis for H. Then $(e_m \otimes \overline{e_n}, m, n \in \mathbb{N})$ is a complete orthonormal basis for $H \otimes H = L^2(S \times S, \Sigma \otimes \Sigma, \mu \times \mu)$. Given $k \in H \otimes H$, we may assert that there exists $(c_{m,n}, m, n \in \mathbb{N}) \in l^2(\mathbb{N}^2)$ so that

$$k = \sum_{m,n=1}^{\infty} c_{m,n} e_m \otimes \overline{e_n}.$$

Then for all $k \in \mathbb{N}, x \in S$,

$$Te_k(x) = \int_S \left(\sum_{m,n=1}^{\infty} c_{m,n} e_m(x) \overline{e_n(y)} \right) e_k(y)\mu(dy) = \sum_{m=1}^{\infty} c_{m,k} e_m(x).$$

Hence

$$\sum_{m=1}^{\infty} ||Te_k||^2 = \sum_{m,k=1}^{\infty} |c_{m,k}|^2 = ||k||_{H \otimes H}^2 < \infty,$$

and so $T \in \mathcal{I}_2(H)$. For necessity it can be shown that $T \in \mathcal{I}_2(H)$ has a kernel of the form $k = \sum_{m,n=1}^{\infty} c_{m,n} e_m \otimes \overline{e_n}$ by making an approximation with finite-rank operators. □

A.6.3 Trace-Class Operators

If $T \in \mathcal{L}(H)$ the operator T^*T is positive and self-adjoint. Hence by spectral theory, it has a positive self-adjoint square root which we denote as $|T| = (T^*T)^{\frac{1}{2}}$. We say that T is *trace class* if

$$\mathrm{Tr}(|T|) := \sum_{n=1}^{\infty} \langle |T|e_n, e_n\rangle < \infty,$$

for some (and hence all) orthonormal basis $(e_n, n \in \mathbb{N})$ of H. The linear space of all trace class operators on H is denoted $\mathcal{I}_1(H)$. It is a Banach space with respect to the norm $||\cdot||_1$, where $||T||_1 := \mathrm{Tr}(|T|)$ for $T \in \mathcal{I}_1(H)$.

1. $\mathcal{I}_1(H)$ is a (two-sided) *-ideal of $\mathcal{L}(H)$.
2. $\mathcal{I}_1(H) \subseteq \mathcal{C}(H)$ and if $T \in \mathcal{C}(H)$, then $T \in \mathcal{I}_1(H)$ if and only if $\sum_{n=1}^{\infty} |c_n| < \infty$ in (A.6).
3. $\mathcal{F}(H)$ is $||\cdot||_1$-dense in $\mathcal{I}_1(H)$.

Note also that if T is of the form (A.6), then $|T| = \sum_{n=1}^{\infty} |c_n| \langle \cdot, e_n\rangle e_n$ and

$$||T||_1 = \sum_{n=1}^{\infty} |c_n|. \tag{A.9}$$

Theorem A.0.8 *$\mathcal{I}_1(H)$ is continuously embedded into $\mathcal{I}_2(H)$. Indeed if $T \in \mathcal{I}_1(H)$, then $||T||_2 \leq ||T||_1$.*

Proof This follows from comparing (A.9) with (A.8). □

Define $\mathcal{I}_1'(H) := \{T \in \mathcal{I}_2(H); ||T||_1' < \infty\}$, where the norm

$$||T||_1' := \sup\{\langle T, X\rangle_2, X \in \mathcal{L}(H), ||X|| \leq 1\}.$$

Theorem A.0.9 *$\mathcal{I}_1'(H) = \mathcal{I}_1(H)$ and $||\cdot||_1 = ||\cdot||_1'$.*

Proof We follow Walter [213]. Let $T \in \mathcal{I}_1(H)$ with $T = \sum_{n=1}^{\infty} c_n \langle \cdot, e_n\rangle f_n$. Then if $X \in \mathcal{L}(H)$ with $||X|| \leq 1$ we have

$$
\begin{aligned}
|\langle T, X\rangle_2| &= \left|\sum_{m=1}^{\infty}\sum_{n=1}^{\infty} c_n\langle e_m, e_n\rangle\langle f_n, Xe_m\rangle\right| \\
&\leq \sum_{n=1}^{\infty} |c_n|||e_n||.||X||.||f_n|| \\
&\leq \sum_{n=1}^{\infty} |c_n|,
\end{aligned}
$$

and so $T \in \mathcal{I}_1'(H)$ with $||T||_1' \leq ||T||_1$. Conversely, if $T \in \mathcal{I}_1'(H)$ choose $d_n \in \mathbb{C}$ so that $|d_n| = 1$ and $|c_n| = c_n d_n$ for all $n \in \mathbb{N}$ and consider the finite-rank operator $X_N := \sum_{n=1}^{N} d_n\langle\cdot, e_n\rangle f_n$. We easily check that $||X_N|| \leq 1$, and then

$$
\begin{aligned}
||T||_1' &\geq |\langle T, X_N\rangle_2| \\
&= \left|\sum_{n=1}^{\infty}\langle Te_n, X_N e_n\rangle\right| \\
&= \left|\sum_{n=1}^{N} c_n d_n\right| \\
&= \sum_{n=1}^{N} |c_n| \\
&\to ||T||_1 \text{ as } N \to \infty,
\end{aligned}
$$

where we have used (A.9). Hence $T \in \mathcal{I}_1(H)$ and the required result follows. □

The main purpose of introducing the norm $||\cdot||_1'$ is for ease of proving the next theorem.

Theorem A.0.10 *$T \in \mathcal{I}_1(H)$ if and only if there exist $S_1, S_2 \in \mathcal{I}_2(H)$ so that $T = S_1 S_2$.*

Proof We again follow Walter [213].
(Necessity) If T has the form (A.6), then just choose $S_1 = \sum_{n=1}^{\infty}\sqrt{c_n}\langle\cdot, e_n\rangle f_n$ and $S_2 = \sum_{n=1}^{\infty}\sqrt{c_n}\langle\cdot, e_n\rangle e_n$.

(Sufficiency) Let $S_1, S_2 \in \mathcal{I}_2(H)$ and $X \in \mathcal{I}_2(H)$ with $||X|| \leq 1$. Then

$$
\begin{aligned}
|\langle S_1 S_2, X\rangle_2| &= |\langle S_2, S_1^* X\rangle_2| \\
&\leq ||S_2||_2||S_1^* X||_2 \\
&= ||S_2||_2||X^* S_1||_2 \\
&\leq ||S_2||_2||X||.||S_1||_2 \\
&\leq ||S_1||_2.||S_2||_2.
\end{aligned}
$$

We then conclude that $S_1 S_2 \in \mathcal{I}_1(H)$ with $||S_1 S_2||_1 \leq ||S_1||_2.||S_2||_2$. □

A.7 Appendix 7: Semigroups of Linear Operators

In this brief appendix we describe some of the key basic ideas of the analytic theory of semigroups. Standard references are Davies [53] which is revised and updated within Chaps. 6–8 of Davies [54], and Pazy [158].

A *C_0-semigroup* acting in a (real, or complex) Banach space E is a family $(T_t, t \geq 0)$ of bounded linear operators acting on E satisfying the conditions:

(i) (Semigroup condition) $T_{s+t} = T_s T_t$ for all $s, t \geq 0$,
(ii) $T_0 = I$,
(iii) (Strong continuity) The mapping $t \to T_t\psi$ is continuous from $\mathbb{R}^+$ to E for all $\psi \in E$.

A C_0 semigroup is said to be a *contraction semigroup* if T_t is a contraction for all $t \geq 0$. We define a subset $\mathcal{D}_L$ of E by the prescription

$$\mathcal{D}_L := \left\{\psi \in E, \text{ there exists } \phi_\psi \in E \text{ so that } \lim_{t\to 0}\left|\left|\frac{T_t\psi - \psi}{t} - \phi_\psi\right|\right| = 0\right\}.$$

Then $\mathcal{D}_L$ is a linear subspace of E and the mapping $L : \mathcal{D}_L \to E$ defined by $L\psi = \phi_\psi$ is well-defined and linear. We call L the *infinitesimal generator* of the semigroup $(T_t, t \geq 0)$. Then $\mathcal{D}_L$ is the domain of L, and we have

$$L\psi = \lim_{t\to 0}\frac{T_t\psi - \psi}{t},$$

for all $\psi \in \mathcal{D}_L$, i.e. $L\psi = \left.\frac{dT_t\psi}{dt}\right|_{t=0}$ in the sense of the strong derivative on Banach space.

Theorem A.0.11 *The infinitesimal generator of a C_0-semigroup is densely defined.*

Proof We must show that D_L is dense in E. Let $\psi \in E$ be arbitrary and fix $u > 0$. Define $\phi_u = \int_0^u T_t\psi dt$. Then $\phi_u \in \mathcal{D}_L$ since for all $s > 0$,

$$\begin{aligned}
\frac{T_s\phi_u - \phi_u}{s} &= \frac{1}{s}\left\{\int_0^u T_{s+t}\psi dt - \int_0^u T_t\psi dt\right\}\\
&= \frac{1}{s}\left\{\int_s^{s+u} T_t\psi dt - \int_0^u T_t\psi dt\right\}\\
&= \frac{1}{s}\left\{\int_u^{s+u} T_t\psi dt - \int_0^s T_t\psi dt\right\}\\
&\to T_u\psi - \psi, \text{ as } s \to 0.
\end{aligned}$$

Now $\psi = \lim_{u\to 0}\frac{1}{u}\int_0^u T_t\psi dt$ and the result follows. □

Theorem A.0.12 *For all* $t \geq 0$, $T_t \mathcal{D}_L \subseteq \mathcal{D}_L$ *and if* $\psi \in \mathcal{D}_L$

(i) $T_t L\psi = LT_t\psi$,

(ii) $\frac{d}{dt}T_t\psi = LT_t\psi$,

(iii) $T_t\psi - \psi = \int_0^t T_s L\psi ds$.

Proof (Sketch) Everything follows from the following

$$\begin{aligned} LT_t\psi &= \lim_{s\to 0}\frac{T_{s+t}\psi - T_t\psi}{s} \\ &= T_t \lim_{s\to 0}\frac{T_s\psi - \psi}{s} \end{aligned}$$

□

We can interpret Theorem A.7.2 (ii) as telling us that $u(t) = T_t\psi$ is the unique solution of the abstract E-valued differential equation $\frac{du(t)}{dt} = Lu(t)$ with initial condition $u(0) = \psi$. Many examples of this equation where $E = C_0(G, \mathbb{R})$ are studied in Chap. 5, where G is a Lie group and $\mathcal{L}$ is the Hunt generator of a convolution semigroup. Another much-studied example is the *heat semigroup* on $C_0(M, \mathbb{R})$, where M is a Riemannian manifold and L is the Laplace-Beltrami operator Δ. If M is compact, there is a *heat kernel* $k \in C^\infty((0,\infty) \times M \times M, \mathbb{R})$ (i.e. a fundamental solution of the heat equation $\frac{\partial u}{\partial t} = \Delta u$) so that

$$T_t f(x) = \int_M f(y)k_t(x,u)\mu(dy),$$

for all $t > 0$, $f \in C(M, \mathbb{R})$, $x \in M$, where μ is the volume measure on M. Another important class of semigroups are those generated by the Schrödinger operators $L = -\Delta + V$, where V is a suitable multiplication operator acting on $L^2(M, \mu)$.

Theorem A.0.13 *L is a closed operator.*

Proof Let $(\phi_n, n \in \mathbb{N})$ be a sequence of vectors in E with $\phi_n \in D_L$ for all $n \in \mathbb{N}$ which converges to $\phi \in E$ and suppose that we also have $L\phi_n \to \psi \in E$. We are done if we can show that $\phi \in D_L$ and that $\psi = L\phi$. To establish this, observe that by Theorem A.7.2 (iii), for all $n \in \mathbb{N}, t > 0$,

$$\frac{1}{t}(T_t\phi_n - \phi_n) = \frac{1}{t}\int_0^t T_s L\phi_n ds.$$

Now take limits as $n \to \infty$ to obtain

$$\frac{1}{t}(T_t\phi - \phi) = \frac{1}{t}\int_0^t T_s \psi ds.$$

Finally we take limits as $t \to 0$ to find that

$$\lim_{t \to 0} \frac{1}{t}(T_t\phi - \phi) = \psi,$$

and the result follows. □

A C_0-semigroup becomes a *one-parameter group* of operators if it can be extended to a family of bounded operators $(T_t, t \in \mathbb{R})$ on E so that $T_{s+t} = T_s T_t$ for all $s, t \in \mathbb{R}$. An example of great importance arises when E is a complex Hilbert space (which we henceforth denote as $\mathcal{H}$), and each T_t is a unitary operator (which from now on, we write as U_t) acting on $\mathcal{H}$. We then call $(U_t, t \in \mathbb{R})$ a *one-parameter unitary group*. A deep insight into these groups is given by the following:

Theorem A.0.14 (Stone's theorem) *If $(U_t, t \in \mathbb{R})$ is a one-parameter group acting in a complex, separable Hilbert space $\mathcal{H}$, then there exists a self-adjoint operator H acting in $\mathcal{H}$ so that for all $t \in \mathbb{R}$,*

$$U_t = e^{itH}. \tag{A.10}$$

For a proof, see e.g. Reed and Simon [166] Theorem VIII.8, pp. 266–268. Note that the operator H is typically unbounded (unless $\mathcal{H}$ is finite-dimensional), and the exponential in (A.10) is defined via spectral theory.

Stone's theorem plays an important role in quantum mechanics. Fix $\psi \in \mathcal{H}$ and define $\psi(t) = U(t)\psi$, for $t \in \mathbb{R}$. Then the mapping $t \to \psi(t)$ is strongly differentiable, and formally differentiating in (A.10) yields an abstract form of *Schrödinger's equation*:

$$\frac{d\psi(t)}{dt} = iH\psi(t),$$

in which the operator H is playing the role of the Hamiltonian. Stone's theorem may also be utilised to give a delightful proof of Bochner's theorem, see Reed and Simon [166] Theorem IX.9, pp. 330–331.

A.8 Appendix 8: Cores of Closed Linear Operators on Banach Spaces

Let E be a real or complex separable Banach space and $T : E \to E$ be a densely defined closed linear operator with domain Dom(T). A linear manifold $\mathcal{D} \subset$ Dom(T) is said to be a *core* for T if $\overline{T|_{\mathcal{D}}} = T$. Equivalently, given any $g \in$ Dom(T) there exists a sequence $(g_n, n \in \mathbb{N})$, where $g_n \in \mathcal{D}$ for all $n \in \mathbb{N}$, such that $\lim_{n\to\infty} g_n = g$ and $\lim_{n\to\infty} Tg_n = Tg$ (see e.g. Davies [54]). Since T is closed, its *resolvent set* $\rho_T := \{\lambda \in \mathbb{C} : \lambda I - T \text{ is invertible}\}$ is a non-empty open set in $\mathbb{C}$. For each

$\lambda \in \rho(T)$, the *resolvent* $R_\lambda := (\lambda I - T)^{-1}$ is a bounded linear operator from E onto Dom(T). Hence $\lambda I - T$ maps Dom(T) onto E.

Our main result is the following (see also the first part of the proof of Proposition 1.1 in Breuillard [35]):

Theorem A.0.15 *The following are equivalent.*

1. *$\mathcal{D}$ is a core for T.*
2. *$\lambda I - T$ maps $\mathcal{D}$ to a dense linear manifold in E for some $\lambda \in \rho_T$.*
3. *$\lambda I - T$ maps $\mathcal{D}$ to a dense linear manifold in E for all $\lambda \in \rho_T$.*

Proof (2) $\Rightarrow$ (1). Let $g \in$ Dom(T) be arbitrary. By assumption there exists $\lambda \in \rho_T$ and a sequence $(g_n, n \in \mathbb{N})$ where $g_n \in \mathcal{D}$ for all n such that $\lim_{n\to\infty}(\lambda g_n - Tg_n) = \lambda g - Tg$. Applying the bounded operator R_λ, we find that $\lim_{n\to\infty} g_n = g$. Then $\lim_{n\to\infty} Tg_n = Tg$ follows easily.

(1) $\Rightarrow$ (3) Conversely, let $\mathcal{D}$ be a core for T. Given any $\lambda \in \rho_T$ we know that $\lambda I - T$ maps Dom(T) onto E. Hence given any $f \in E$ we can find $g \in$ Dom(T) so that $f = \lambda g - Tg$. But by the core property, given any $\epsilon > 0$ there exists $g' \in \mathcal{D}$ so that $||(\lambda I - T)(g - g')|| < \epsilon$. The result follows.

(3) $\Rightarrow$ (2) is obvious. □

References

1. S. Albeverio, M. Gordina, Lévy processes and their subordination in matrix Lie groups. Bull. Sci. Math. **131**, 738–760 (2007)
2. S.I. Amari, *Methods of Information Geometry, Translations of Mathematical Monographs* (American Mathematical Society, Providence, 2000)
3. G.W. Anderson, A. Guionnet, *An Introduction to Random Matrices* (Cambridge University Press, Zeitouni, 2010)
4. D. Applebaum, Compound Poisson processes and Lévy processes in groups and symmetric spaces. J. Theor. Prob. **13**, 383–425 (2000)
5. D. Applebaum, Operator-valued stochastic differential equations arising from unitary group representations. J. Theor. Prob. **14**, 61–76 (2001)
6. D. Applebaum, in *Lévy Processes in Stochastic Differential Geometry in Lévy Processes: Theory and Applications*, ed. by O. Barndorff-Nielsen, T. Mikosch, S. Resnick (Birkhäuser, Boston, Basel, Berlin, 2001), pp. 111–139
7. D. Applebaum, *Lévy Processes and Stochastic Calculus*, 2nd edn. (Cambridge University Press, Cambridge, 2009)
8. D. Applebaum, Probability measures on compact groups which have square-integrable densities. Bull. Lond. Math. Sci. **40**,1038–1044 (2008) (Corrigendum 42, 948, 2010)
9. D. Applebaum, Some L^2 properties of semigroups of measures on Lie groups. Semigroup Forum **79**, 217–228 (2009)
10. D. Applebaum, Infinitely divisible central probability measures on compact Lie groups—regularity, semigroups and transition kernels. Ann. Prob. **39**, 2474–2496 (2011)
11. D. Applebaum, Pseudo differential operators and Markov semigroups on compact Lie groups. J. Math. Anal. Appl. **384**, 331–348 (2011)
12. D. Applebaum, Smoothness of densities on compact Lie groups, to appear in Rendiconti del Seminario Matematico (2013)
13. D. Applebaum, R.Bañuelos, Martingale transform and Lévy Processes on Lie Groups, to appear in Indiana Univ. Math. J.

14. D. Applebaum, A.Dooley, A generalised Gangolli-Lévy-Khintchine formula for infinitely divisible measures and Lévy Processes on semi-simple Lie groups and symmetric spaces, to appear in Ann. d'Inst. Henri Poincaré (Prob. Stat.)
15. D. Applebaum, H. Kunita, Lévy flows on manifolds and Lévy processes on Lie groups. J. Math. Kyoto Univ. **33**, 1103–1123 (1993)
16. J.C. Baez, The octonions. Bull. Am. Math. Soc. **39**, 145–207 (2002)
17. J.C. Baez, H. Huerta, The algebra of grand unified theories. Bull. Am. Math. Soc. **47**, 483–502 (2010)
18. R. Bañuelos, F. Baudoin, Trace and heat kernel asymptotics for subordinate semigroups on manifolds, arXiv:1308.4944v1 (2013)
19. O.E. Barndorff-Nielsen, D.R. Cox, N. Reid, The role of differential geometry in statistical theory. Int. Stat. Rev. **54**, 83–96 (1986)
20. F. Baudoin, M. Hairer, J. Teichmann, Ornstein-Uhlenbeck processes on Lie groups. J. Funct. Anal. **255**, 877–890 (2008)
21. H. Bauer, *Measure and Integration Theory* (Walter de Gruyter, Berlin, New York, 2001)
22. R. Beals, R. Wong, *Special Functions: A Graduate Text* (Cambridge University Press, Cambridge, 2010)
23. A. Bendikov, L. Saloff-Coste, Gaussian bounds for derivatives of central Gaussian semigroups on compact groups. Trans. Am. Math. Soc. **354**, 1279–1298 (2001)
24. A. Bendikov, L. Saloff-Coste, Central Gaussian semigroups of measures with continuous density. J. Funct. Anal. **186**, 206–286 (2001)
25. A. Bendikov, L. Saloff-Coste, On the sample paths of Brownian motions on compact infinite dimensional groups. Ann. Prob. **31**, 1464–1494 (2003)
26. C. Berg, G. Forst, *Potential Theory on Locally Compact Abelian Groups* (Springer, New York, 1975)
27. J. Bigot, C. Christophe, S. Gedat, Random action of compact Lie groups and minimax estimation of a mean pattern. IEEE Trans. Inf. Theory **58**, 3509–3520 (2012)
28. M. Bingham, Central limit theory on locally compact abelian groups. in *Probability Measures on Groups and Related Structures*, XI, 1994 (World Scientific Publishing, Oberwolfach, River Edge, NJ, 1995) pp. 14–37
29. N.H. Bingham, Random walks on spheres. Z.Wahrscheinlichkeitstheorie verw. Geb. **22**, 169–192 (1972)
30. N.H. Bingham, C.M. Goldie, J.L. Teugels, *Regular Variation* (Cambridge University Press, Cambridge, 1987)
31. W.R. Bloom, H. Heyer, *Harmonic Analysis of Probability Measures on Hypergroups* (de Gruyter, Berlin, New York, 1995)
32. R.M. Blumenthal, R.K. Getoor, The asymptotic distribution of the eigenvalues for a class of Markov operators. Pac. J. Math. **9**, 399–408 (1959)
33. L. Boi, Clifford geometric algebras, spin manifolds and group actions in mathematics and physics. Adv. Appl. Clifford Algebras. **19**, 611–656 (2009)
34. E. Born, An explicit Lévy-Hinčin formula for convolution semigroups on locally compact groups. J. Theor. Prob. **2**, 325–42 (1989)
35. E. Breuillard, Random walks on Lie groups, preprint (2004), http://www.math.u-psud.fr/~breuilla/part0gb.pdf
36. T. Bröcker, T. tom Dieck, Representations of Compact Lie Groups (Springer, New York, 1985)
37. R.G.M. Brummelhuis, An F. and M. Riesz theorem for bounded symmetric domains. Ann. Inst. Fourier **37**, 139–150 (1987)
38. R.G.M. Brummelhuis, An F. and M. Riesz theorem for compact groups. Math. Scand. **64**, 226–32 (1989)
39. J.M. Burns, An elementary proof of the 'strange formula' of Freudenthal and de Vries. Q. J. Math. **51**, 295–297 (2000)
40. R. Carter, G. Segal, I. Macdonald, *Lectures on Lie Groups and Lie Algebras*, London Mathematical Society 32 (Cambridge University Press, Cambridge, 1995)

41. P. Cartier, *A Primer of Hopf Algebras*, Frontiers in Number Theory, Physics, and Geometry II (Springer, Berlin, 2007), pp. 537–615
42. F. Chamizo, H. Iwaniec, On the sphere problem. Rev. Mat. Iberoamericana **11**, 417–429 (1995)
43. I. Chavel, *Riemannian Geometry: A Modern Introduction*, Cambridge Tracts in Mathematics 108 (Cambridge University Press, Cambridge, 1993)
44. I. Chavel, *Eigenvalues in Riemannian Geometry* (Academic Press, London, 1984)
45. C. Chevalley, *Theory of Lie Groups I* (Princeton University Press, Princeton, 1946)
46. G.S. Chirikjian, *Stochastic Models, Information Theory and Lie Groups, Volume 1: Classical results and Geometric Methods* (Birkhäuser, Boston, Basel, Berlin, 2009)
47. Gs Chirikjian, *Stochastic Models, Information Theory and Lie Groups, Volume 2: Analytical Methods and Modern Applications* (Birkhäuser, Boston, Basel, Berlin, 2012)
48. G.S. Chirikjian, A.B. Kyatkin, *Engineering Applications of Noncommutative Harmonic Analysis* (CRC Press LLC, Boca Raton, 2001)
49. S. Cohen, Some Markov properties of stochastic differential equations with jumps, in Séminaire de Probabilités XXIX, in *Lecture Notes in Math*, ed. by J. Azéma, M. Emery, P.A. Meyer, M. Yor (Springer, Berlin, Heidelberg, 1995), pp. 181–194
50. D.L. Cohn, *Measure Theory* (Birkhaüser, Boston, 1980)
51. S.G. Dani, M. McCrudden, Embeddability of infinitely divisible distributions on linear Lie groups. Invent. Math. **110**, 237–61 (1992)
52. S.G. Dani, M. McCrudden, Convolution roots and embedding of probability measures on Lie groups. Adv. Math. **209**, 198–211 (2007)
53. E.B. Davies, *One-Parameter Semigroups* (Academic Press, London, 1980)
54. E.B. Davies, *Linear Operators and their Spectra* (Cambridge University Press, Cambridge, 2007)
55. L. Debnath, P. Mikusiński, *Introduction to Hilbert Spaces with Applications*, 3rd edn. (Academic Press, London, 2005)
56. P. Diaconis, *Group Representations in Probability and Statistics*, Lecture Notes-Monograph Series (Institute of Mathematical Statistics, Hayward, California, 1988)
57. P. Diaconis, M. Shahshahani, On the eigenvalues of random matrices. J. Appl. Prob. **31**, 49–62 (1994)
58. A.J. Dragt, The symplectic group and classical mechanics. Ann. N.Y. Acad. Sci. **1045**, 291–307 (2005)
59. B.K. Driver, Integration by parts and quasi-invariance for heat kernel measures on loop groups, J. Funct. Anal. **149**, 470–547 (1997) (corrigendum 155, 297–301, 1998)
60. B.K. Driver, T. Lohrenz, Logarithmic Sobolev inequalities for pinned loop groups. J. Funct. Anal. **140**, 381–448 (1996)
61. R.E. Edwards (ed.), *Integration and Harmonic Analysis on Compact Groups*, London Mathematical Society Lecture Note Series 8 (Cambridge University Press, Cambridge, 1972)
62. K.D. Elworthy, *Geometric aspects of diffusions on manifolds in École d'Été de Probabilitès de Saint-Flour XV-XVII, 1985–1987, 277–425*, vol. 1362. Lecture Notes in Math (Springer, Berlin, 1988)
63. J. Faraut, *Analysis on Lie Groups* (Cambridge University Press, Cambridge 2008)
64. H.D. Fegan, The heat equation and modular forms. J. Differ. Geom. **13**, 589–602 (1978)
65. H.D. Fegan, The fundamental solution of the heat equation on a compact Lie group. J. Differ. Geom. **18**, 659–668 (1983)
66. H.D. Fegan, *Introduction to Compact Lie Groups* (World Scientific Publishing, River Edge, 1991)
67. P. Feinsilver, Processes with independent increments on a Lie group. Trans. Am. Math. Soc. **242**, 73–121 (1978)
68. G.B. Folland, *A Course in Abstract Harmonic Analysis* (CRC Press, Inc., Boca Raton, 1995)
69. M. Fukushima, Y. Oshima, M. Takeda, *Dirichlet Forms and Symmetric Markov Processes* (de Gruyter, Berlin, 1994)

70. R. Gangolli, Sample functions of certain differential processes on symmetric spaces Pacific. J. Math. **15**, 477–496 (1965)
71. B. Gelbaum, G.K. Kalisch, J.M.H. Olmsted, On the embedding of topological semigroups and integral domains. Proc. Am. Math. Soc. **2**, 807–821 (1951)
72. M. Gordina, J. Haga, Lévy processes in a step 3 nilpotent group, Preprint (2012)
73. U. Grenander, *Probabilities on Algebraic Structures* (Wiley, New York, 1963)
74. A. Grigor'yan, *Heat Kernel and Analysis on Manifolds, AMS/IP Studies in Advanced Mathematics 47* (American Mathematical Society, Providence, 2009)
75. Y. Guivarc'h, M. Keane, B. Roynette, *Marches Aléatores sur les Groups de Lie*, Lecture Notes in Mathematics volume 624 (Spinger, Berlin, Heidelberg, New York, 1977)
76. D. Gurarie, *Symmetries and Laplacians* (Dover Publications Inc., Mineola, New York, 1992, 2008)
77. B.C. Hall, *Lie Groups, Lie algebras and Representations—An Elementary Introduction* (Springer, New York, 2003)
78. E.J. Hannan, Group representations and applied probability. J. Appl. Prob. **2**, 1–68 (1965)
79. K.E. Hare, The size of characters of compact Lie groups. Stud. Math. **129**, 1–18 (1998)
80. Harish-Chandra, Harmonic analysis on semisimple Lie groups. Bull. Am. Math. Soc. **76**, 529–551 (1970)
81. P. Harremoës, Maximum entropy on compact groups. Entropy **11**, 222–237 (2009)
82. J. Hawkes, Potential theory of Lévy processes. Proc. London Math. Soc. **38**, 335–352 (1979)
83. T. Hawkins, *Emergence of the Theory of Lie Groups, An Essay in the History of Mathematics 1869–1926*, Sources and Studies in the History of Mathematics and Physical Sciences (Springer, New York, 2000)
84. W. Hazod, *Stetige Faltungshalbgruppen von Wahrscheinlichkeitsmassen und Erzeugende Distributionen*, Lecture Notes in Mathematics, vol. 595 (Springer, New York, 1977)
85. W. Hazod, E. Siebert, *Stable Probability Measures on Euclidean Spaces and on Locally Compact Groups. Structural Properties and Limit Theorems*, Mathematics and its Applications, vol. 531 (Kluwer Academic Publishers, Dordrecht, 2001)
86. D.M. Healy Jr, H. Hendriks, P.T. Kim, Spherical deconvolution. J. Multivar. Anal. **67**, 1–22 (1998)
87. E. Hebay, *Sobolev Spaces on Riemannian Manifolds*, Lecture Notes in Mathematics vol. 1635 (Springer, Berlin, Heidelberg, New York, 1996)
88. S. Helgason, *Differential Geometry, Lie Groups and Symmetric Spaces* (Academic Press, New York, 1978) reprinted with corrections American Mathematical Society (2001)
89. S. Helgason, *Groups and Geometric Analysis* (Academic Press, New York, 1984), reprinted with corrections by the American Mathematical Society (2000)
90. H. Hendriks, Nonparametric estimation of a probability density on a Riemannian manifold using Fourier expansions. Ann. Stat. **18**, 832–849 (1990)
91. E. Hewitt, K.A. Ross, *Abstract Harmonic Analysis Volume 1: Structure of Topological Groups, Integration Theory, Group Representations* (Springer, Berlin, Gottingen, Heidelberg, 1963)
92. E. Hewitt, K.A. Ross, *Abstract Harmonic Analysis Volume 2: Structure and Analysis for Compact Groups, Analysis on Locally Compact Abelian Groups* (Springer, New York, Heidelberg, Berlin, 1970)
93. H. Heyer, L'analyse de Fourier non-commutative et applications à la théorie des probabilités, Ann. Inst. Henri Poincaré (Prob.Stat.) **4**, 143–168 (1968)
94. H. Heyer, *Infinitely Divisible Probability Measures on Compact Groups, in Lectures on Operator Algebras*. Lecture Notes in Math, vol. 247 (Springer, Berlin, Heidelberg, New York, 1972) pp. 55–249
95. H. Heyer, *Probability Measures on Locally Compact Groups* (Springer, Berlin, Heidelberg, 1977)
96. H. Heyer, *Structural Aspects in the Theory of Probability*, 2nd enlarged edn. (World Scientific Publishing, Singapore, 2010)
97. H. Heyer, A Bochner-type representation of positive definite mappings on the dual of a compact group, to appear in Commun. Stoch. Anal. (2013)

98. P.J. Higgins, *An Introduction to Topological Groups*, London Mathematical Society Lecture Note Series 15 (Cambridge University Press, Cambridge, 1974)
99. K.H. Hofmann, S.A. Morris, *The Structure of Compact Groups*, de Gruyter Studies in Mathematics 25 (2006)
100. A.S. Holevo, *An analog of the Itô decomposition for multiplicative processes with values in a Lie group, Quantum Probability and Applications V*, ed. by L. Accardi, W. von Waldenfels. Springer Lectures Nots in Math, vol. 1442 (1990) pp. 211–215
101. L. Hörmander, Hypoelliptic second order differential equations. Acta Math. **119**, 147–171 (1967)
102. F. Hirsch, J.P. Roth, Opérateurs dissipatifs et codissipatifs invariants sur un espace homogène, Théorie du Potentiel et Analyse Harmonique, Strasbourg, 1973, pp. 229–245. Lecture Notes in Math, vol. 404 (Springer, New York, 1974)
103. G.A. Hunt, Semigroups of measures on Lie groups. Trans. Am. Math. Soc. **81**, 264–293 (1956)
104. S. Hurst, The characteristic function of the Student t distribution, Centre for Mathematics and its Applications, School of Mathematical Sciences, ANU. http://maths-old.anu.edu.au/research.reports/srr/95/044/SRR95-044-scan.pdf. Accessed 1995
105. K. Itô, Brownian motions in a Lie groups. Proc. Jpn. Acad. **26**, 4–10 (1950)
106. N. Jacob, *Pseudo-Differential Operators and Markov Processes*, Mathematical Research 94 (Akademie-Verlag, Berlin, 1996)
107. N. Jacob, *Pseudo-Differential Operators and Markov Processes: 3*, Markov Processes and Applications (World Scientific Publishing, Singapore, 2005)
108. N. Jacobson, *Lie Algebras* (Wiley, New York, 1962)
109. O. Johnson, Y. Suhov, Entropy and convergence on compact groups. J. Theor. Prob. **13**, 843–857 (2000)
110. H.F. Jones, *Groups, Representations and Physics*, 2nd edn. (Taylor and Francis, UK, 1998)
111. M. Kac, On some connections between probability theory and differential and integral equations. in *Proceedings of the Second Berkeley Symposium on Mathematical Statistics and Probability 1950*, pp. 189–215. University of California Press, Berkeley and Los Angeles, 1951
112. I. Kaplansky, *Lie Algebras and Locally Compact Groups* (The University of Chicago Press, Chicago, London, 1971)
113. Y. Katznelson, *An Introduction to Harmonic Analysis*, 3rd edn. (Cambridge University Press, Cambridge, 1994)
114. Y. Kawada, K. Itô, On the probability distribution on a compact group I. Proc. Phys. Mat. Soc. Jpn. **22**, 977–998 (1940)
115. J.P. Keating, N.C. Snaith, Random matrix theory and $\zeta(1/2+it)$. Commun. Math. Phys. **214**, 57–89 (2000)
116. J.L. Kelley, *General Topology* (Springer, New York, Berlin, Heidelberg, Tokyo, 1955)
117. J.T. Kent, The infinite divisibility of the von Mises-Fisher distribution for all values of the parameter in all dimensions. Proc. Lond. Math. Soc **35**, 359–384 (1979)
118. P.T. Kim, D.S. Richards, Deconvolution density estimators on compact Lie groups. Contemp. Math. **287**, 155–171 (2001)
119. A.W. Knapp, *Representation Theory of Semisimple Groups* (Princeton, New Jersey, 1986)
120. A.W. Knapp, *Lie Groups Beyond an Introduction*, 2nd edn. (Birkhäuser, Boston, 1996, 2002)
121. A.W. Knapp, *Basic Real Analysis* (Birkhäuser, Boston, Basel, Berlin, 2005)
122. A.W. Knapp, *Advanced Real Analysis* (Birkhäuser, Boston, Basel, Berlin, 2005)
123. J.-Y. Koo, P.T. Kim, Asymptotic minimax bounds for stochastic deconvolution over groups. IEEE Trans. Inf. Theory **54**, 289–298 (2008)
124. C. Kosniowski, *A First Course in Algebraic Topology* (Cambridge University Press, Cambridge, 1980)
125. H. Kunita, Stable limit distributions over a nilpotent Lie group. Proc. Jpn. Acad. Ser. A **71**, 1–5 (1995)

126. H. Kunita, Convolution semigroups of stable distributions over a nilpotent Lie group. Proc. Jpn. Acad. Ser. A **70**, 305–310 (1994)
127. H. Kunita, Stable Lévy processes on nilpotent Lie groups. Stochast. Anal. Infinite Dimension. Spaces Pitman Res. Notes **310**, 167–182 (1994)
128. H. Kunita, Analyticity and injectivity of convolution semigroups on Lie groups. J. Funct. Anal. **165**, 80–100 (1999)
129. H. Kunita, Fundamental solutions and their short time estimates for jump-diffusions on manifolds, in preparation (2013)
130. H.H. Kuo, *Gaussian measures in Banach Spaces*, Lecture Notes in Mathematics, vol. 463 (Springer, Berlin, New York, 1975)
131. M. Liao, Lévy processes and Fourier analysis on compact Lie groups. Ann. Prob. **32**, 1553–1573 (2004)
132. M. Liao, *Lévy Processes in Lie Groups* (Cambridge University Press, Cambridge, 2004)
133. M. Liao, L. Wang, Lévy-Khinchine formula and existence of densities for convolution semigroups on symmetric spaces. Potential Anal. **27**, 133–150 (2007)
134. E.H. Lieb, H.-T. Yau, The stability and instability of relativistic matter. Commun. Math. Phys. **118**, 117–213 (1988)
135. W. Linde, *Probability in Banach Spaces—Stable and Infinitely Divisible Distributions* (Wiley-Interscience, New York, 1986)
136. J.T.-H. Lo, S.-K. Ng, Characterizing Fourier series representations of probability distributions on compact Lie groups. Siam J. Appl. Math. **48** 222–228 (1988)
137. A. Lozano-Robledo, *Elliptic Curves, Modular Forms and Their L-Functions*, Student Mathematical Library, IAS/Park City Mathematical Subseries (American Mathematical Society, Providence, 2011)
138. Z.M. Luo, P.T. Kim, T.Y. Kim, J.Y. Koo, Deconvolution on the Euclidean motion group SE(3). Inverse Prob. **27**, 1–30 (2011)
139. G.W. Mackey, *The Theory of Unitary Group Representations* (The University of Chicago Press, Chicago, London, 1976)
140. G.W. Mackey, *Unitary Group Representations in Physics, Probability, and Number Theory*, Mathematics Lecture Note Series 55 (Benjamin/Cummings Publishing Co., Inc, Reading, Mass., Redwood City, 1978)
141. G.W. Mackey, *The Scope and History of Commutative and Noncommutative Harmonic Analysis, History of Mathematics*, vol. 5 (American Mathematical Society, London Mathematical Society, Providence, 1992)
142. P. Major, S.R. Shlosman, A local limit theorem for the convolution of probability measures on a compact connected group. Z. Wahrsch. verv. Geb. **50**, 137–148 (1979)
143. M.P. Malliavin, P. Malliavin, *Factorisations et lois limites de la diffusion horizontale audessus d'un espace Riemmanian symetrique, in Théorie du Potentiel et Analyse Harmonique*, Lecture Notes Math, vol. 404 (Springer, Berlin, 1974) pp. 164–217
144. P. Malliavin (with H. Aurault, L. Kay, G. Letac), *Integration and Probability* (Springer, New York, 1995)
145. P. Malliavin, *Stochastic Analysis* (Springer, Berlin, Heidelberg, 1997)
146. A. Malyarenko, *Invariant Random Fields on Spaces with a Group Action* (Springer, Berlin, Heidelberg, 2013)
147. D. Marinucci, G. Peccati, *Random Fields on the Sphere—Representation, Limit Theorems and Cosmological Applications*, London Mathematical Society Lecture Note Series 389 (Cambridge University Press, Cambridge, 2011)
148. P.A. Mello, Central-limit theorems on groups. J. Math. Phys. **27**, 2876–2891 (1986)
149. W. Miller Jr, *Lie Theory and Special Functions* (Academic Press, New York, 1968)
150. S. Minakshisundaram, A. Pleijel, Some properties of the eigenfunctions of the Laplace operator on Riemannian manifolds. Canad. J. Math. **1**, 242–256 (1949)
151. D. Montgomery, L. Zippin, *Topological Transformation Groups* (Interscience Publishers, New York, London, 1955)

152. K. Nagami, Baire sets, Borel sets and some typical semi-continuous functions. Nagoya Math. J. **7**, 85–93 (1954)
153. D. Neuenschwander, *Probabilities on the Heisenberg Group—Limit Theorems and Brownian Motion* (Springer, Berlin, Heidelberg, 1996)
154. D. Ornstein, B. Weiss, Entropy and isomorphism theorems for actions of amenable groups. J. Anal. Math. **48**, 1–141 (1987)
155. K.R. Parthasarathy, *Probability Measures on Metric Spaces* (Academic Press, New York, 1967)
156. K.R. Parthasarathy, On the embedding of an infinitely divisible probability distribution in a one-parameter convolution semigroup. Theor. Prob. Appl. **12**, 373–380 (1967)
157. K.R. Parthasarathy, On the embedding of an infinitely divisible probability distribution in a one-parameter convolution semigroup. Sankhyā Ser. A **35**, 124–132 (1973)
158. A. Pazy, *Semigroups of Linear Operators and Applications to Partial Differential Equations* (Springer, New York, 1983)
159. F. Perrin, Étude mathématique du mouvement brownien de rotation. Ann. Sci. École Norm. Sup. **3**, 1–51 (1928)
160. S. Peszat, S. Tindel, Stochastic heat and wave equations on a Lie group. Stoch. Anal. Appl. **28**, 662–695 (2010)
161. J. Picard, C. Savona, Smoothness of the law of manifold-valued Markov processes with jumps (2011). arXiv:1106.4721v1
162. M. Pontier, A.S. Üstünel, Analyse stochastique sur l'espace de Lie-Wiener. C.R. Acad. Sci. Paris **313**(Série I), 313–316 (1991)
163. D. Ragozin, Central measures on compact simple Lie groups. J. Funct. Anal. **10**, 212–229 (1972)
164. D.A. Raikov, On absolutely continuous set functions. Doklady Acad. Nauk SSR (N.S.) **34**, 239–241 (1942)
165. S. Ramaswami, Semigroups of measures on Lie groups. J. Indian Math. Soc. **38**, 175–189 (1974)
166. M. Reed, B. Simon, *Methods of Modern Mathematical Physics, Functional Analysis*, revised and enlarged edn. vol. 1 (Academic Press, New York, 1980)
167. D. Revuz, *Markov Chains*, 2nd edn. vol. 11 (North Holland Mathematical Library, Elsevier Science Publishers B.V., 1984) (1st edn. (1975))
168. J. Roe, *Elliptic Operators, Topology and Asymptotic Methods*, 2nd edn. (Chapman and Hall, CRC, NJ, 1998)
169. S. Rosenberg, *The Laplacian on a Riemannian Manifold* (Cambridge University Press, Cambridge, 1997)
170. J.S. Rosenthal, Random rotations: characters and random walks on SO(n). Ann. Prob. **22**, 398–423 (1994)
171. W. Rudin, *Fourier Analysis on Groups* (Wiley Interscience, New York, 1962)
172. M. Ruzhansky, V. Turunen, *Pseudo-differential Operators and Symmetries: Background Analysis and Advanced Topics* (Birkhäuser, Basel, 2010)
173. S. Said, C. Lageman, N. LeBihan, J.H. Manton, Decompounding on compact Lie groups. IEEE Trans. Inf. Theory **56**, 2766–2777 (2010)
174. S. Said, J.H. Manton, Extrinsic mean of Brownian distributions on compact Lie groups. IEEE Trans. Inf. Theory **58**, 3521–3535 (2012)
175. L. Saloff-Coste, Analysis on compact Lie groups of large dimension and on connected compact groups. Colloq. Math. **118**, 183–199 (2010)
176. L. Saloff-Coste, *The heat kernel and its estimates, in Probabilistic approach to geometry*, Advanced Studies in Pure Mathematics 57 (Mathematical Society Japan, Tokyo, 2010) pp. 405–436
177. K.-I. Sato, *Lévy Processes and Infinite Divisibility* (Cambridge University Press, Cambridge, 1999)
178. R.L. Schilling, R. Song, Z. Vondraček, *Bernstein Functions, Theory and Applications*, Studies in Mathematics 37 (De Gruyter, Berlin, 2010)

179. S.B. Shlosman, Limit theorems of probability theory on compact topological groups. Theory Prob. App. **25**, 604–609 (1980)
180. S.B. Shlosman, A local limit theorem for the convolution of probability measures on a compact connected group. Z. Wahrsch. verv. Geb. **65**, 627–636 (1984)
181. M. Schürmann, *White Noise on Bialgebras, Lecture Notes in Mathematics* vol. 1544 (Springer, Berlin, 1991)
182. E. Seneta, Fitting the variance-gamma model to financial data, J. Appl. Prob. (Special, Vol.)41A, 177–187 (2004)
183. M.R. Sepanski, *Compact Lie Groups, Graduate Texts in Mathematics*, vol. 235 (Springer, Berlin, 2007)
184. E. Siebert, Absolut-Stetigkeit und Träger von Gauss-Verteilungen auf lokalkompakten Gruppen, Math. Ann. **210**, 129–147 (1974)
185. E. Siebert, Fourier analysis and limit theorems for convolution semigroups on a locally compact group. Adv. Math. **39**, 111–154 (1981)
186. B.W. Silverman, *Density Estimation for Statistics and Data Analysis, Monographs on Statistics and Applied Probability*, vol. 26 (Chapman and Hall, London, 1986)
187. G.F. Simmons, *Introduction to Topology and Modern Analysis* (McGraw Hill Book Company, Inc. New York, 1963)
188. B. Simon, *Representations of Finite and Compact Groups, Graduate Studies in Math*, vol. 10 (American Mathematical Society, Rhode Island, 1996)
189. M. Spivak, *Calculus on Manifolds* (W.A.Benjamin Inc. New York, 1965)
190. R.J. Stanton, Convergence of fourier series on compact Lie groups. Trans. Am. Math. Soc. **218**, 61–87 (1976)
191. R.J. Stanton, P.A. Tomas, Polyhedral summability of fourier series on compact lie groups. Am. J. Math. **100**, 477–493 (1978)
192. E.M. Stein, *Topics in Harmonic Analysis Related to the Littlewood-Paley Theory* (Princeton University Press and the University of Tokyo Press, Princeton, 1970)
193. S. Sternberg, *Group Theory and Physics* (Cambridge University Press, Cambridge, 1994)
194. I. Stewart, *Why Beauty Is Truth* (Basic Books, New York, 2007)
195. C.J. Stone, Optimal rates of convergence for nonparametric estimators. Ann. Stat. **8**, 1348–1360 (1980)
196. D.W. Stroock, S.R.S. Varadhan, Limit theorems for random walks on Lie groups, Sankhyā. Series A **35**, 27–93 (1973)
197. K. Stromberg, A note on the convolution of regular measures. Math. Scand. **7**, 347–52 (1959)
198. K. Stromberg, Probabilities on a compact group, Trans. Am. Math. Soc. **94** 295–309 (1960)
199. M. Stroppel, *Locally Compact Groups, EMS Textbooks in Mathematics* (European Mathematical Society, Zurich, 2006)
200. M. Sugiura, Fourier series of smooth functions on compact Lie groups. Osaka J. Math. **8**, 33–47 (1971)
201. M.Sugiura, *Unitary Representations and Harmonic Analysis* 2nd edn. North Holland, Amsterdam 1975, 1990)
202. J.D. Talman, *Special Functions, A Group Theoretic Approach* (Benjamin Inc, W.A, 1968)
203. M.E. Taylor *Noncommutative Harmonic Analysis, Mathematical Surveys and Monographs*, vol. 22 (American Mathematical Society, Rhode Island, 1986)
204. S. Tindel, F. Viens, On space-time regularity for the stochastic heat equation on Lie groups. J. Funct. Anal. **169**, 559–603 (1999)
205. F. Treves, *Linear Partial Differential Equations* (Academic Press, New York 1975)
206. V.S. Varadarajan, *Lie Groups, Lie Algebras and their Representations* (Springer, New York, 1984) (first published by Prentice-Hall (1974)
207. V.S. Varadarajan, *An Introduction to Harmonic Analysis on Semisimple Lie Groups, Cambridge Studies in Advanced Mathematics* vol. 16 (Cambridge University Press, Cambridge 1989)
208. V.S. Varadarajan, Historical review of Lie theory. http://www.math.ucla.edu/~vsv/liegroups2007/liegroups2007.html

209. NTh Varopoulos, L. Saloff-Coste, T. Coulhon, *Analysis and Geometry on Groups* (Cambridge University Press, Cambridge, 1992)
210. N.Ja. Vilenkin, *Special Functions and the Theory of Group Representations, Translated from the Russian by V. N. Singh.* Translations of Mathematical Monographs, vol. 22 (American Mathematical Society, Providence, Rhode Island, 1968)
211. N.Ja. Vilenkin, A.U. Klimyk, *Representations of Lie Groups and Special Functions Translated from the Russian by V. A. Groza and A. A* vol. 1, 2, 3 (Groza. Kluwer Academic Publishers Group, Dordrecht, 1991, 1992, 1993)
212. M. Voit, Martingale characterizations of stochastic processes on compact groups. Prob. Math. Stat. **19**, 389–405 (1999)
213. M. Walter, Hilbert-Schmidt and trace class operators. http://www.leetspeak.org/math/wiko/hilbert_schmidt_nuclear.pdf
214. F.W. Warner, *Foundations of Differentiable Manifolds and Lie Groups* (Springer, New York, 1983)
215. G. N. Watson, *A Treatise on the Theory of Bessel Functions* (Cambridge University Press, Cambridge, 1922)
216. D. Wehn, Some remarks on Gaussian distributions on a Lie group, Z.Wahrscheinlichkeitstheorie verw. Gebiete **30**, 255–63 (1974)
217. J.G. Wendel, Haar measure and the semigroup of measures on a compact group. Proc. Am. Math. Soc. **5**, 923–929 (1954)
218. I.M. Yaglom, *Felix Klein and Sophus Lie* (Birkhäuser, Boston, 1988)
219. B. Yazici, Stochastic deconvolution over groups. IEEE Trans. Inf. Theory **50**, 494–510 (2004)
220. D.P. Želobenko, *Compact Lie Groups and their Representations, Translations of Mathematical Monographs*, vol. 40 (American Mathematical Society, Providence, Rhode Island, 1973)

Index

D. Applebaum, *Probability on Compact Lie Groups*, Probability Theory and Stochastic Modelling 70, DOI: 10.1007/978-3-319-07842-7,

Printed in the United States
By Bookmasters

Printed in the United States
By Bookmasters